Bootstrapping Service Mesh Implementations with Istio

Build reliable, scalable, and secure microservices on Kubernetes with Service Mesh

Anand Rai

BIRMINGHAM—MUMBAI

Bootstrapping Service Mesh Implementations with Istio

Copyright © 2023 Packt Publishing

All rights reserved. No part of this book may be reproduced, stored in a retrieval system, or transmitted in any form or by any means, without the prior written permission of the publisher, except in the case of brief quotations embedded in critical articles or reviews.

Every effort has been made in the preparation of this book to ensure the accuracy of the information presented. However, the information contained in this book is sold without warranty, either express or implied. Neither the author, nor Packt Publishing or its dealers and distributors, will be held liable for any damages caused or alleged to have been caused directly or indirectly by this book.

Packt Publishing has endeavored to provide trademark information about all of the companies and products mentioned in this book by the appropriate use of capitals. However, Packt Publishing cannot guarantee the accuracy of this information.

Group Product Manager: Mohd Riyan Khan
Publishing Product Manager: Surbhi Suman
Senior Editor: Tanya D'cruz
Technical Editor: Nithik Cheruvakodan
Copy Editor: Safis Editing
Project Coordinator: Ashwin Kharwa
Proofreader: Safis Editing
Indexer: Hemangini Bari
Production Designer: Ponraj Dhandapani
Marketing Coordinator: Agnes D'souza

First published: April 2023

Production reference: 1230323

Published by Packt Publishing Ltd.
Livery Place
35 Livery Street
Birmingham
B3 2PB, UK.

ISBN 978-1-80324-681-9

www.packtpub.com

I am grateful to my kids, Yashasvi and Agastya, who sacrificed their playtime with me so I could write this book, and to Pooja, my loving wife, who supported me and kept me committed to this venture. This book, like everything else in my life, would not be possible without the blessings, love, and care of my beloved father, Mr. Jitendra Rai, and my loving mother, Mrs. Prem Lata Rai. This book is the result of the inspiration I get from my uncles, Mr. Pradeep Kumar Rai and Mr. Awadhesh Rai, who have been my pillars of support, my mentors and coaches. It was their motivation and guidance that led me to pursue computer science as a hobby and a career.

– Anand Rai

Contributors

About the author

Anand Rai has over 18 years' experience of working in information technology for various organizations including technology providers and consumers. He has held a variety of executive and senior roles in those organizations but has always taken a hands-on approach to technology. This experience has given him perspectives on how development in information technology drives productivity and improvements in our daily lives. His areas of specialization are application integration, API management, microservices architectures, the cloud, DevOps, and Kubernetes. He loves solving problems, visualizing new solutions, and helping organizations to use technology as an enabler for achieving business outcomes.

About the reviewers

Andres Sacco is a technical leader at TravelX and has experience with many different languages, including Java, PHP, and Node.js. In his previous job, Andres helped find alternative ways to optimize data transfer between microservices, which reduced the infrastructure cost by 55%. Before introducing these optimizations, he investigated different testing methods for better coverage of the microservices than the unit tests provided. He has also dictated internal courses about new technologies and has written articles on Medium. Andres is a co-author of the book *Beginning Scala 3*, published by Apress.

Ajay Reddy Yeruva currently works as a senior software engineer in the IP-DevOps team at Ritchie Bros. Auctioneers, along with volunteering as vice president at AAITP. He has an IT career that spans around 10 years. Before moving to Ritchie Bros. Auctioneers, he worked at Cisco Systems and Infosys Limited as a systems/software engineering consultant, helping to set up and maintain production applications, infrastructure, CI/CD pipelines, and operations. He is an active member of DevSecOps, AIOps, GitOps, and DataOps communities on different forums. When it comes to the best monitoring and observability talents Ajay tops the global list!

Table of Contents

3

Understanding Istio Control and Data Planes 61

Part 2: Istio in Practice

4

Managing Application Traffic 89

5

Managing Application Resiliency 127

6

Securing Microservices Communication 153

7

Service Mesh Observability 179

Part 3: Scaling, Extending, and Optimizing

8

Scaling Istio to Multi-Cluster Deployments Across Kubernetes 211

9

Extending Istio Data Plane 241

10

Deploying Istio Service Mesh for Non-Kubernetes Workloads 265

11

Troubleshooting and Operating Istio 285

12

Summarizing What We Have Learned and the Next Steps 315

Appendix – Other Service Mesh Technologies 341

Preface

Istio is one of the most widely adopted Service Mesh technologies. It is used to manage application networking to provide security and operational efficiency to microservices. This book explores Istio layer by layer to explain how it is used to manage application networking, resiliency, observability, and security. Using various hands-on examples, you'll learn about Istio Service Mesh installation, its architecture, and its various components. You will perform a multi-cluster installation of Istio along with integrating legacy workloads deployed on virtual machines. You'll learn how to extend the Istio data plane using **WebAssembly** (**WASM**), as well as covering Envoy and why it is used as the data plane for Istio. You'll see how OPA Gatekeeper can be used to automate best practices for Istio. You'll learn how to observe and operate Istio using Kiali, Prometheus, Grafana, and Jaeger. You'll also explore other Service Mesh technologies such as Linkerd, Consul, Kuma, and Gloo Mesh. The easy-to-follow hands-on examples built using lightweight applications throughout the book will help you to focus on implementing and deploying Istio to cloud and production environments instead of having to deal with complex demo applications.

After reading this book, you'll be able to perform reliable and zero-trust communication between applications, solve application networking challenges, and build resilience in distributed applications using Istio.

Who this book is for

Software developers, architects, and DevOps engineers with experience in using microservices in Kubernetes-based environments and who want to solve application networking challenges that arise in microservice communications will benefit from this book. To get the most out of this book, you will need to have some experience in working with the cloud, microservices, and Kubernetes.

What this book covers

Chapter 1, Introducing Service Meshes, covers the fundamentals of cloud computing, microservices architecture, and Kubernetes. It then outlines the context as to why a Service Mesh is required and what value it delivers. If you don't have hands-on experience in dealing with large-scale deployment architecture using Kubernetes, the cloud, and microservices architecture, then this chapter will familiarize you with these concepts and give you a good foundation for understanding the more complex subjects in the subsequent chapters.

Chapter 2, Getting Started with Istio, describes why Istio has experienced viral popularity among the Service Mesh technologies available. The chapter then provides instructions to install and run Istio and walks you through Istio's architecture and its various components. Once installed, you will then enable Istio sidecar injection in an example application packaged with the Istio installation. The chapter provides a step-by-step look at the pre- and post-enablement of Istio in the example application to give you an idea of how Istio works.

Chapter 3, Understanding Istio Control and Data Planes, dives deeper into Istio's control plane and data plane. This chapter will help you understand the Istio control plane so you can plan the installation of control planes in a production environment. After reading this chapter, you should be able to identify the various components of the Istio control plane including istiod, along with the functionality they each deliver in the overall working of Istio. The chapter will also familiarize you with Envoy, its architecture, and how to use Envoy as a standalone proxy.

Chapter 4, Managing Application Traffic, provides details on how to manage application traffic using Istio. The chapter is full of hands-on examples, exploring the management of Ingress traffic using the Kubernetes Ingress resource and then showing how to do this using Istio Gateway, along with securely exposing Ingress over HTTPS. The chapter provides examples of canary releases, traffic mirroring, and routing traffic to services outside the mesh. Finally, we'll see how to manage traffic egressing from the mesh.

Chapter 5, Managing Application Resiliency, provides details on how to make use of Istio to increase the application resiliency of microservices. The chapter discusses various aspects of application resiliency including fault injection, timeout and retries, load balancing, rate limiting, circuit breakers, and outlier detection, and how each of these is addressed by Istio.

Chapter 6, Securing Microservices Communication, dives deeper into advanced topics on security. The chapter starts with explaining Istio's security architecture, followed by implementing mutual TLS for service communication both with other services in the mesh and with downstream clients outside the mesh. The chapter will walk you through various hands-on exercises to create custom security policies for authentication and authorization.

Chapter 7, Service Mesh Observability, provides insight into why observability is important, how to collect telemetry information from Istio, the different types of metrics available and how to fetch them via APIs, and how to enable distributed tracing for applications deployed in the mesh.

Chapter 8, Scaling Istio to Multi-Cluster Deployments Across Kubernetes, walks you through how Istio can be used to provide seamless connectivity between applications deployed across multiple Kubernetes clusters. The chapter also covers multiple installation options for Istio to achieve high availability and continuity with the Service Mesh. The chapter covers advanced topics of Istio installation and familiarizes you with how to set up Istio in a primary-remote configuration on multiple networks, primary-remote configuration on a single network, multi-primary configuration on different networks, and multi-primary configuration on a single network.

Chapter 9, Extending Istio Data Plane, provides various options to extend the Istio data plane. The chapter discusses EnvoyFilter and WebAssembly in great detail and examines how they can be used to extend the functionality of the Istio data plane beyond what is offered out of the box.

Chapter 10, Deploying the Istio Service Mesh for Non-Kubernetes Workloads, provides a background as to why organizations have a significant number of workloads still deployed on virtual machines. The chapter then introduces the concept of hybrid architecture, a combination of modern and legacy architecture, followed by showing how Istio helps to marry these two worlds of legacy and modern technologies and how you can extend Istio beyond Kubernetes to virtual machines.

Chapter 11, Troubleshooting and Operating Istio, provides details of common problems you will encounter when operating Istio and how to distinguish and isolate them from other issues. The chapter then covers various techniques to analyze and troubleshoot the day-2 problems often faced by operations and reliability engineering teams. The chapter provides various best practices for deploying and operating Istio and shows how to automate the enforcement of best practices using OPA Gatekeeper.

Chapter 12, Summarizing What We Have Learned and the Next Steps, helps you revise what you've learned from this book by putting it to use to deploy and configure an open source application, helping you gain confidence in employing your learning in real-world applications. The chapter also provides various resources you can explore to advance your learning and expertise in Istio. Finally, the chapter introduces eBPF, an advanced technology poised to make a positive impact on service meshes.

Appendix – Other Service Mesh Technologies, introduces other Service Mesh technologies including Linkerd, Gloo Mesh, and Consul Connect, which are gaining popularity, recognition, and adoption by organizations. The information provided in this appendix is not exhaustive, but rather aims to make you familiar with the alternatives to Istio and help you form an opinion on how these technologies fare in comparison to Istio.

To get the most out of this book

Readers will need hands-on experience of using and deploying microservices on Kubernetes-based environments. Readers need to be familiar with using YAML and JSON and performing basic operations of Kubernetes. As the book makes heavy usage of various cloud provider services, it is helpful to have some experience of using various cloud platforms.

Software/hardware covered in the book	Operating system requirements
A workstation with a quad-core processor and 16 GB RAM at a minimum	macOS or Linux
Access to AWS, Azure, and Google Cloud subscriptions	N/A
Visual Studio Code or similar	N/A
minikube, Terraform	N/A

If you are using the digital version of this book, we advise you to type the code yourself or access the code from the book's GitHub repository (a link is available in the next section). Doing so will help you avoid any potential errors related to the copying and pasting of code.

Download the example code files

You can download the example code files for this book from GitHub at `https://github.com/PacktPublishing/Bootstrap-Service-Mesh-Implementations-with-Istio`. If there's an update to the code, it will be updated in the GitHub repository.

We also have other code bundles from our rich catalog of books and videos available at `https://github.com/PacktPublishing/`. Check them out!

Download the color images

We also provide a PDF file that has color images of the screenshots and diagrams used in this book. You can download it here: `https://packt.link/DW41O`.

Conventions used

There are a number of text conventions used throughout this book.

`Code in text`: Indicates code words in text, database table names, folder names, filenames, file extensions, pathnames, dummy URLs, user input, and Twitter handles. Here is an example: "The configuration patch is applied to `HTTP_FILTER` and in particular to the HTTP router filter of the `http_connection_manager` network filter."

A block of code is set as follows:

```
"filterChainMatch": {
                     "destinationPort": 80,
                     "transportProtocol": "raw_buffer"
                 },
```

When we wish to draw your attention to a particular part of a code block, the relevant lines or items are set in bold:

```
"filterChainMatch": {
                     "destinationPort": 80,
                     "transportProtocol": "raw_buffer"
                 },
```

Any command-line input or output is written as follows:

```
% curl -H "Host:httpbin.org" http://
a816bb2638a5e4a8c990ce790b47d429-1565783620.us-east-1.elb.
amazonaws.com/get
```

Bold: Indicates a new term, an important word, or words that you see onscreen. For instance, words in menus or dialog boxes appear in **bold**. Here is an example: "**Cloud computing** is utility-style computing with a business model similar to what is provided by businesses selling utilities such as LPG and electricity to our homes"

> **Tips or important notes**
> Appear like this.

Get in touch

Feedback from our readers is always welcome.

General feedback: If you have questions about any aspect of this book, email us at customercare@packtpub.com and mention the book title in the subject of your message.

Errata: Although we have taken every care to ensure the accuracy of our content, mistakes do happen. If you have found a mistake in this book, we would be grateful if you would report this to us. Please visit www.packtpub.com/support/errata and fill in the form.

Piracy: If you come across any illegal copies of our works in any form on the internet, we would be grateful if you would provide us with the location address or website name. Please contact us at copyright@packt.com with a link to the material.

If you are interested in becoming an author: If there is a topic that you have expertise in and you are interested in either writing or contributing to a book, please visit authors.packtpub.com.

Share Your Thoughts

Once you've read *Bootstrapping Service Mesh Implementations with Istio*, we'd love to hear your thoughts! Scan the QR code below to go straight to the Amazon review page for this book and share your feedback.

`https://packt.link/r/1803246812`

Your review is important to us and the tech community and will help us make sure we're delivering excellent quality content.

Download a free PDF copy of this book

Thanks for purchasing this book!

Do you like to read on the go but are unable to carry your print books everywhere?

Is your eBook purchase not compatible with the device of your choice?

Don't worry, now with every Packt book you get a DRM-free PDF version of that book at no cost.

Read anywhere, any place, on any device. Search, copy, and paste code from your favorite technical books directly into your application.

The perks don't stop there, you can get exclusive access to discounts, newsletters, and great free content in your inbox daily

Follow these simple steps to get the benefits:

1. Scan the QR code or visit the link below

`https://packt.link/free-ebook/9781803246819`

2. Submit your proof of purchase
3. That's it! We'll send your free PDF and other benefits to your email directly

Part 1: The Fundamentals

In this part of the book, we will cover the fundamentals of the Service Mesh, why it is needed, and what type of applications need it. You will understand the difference between Istio and other Service Mesh implementations. This part will also walk you through the steps to configure and set up your environment and install Istio, and while doing so, it will unravel the Istio control plane and data plane, how they operate, and their roles in the Service Mesh.

This part contains the following chapters:

- *Chapter 1, Introducing Service Meshes*
- *Chapter 2, Getting Started with Istio*
- *Chapter 3, Understanding Istio Control and Data Planes*

1

Introducing Service Meshes

Service Mesh are an advanced and complex topic. If you have experience using the cloud, Kubernetes, and developing and building an application using microservices architecture, then certain benefits of Service Mesh will be obvious to you. In this chapter, we will familiarize ourselves with and refresh some key concepts without going into too much detail. We will look at the problems you experience when you are deploying and operating applications built using microservices architecture and deployed on containers in the cloud, or even traditional data centers. Subsequent chapters will focus on Istio, so it is good to take some time to read through this chapter to prepare yourself for the learning ahead.

In this chapter, we're going to cover the following main topics:

- Cloud computing and its advantages
- Microservices architecture
- Kubernetes and how it influences design thinking
- An introduction to Service Mesh

The concepts in the chapter will help you build an understanding of Service Mesh and why they are needed. It will also provide you guidance on identifying some of the signals and symptoms in your IT environment that indicate you need to implement Service Mesh. If you don't have hands-on experience in dealing with large-scale deployment architecture using Kubernetes, cloud, and microservices architecture, then this chapter will familiarize you with these concepts and give you a good start toward understanding more complex subjects in subsequent chapters. Even if you are already familiar with these concepts, it will still be a good idea to read this chapter to refresh your memory and experiences.

Revisiting cloud computing

In this section, we will look at what cloud computing is in simple terms, what benefits it provides, how it influences design thinking, as well software development processes.

Cloud computing is utility-style computing with a business model similar to what is provided by businesses selling utilities such as LPG and electricity to our homes. You don't need to manage the production, distribution, or operation of electricity. Instead. you focus on consuming it effectively and efficiently by just plugging in your device to the socket on the wall, using the device, and paying for what you consume. Although this example is very simple, it is still very relevant as an analogy. Cloud computing providers provide access to compute, storage, databases, and a plethora of other services, including **Infrastructure as a Service** (**IaaS**), **Platform as a Service** (**PaaS**), and **Software as a Service** (**SaaS**) over the internet.

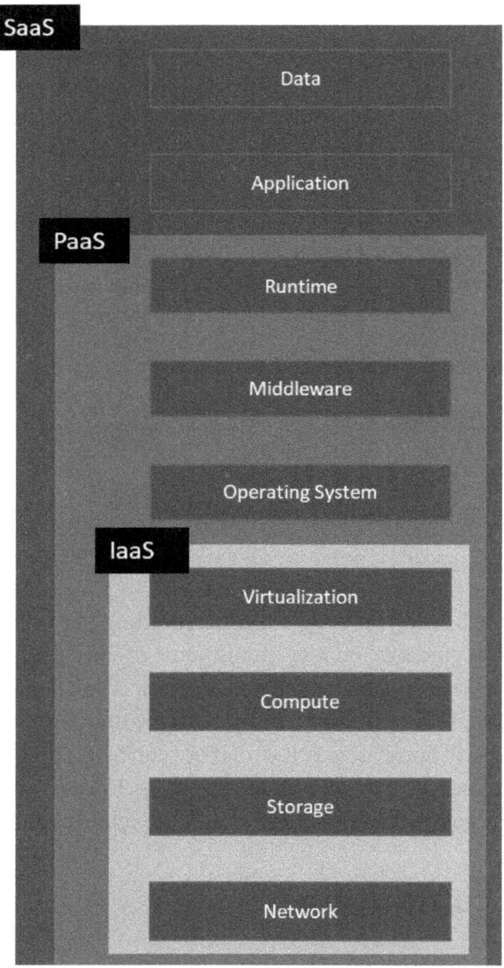

Figure 1.1 – Cloud computing options

Figure 1.1 illustrates the cloud computing options most commonly used:

- **IaaS** provides infrastructure such as networking to connect your application with other systems in your organization, as well as everything else you would like to connect to. IaaS gives you access to computational infrastructure to run your application, equivalent to **Virtual Machines** (**VMs**) or bare-metal servers in traditional data centers. It also provides storage for host data for your applications to run and operate. Some of the most popular IaaS providers are Amazon EC2, Azure virtual machines, Google Compute Engine, Alibaba E-HPC (which is very popular in China and the Greater China region), and VMware vCloud Air.
- **PaaS** is another kind of offering that provides you with the flexibility to focus on building applications rather than worrying about how your application will be deployed, monitored, and so on. PaaS includes all that you get from IaaS but also middleware to deploy your applications, development tools to help you build applications, databases to store data, and so on. PaaS is especially beneficial for companies adopting microservices architecture. When adopting microservices architecture, you also need to build an underlying infrastructure to support microservices. The ecosystem required to support microservices architecture is expensive and complex to build. Making use of PaaS to deploy microservices makes microservices architecture adoption much faster and easier. There are many examples of popular PaaS services from cloud providers. However, we will be using Amazon **Elastic Kubernetes Service** (**EKS**) as a PaaS to deploy the sample application we will explore hands-on with Istio.
- **SaaS** is another kind of offering that provides a complete software solution that you can use as a service. It is easy to get confused between PaaS and SaaS services, so to make things simple, you can think of SaaS as services that you can consume without needing to write or deploy any code. For example, it's highly likely that you are using an email service as SaaS with the likes of Gmail. Moreover, many organizations use productivity software that is SaaS, and popular examples are services such as Microsoft Office 365. Other examples include CRM systems such as Salesforce and **enterprise resource planning** (**ERP**) systems. Salesforce also provides a PaaS offering where Salesforce apps can be built and deployed. Salesforce Essentials for small businesses, Sales Cloud, Marketing Cloud, and Service Cloud are SaaS offerings, whereas Salesforce Platform, which is a low-code service for users to build Salesforce applications, is a PaaS offering. Other popular examples of SaaS are Google Maps, Google Analytics, Zoom, and Twilio.

Cloud services providers also provide different kinds of cloud offerings, with varying business models, access methods, and target audiences. Out of many such offerings, the most common are a public cloud, a private cloud, a hybrid cloud, and a community cloud:

- A **public cloud** is the one you most probably are familiar with. This offering is available over the internet and is accessible to anyone and everyone with the ability to subscribe, using a credit card or similar payment mechanism.

- A **private cloud** is a cloud offering that can be accessed over the internet or a restricted private network to a restricted set of users. A private cloud can be an organization providing IaaS or PaaS to its IT users; there are also service providers who provide a private cloud to organizations. The private cloud delivers a high level of security and is widely used by organizations that have highly sensitive data.
- A **hybrid cloud** refers to an environment where public and private clouds are collectively used. Also, a hybrid cloud is commonly used when more than one cloud offering is in use – for example, an organization using both AWS and Azure with applications deployed and data flowing across the two. A hybrid cloud is a good option when there are data and applications that are required to be hosted in a private cloud due to security reasons. Conversely, there may be other applications that don't need to reside in the private cloud and can benefit from the scalability and elasticity features of a public cloud. Rather than restricting yourself to a public or private cloud, or one cloud provider or another, you should reap the benefit of the strengths of various cloud providers and create an IT landscape that is secure, resilient, elastic, and cost-effective.
- A **community cloud** is another cloud offering available to a set of organizations and users. Some good examples are AWS GovCloud in the US, which is a community cloud for the US government. This kind of cloud restricts who can use it – for example, AWS GovCloud can only be used by US government departments and agencies.

Now that you understand the true crux of cloud computing, let's look at some of its key advantages in the following section.

Advantages of cloud computing

Cloud computing enables organizations to easily access all kinds of technologies without going through high upfront investment in expensive hardware and software procurement. By utilizing cloud computing, organizations achieve agility, as they can innovate faster by having access to high-end compute power and infrastructure (such as a load balancer, compute instances, and so on) and also to software services (such as machine learning, analytics, messaging infrastructure, AI, databases, and so on) that can be integrated as building blocks in a plug-and-play style to build software applications.

For example, if you're building a software application, then most probably it will need the following:

- Load balancers
- Databases
- Servers to run and compute servers to host an application
- Storage to host the application binaries, logs, and so on
- A messaging system for asynchronous communication

You will need to procure, set up, and configure this infrastructure in an on-premises data center. This activity, though important for launching and operationalizing your applications in production, does not produce any business differentiators between you and your competition. High availability and resiliency of your software application infrastructure is a requirement that is required to sustain and survive in the digital world. To compete and beat your competition, you need to focus on customer experience and constantly delivering benefits to your consumers.

When deploying on-premises, you need to factor in all upfront costs of procuring infrastructure, which include the following:

- Network devices and bandwidth
- Load balancers
- A firewall
- Servers and storage
- Rack space
- Any new software required to run the application

All the preceding costs will incur **Capital Expenditures** (**CapEx**) for the project. You will also need to factor in the setup cost, which includes the following:

- Network, compute servers, and cabling
- Virtualization, operating systems, and base configuration
- Setup of middleware such as application servers and web servers (if using containerization, then the setup of container platforms, databases, and messaging)
- Logging, auditing, alarming, and monitoring components

All the preceding will incur CapEx for the project but may fall under the organization's **Operating Expenses** (**OpEx**).

On top of the aforementioned additional costs, the most important factor to consider is the time and human resources required to procure, set up, and make the infrastructure ready for use. This significantly impacts your ability to launch features and services on the market (also called *agility* and *time to market*).

When using the cloud, these costs can be procured with a pay-as-you-go model. Where you need compute and storage, it can be procured in the form of IaaS, and where you need middleware, it can be procured in the form of PaaS. You will realize that some of the functionality you need to build might be already available as SaaS. This expedites your software delivery and time to market. On the cost front, some of the costs will still incur CapEx for your project, but your organization can claim it as OpEx, which has certain benefits from a tax point of view. Whereas it previously took months of preparation to set up all that you needed to deploy your application, it can now be done in days or weeks.

Cloud computing also changes the way you design, develop, and operate IT systems. In *Chapter 4*, we will look at cloud-native architecture and how it differs from traditional architecture.

Cloud computing makes it easier to build and ship software applications with low upfront investments. The following section describes microservices architecture and how it is used to build and deliver highly scalable and resilient applications.

Understanding microservices architecture

Before we discuss microservices architecture, let's first discuss **monolithic architecture**. It's highly likely that you will have encountered or probably even participated in building one. To understand it better, let's take a scenario and see how it has been traditionally solved using monolithic architecture.

Let's imagine a book publisher who wants to start an online bookstore. The online bookstore needs to provide the following functionalities to its readers:

- Readers should be able to browse all the books available for purchase.
- Readers should be able to select the books they want to order and save them to a shopping cart. They should also be able to manage their shopping cart.
- Readers should be able to then authorize payment for the book order using a credit card.
- Readers should have the books delivered to their shipping address once payment is complete.
- Readers should be able to sign up, store details including their shipping address, and bookmark favorite books.
- Readers should be able to sign in, check what books they have purchased, download any purchased electronic copies, and update shipping details and any other account information.

There will be many more requirements for an online bookstore, but for the purpose of understanding monolithic architecture, let's try to keep it simple by limiting the scope to these requirements.

It is worth mentioning Conway's law, where he stated that, often, the design of monolithic systems reflects the communication structure of an organization:

> *Any organization that designs a system (defined broadly) will produce a design whose structure is a copy of the organization's communication structure.*
>
> *– Melvin E. Conway*

There are various ways to design this system; we can follow traditional design patterns such as **model-view-controller** (**MVC**), but to do a fair comparison with microservices architecture, let's make use of **hexagonal architecture**. We will also be using hexagonal architecture in microservices architecture.

With a logical view of **hexagonal architecture**, business logic sits in the center. Then, there are adaptors to handle requests coming from outside as well as to send requests outside, which are called inbound and outbound adaptors respectively. The business logic has one or more ports, which are basically a defined set of operations that define how adaptors can interact with business logic as well as how business logic can invoke external systems. The ports through which external systems interact with business logic are called inbound ports, whereas the ports through which business logic interacts with external systems are called outbound ports.

We can summarize the execution flow in a hexagonal architecture in the following two points:

- User interface and REST API adaptors for web and mobile invoke business logic via inbound adaptors
- Business logic invokes external-facing adaptors such as databases and external systems via outbound adaptors

One last but very important point to make about hexagonal architecture is that business logic is made up of modules that are a collection of domain objects. To know more about domain-driven design definitions and patterns, you can read the reference guide written by Eric Evans at `https://domainlanguage.com/wp-content/uploads/2016/05/DDD_Reference_2015-03.pdf`.

Returning to our online bookstore application, the following will be the core modules:

- **Order management**: Managing customer orders, shopping carts, and updates on order progress
- **Customer management**: Managing customer accounts, including sign-up, sign-in, and subscriptions
- **Payment management**: Managing payments
- **Product catalog**: Managing all the products available
- **Shipping**: Managing the delivery of orders
- **Inventory**: Managing up-to-date information on inventory levels

With these in mind, let's draw the hexagonal architecture for this system.

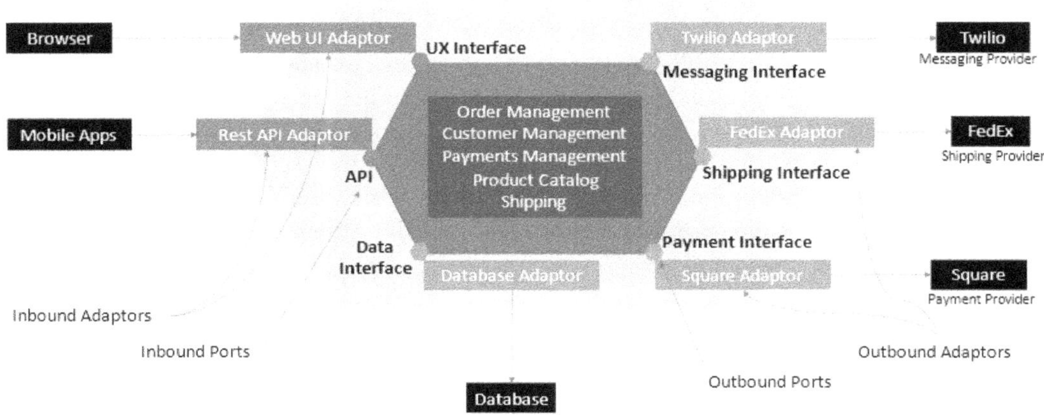

Figure 1.2 – The online book store application monolith

Though the architecture follows hexagonal architecture and some principles of domain-driven design, it is still packaged as one deployable or executable unit, depending on the underlying programming language you are using to write it. For example, if you are using Java, the deployable artifact will be a WAR file, which will then be deployed on an application server.

The monolithic application looks awesome when it's greenfield but nightmarish when it becomes brownfield, in which case it would need to be updated or extended to incorporate new features and changes.

Monolithic architectures are difficult to understand, evolve, and enhance because the code base is big and, with time, gets humongous in size and complexity. This means it takes a long time to make code changes and to ship the code to production. Code changes are expensive and require thorough regression testing. The application is difficult and expensive to scale, and there is no option to allocate dedicated computing resources to individual components of the application. All resources are allocated holistically to the application and are consumed by all parts of it, irrespective of their importance in its execution.

The other issue is lock-in to one technology for the whole code base. What this basically means is that you need to constrain yourself to one or a few technologies to support the whole code base. Code lock-in is detrimental to efficient outcomes, including performance, reliability, as well as the amount of effort required to achieve an outcome. You should be using technologies that are the best fit to solve a problem. For example, you can use TypeScript for the UI, Node.js for the API, Golang for modules needing concurrency or maybe for writing the core modules, and so on. Using a monolithic architecture, you are stuck with technologies you used in the past, which might not be the right fit to solve the current problem.

So, how does *microservices architecture* solve this problem? *Microservices* is an overloaded term, and there are many definitions of it; in other words, there is no single definition of microservices. A few well-known personalities have contributed their own definitions of microservices architecture:

> *The term Microservices architecture has sprung up over the last few years to describe a particular way of designing software applications as suites of independently deployable services. While there is no precise definition of this architectural style, there are certain common characteristics around organization around business capability, automated deployment, intelligence in the endpoints, and decentralized control of languages and data.*
>
> *– Martin Fowler and James Lewis*

The definition was published on `https://martinfowler.com/articles/microservices.html` and is dated March 25, 2014, so you can ignore "sprung up over the last few years" in the description, as microservices architecture has becoming mainstream and pervasive.

Another definition of microservices is from Adam Cockcroft: "*Loosely coupled service-oriented architecture with bounded contexts.*"

In microservices architecture, the term *micro* is a topic of intense debate, and often the question asked is, "*How micro should microservices be?*" or "*How should I decompose my application?*". There is no easy answer to this; you can follow various decomposing strategies by following domain-driven design and decomposing applications into services based on business capability, functionality, the responsibility or concern of each service or module, scalability, bounded context, and blast radius. There are numerous articles and books written on the topic of microservices and decomposition strategies, so I am sure you can find enough to read about strategies for sizing your application in microservices.

Let's get back to the online bookstore application and redesign it using a microservices architecture. The following diagram represents the online bookstore applications built using microservices architecture principles. The individual services are still following hexagonal architecture, and for brevity, we have not represented the inbound and outbound ports and adaptors. You can assume that ports, adaptors, and containers are within the hexagon itself.

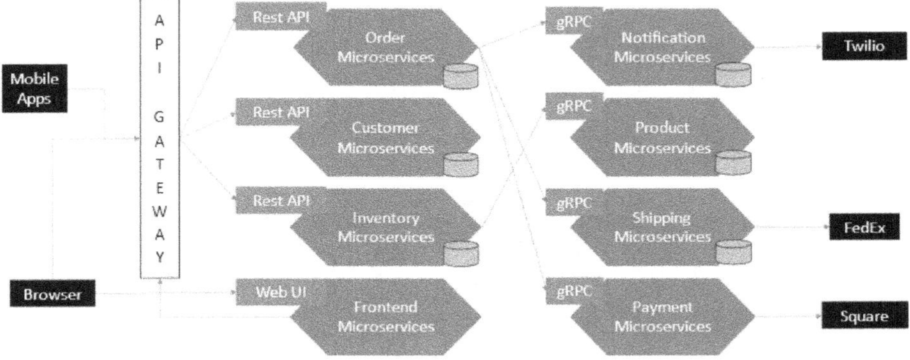

Figure 1.3 – The online bookstore microservices architecture

Microservices architecture provides several benefits over monolithic architecture. Having independent modules segregated based on functionality and decoupled from each other unlocks the monolithic shackles that drag the software development process. Microservices can be built faster at a comparatively lower cost than a monolith and are well adept for continuous deployment processes and, thus, have faster time to production. With microservices architecture, developers can release code to production as frequently as they want. The smaller code base of microservices is easy to understand, and developers only need to understand microservices and not the whole application. Also, multiple developers can work on microservices within the application without any risk of code being overwritten or impacting each other's work. Your application, now made up of microservices, can leverage polyglot programming techniques to deliver performance efficiency, with less effort for more outcomes, and best-of-breed technologies to solve a problem.

Microservices as self-contained independent deployable units provide you with fault isolation and a reduced blast radius – for example, assume that one of the microservices starts experiencing exceptions, performance degradation, memory leakage, and so on. In this case, because the service is deployed as a self-contained unit with its own resource allocation, this problem will not affect other microservices. Other microservices will not get impacted by overconsumption of memory, CPU, storage, network, and I/O.

Microservices are also easier to deploy because you can use varying deployment options, depending on microservices requirements and what is available to you – for example, you can have a set of microservices deployed on a serverless platform and, at the same time, another set on a container platform along with another set on virtual machines. Unlike monolithic applications, you are not bounded by one deployment option.

While microservices provide numerous benefits, they also come with added complexity. This added complexity is because now you have too much to deploy and manage. Not following correct decomposition strategies can also create micro-monoliths that are nightmarish to manage and operate. Another important aspect is communication between microservices. As there will be lots of microservices that need to talk to each other, it is very important that communication between microservices is swift, performant, reliable, resilient, and secure. In the *Getting to know Service Mesh* section, we will dig deeper into what we mean by these terms.

For now, with a good understanding of microservices architecture, it's time to look at Kubernetes, which is also the de facto platform for deploying microservices.

Understanding Kubernetes

When designing and deploying microservices, it is easy to manage a small number of microservices. As the number of microservices grows, so does the complexity of managing them. The following list showcases some of the complexities caused by the adoption of microservices architecture:

- Microservices will have specific deployment requirements in terms of the kind of base operating systems, middleware, database, and compute/memory/storage. Also, the number of microservices will be large, which, in turn, means that you will need to provide resources to every microservice. Moreover, to keep the cost down, you will need to be efficient with the allocation of resources and their utilization.
- Every microservice will have a different deployment frequency. For example, any updates to payment microservices might be on a monthly basis, whereas updates to frontend UI microservices might be on a weekly or daily basis.
- Microservices need to communicate with each other, for which they need to know about each other's existence, and they should have application networking in place to communicate efficiently.
- Developers who are building microservices need to have consistent environments for all stages of the development life cycle so that there are no unknowns, or near-unknowns, about the behavior of microservices when deployed in a production environment.
- There should be a continuous deployment process in place to build and deploy microservices. If you don't have an automated continuous deployment process, then you will need an army of people to support microservices deployments.
- With so many microservices deployed, it is inevitable that there will be failures, but you cannot burden the microservices developer to solve those problems. Cross-cutting concerns such as resiliency, deployment orchestration, and application networking should be easy to implement and should not distract the focus of microservice developers. These cross-cutting concerns should be facilitated by the underlying platform and should not be incorporated into the microservices code.

Kubernetes, also abbreviated as **K8S**, is an open source system that originated from Google. Kubernetes provides automated deployment, scaling, and management of containerized applications. It provides scalability without you needing to hire an army of DevOps engineers. It fits and suits all kinds of complexities – that is, it works on a small scale as well as an enterprise scale. Google, as well as many other organizations, runs a huge number of containers on the Kubernetes platform.

> **Important note**
>
> A **container** is a self-contained deployment unit that contains all code and associated dependencies, including operating system, system, and application libraries packaged together. Containers are instantiated from images, which are lightweight executable packages. A **Pod** is a deployable unit in Kubernetes and is comprised of one or more containers, with each one in the Pod sharing the resources, such as storage and network. A Pod's contents are always co-located and co-scheduled and run in a shared context.

The following are some of the benefits of the Kubernetes platform:

- Kubernetes provides automated and reliable deployments by taking care of rollouts and rollbacks. During deployments, Kubernetes progressively rolls out changes while monitoring microservices' health to ensure that there is no disruption to the processing of a request. If there is a risk to the overall health of microservices, then Kubernetes will roll back the changes to bring the microservices back to a healthy state.
- If you are using the cloud, then different cloud providers have different storage types. When running in data centers, you will be using various network storage types. When using Kubernetes, you don't need to worry about underlying storage, as it takes care of it. It abstracts the complexity of underlying storage types and provides an API-driven mechanism for developers to allocate storage to the containers.
- Kubernetes takes care of DNS and IP allocation for the Pods; it also provides a mechanism for microservices to discover each other using simple DNS conventions. When more than one copy of services is running, then Kubernetes also takes care of load balancing between them.
- Kubernetes automatically takes care of the scalability requirements of Pods. Depending on resource utilization, Pods are automatically scaled up, which means that the number of running Pods is increased, or scaled down, which means that the number of running Pods is reduced. Developers don't have to worry about how to implement scalability. Instead, they just need average utilization of CPU, memory, and various other custom metrics along with scalability limits.
- In a distributed system, failures are bound to happen. Similarly, in microservices deployments, Pods and containers will become unhealthy and unresponsive. Such scenarios are handled by Kubernetes by restarting the failed containers, rescheduling containers to other worker nodes if underlying nodes are having issues, and replacing containers that have become unhealthy.
- As discussed earlier, microservices architecture being resource-hungry is one of its challenges, and a resource should be allocated efficiently and effectively. Kubernetes takes care of that responsibility by maximizing the allocation of resources without impairing availability or sacrificing the performance of containers.

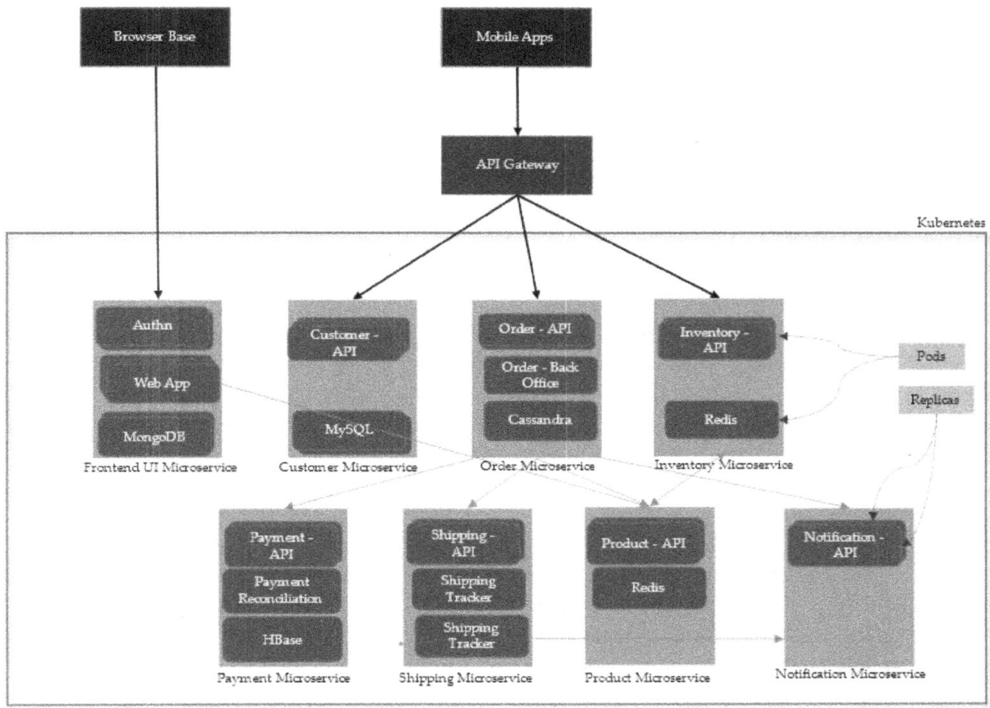

Figure 1.4 – The online bookstore microservice deployed on Kubernetes

The preceding diagram is a visualization of the online bookstore application built using microservices architecture and deployed on Kubernetes.

Getting to know Service Mesh

In the previous section, we read about monolithic architecture, its advantages, and disadvantages. We also read about how microservices solve the problem of scalability and provide flexibility to rapidly deploy and push software changes to production. The cloud makes it easier for an organization to focus on innovation without worrying about expensive and lengthy hardware procurement and expensive CapEx cost. The cloud also facilitates microservices architecture not only by facilitating on-demand infrastructure but also by providing various ready-to-use platforms and building blocks, such as PaaS and SaaS. When organizations are building applications, they don't need to reinvent the wheel every time; instead, they can leverage ready-to-use databases, various platforms including Kubernetes, and **Middleware as a Service** (**MWaaS**).

In addition to the cloud, microservice developers also leverage containers, which makes microservices development much easier by providing a consistent environment and compartmentalization to help achieve modular and self-contained architecture of microservices. On top of containers, the developer should also use a container orchestration platform such as Kubernetes, which simplifies the management of containers and takes care of concerns such as networking, resource allocation, scalability, reliability, and resilience. Kubernetes also helps to optimize the infrastructure cost by providing better utilization of underlying hardware. When you combine the cloud, Kubernetes, and microservices architecture, you have all the ingredients you need to deliver potent software applications that not only do the job you want them to do but also do it cost-effectively.

So, the question on your mind must be, "*Why do I need a Service Mesh?*" or "*Why do I need Service Mesh if I am using the cloud, Kubernetes, and microservices?*" It is a great question to ask and think about, and the answer becomes evident once you have reached a stage where you are confidently deploying microservices on Kubernetes, and then you reach a certain tipping point where networking between microservices just becomes too complex to address by using Kubernetes' native features.

Fallacies of distributed computing

Fallacies of a distributed system are a set of eight assertions made by L Peter Deutsch and others at Sun Microsystems. These assertions are false assumptions often made by software developers when designing distributed applications. The assumptions are that a network is reliable, latency is zero, bandwidth is infinite, the network is secure, the topology doesn't change, there is one administrator, the transport cost is zero, and the network is homogenous.

At the beginning of the *Understanding Kubernetes* section, we looked at the challenges developers face when implementing microservices architecture. Kubernetes provides various features for the deployment of containerized microservices as well as container/Pod life cycle management through declarative configuration, but it falls short of solving communication challenges between microservices. When talking about the challenges of microservices, we used terms such as *application networking* to describe communication challenges. So, let's try to first understand what application networking is and why it is so important for the successful operations of microservices.

Application networking is also a loosely used term; there are various interpretations of it depending on the context it is being used in. In the context of microservices, we refer to application networking as the enabler of distributed communication between microservices. The microservice can be deployed in one Kubernetes cluster or multiple clusters over any kind of underlying infrastructure. A microservice can also be deployed in a non-Kubernetes environment in the cloud, on-premises, or both. For now, let's keep our focus on Kubernetes and application networking within Kubernetes.

Irrespective of where microservices are deployed, you need a robust application network in place for microservices to talk to each other. The underlying platform should not just facilitate communication but also resilient communication. By resilient communication, we mean the kind of communication where it has a large probability of being successful even when the ecosystem around it is in adverse conditions.

Apart from the application network, you also need visibility of the communication happening between microservices; this is also called observability. Observability is important in microservices communication in knowing how the microservices are interacting with each other. It is also important that microservices communicate securely with each other. The communication should be encrypted and defended against man-in-the-middle attacks. Every microservice should have an identity and be able to prove that they are authorized to communicate with other microservices.

So, why Service Meshes? Why can't these requirements be addressed in Kubernetes? The answer lies in Kubernetes architecture and what it was designed to do. As mentioned before, Kubernetes is application life cycle management software. It provides application networking, observability, and security, but at a very basic level that is not sufficient to meet the requirements of modern and dynamic microservices architecture. This doesn't mean that Kubernetes is not modern software. Indeed, it is a very sophisticated and cutting-edge technology, but only for serving container orchestration.

Traffic management in Kubernetes is handled by the Kubernetes network proxy, also called kube-proxy. kube-proxy runs on each node in the Kubernetes cluster. kube-proxy communicates with the Kubernetes API server and gets information about Kubernetes services. Kubernetes services are another level of abstraction to expose a set of Pods as a network service. kube-proxy implements a form of virtual IP for services that sets iptables rules, defining how any traffic for that service will be routed to the endpoints, which are essentially the underlying Pods hosting the application.

To understand it better, let's look at the following example. To run this example, you will need **minikube** and **kubectl** on your computing device. If you don't have this software installed, then I suggest you hold off from installing it, as we will be going through the installation steps in *Chapter 2*.

We will create a Kubernetes deployment and service by following the example in `https://minikube.sigs.k8s.io/docs/start/`:

```
$ kubectl create deployment hello-minikube --image=k8s.gcr.io/
echoserver:1.4
deployment.apps/hello-minikube created
```

We just created a deployment object named `hello-minikube`. Let's execute the `kubectl describe` command:

```
$ kubectl describe deployment/hello-minikube
Name:                   hello-minikube
…….
Selector:               app=hello-minikube
…….
Pod Template:
  Labels:  app=hello-minikube
  Containers:
```

```
    echoserver:
     Image:        k8s.gcr.io/echoserver:1.4
     ..
```

From the preceding code block, you can see that a Pod has been created, containing a container instantiated from the `k8s.gcr.io/echoserver:1.4` image. Let's now check the Pods:

```
$ kubectl get po
hello-minikube-6ddfcc9757-
lq66b   1/1     Running   0          7m45s
```

The preceding output confirms that a Pod has been created. Now, let's create a service and expose it so that it is accessible on a cluster-internal IP on a static port, also called `NodePort`:

```
$ kubectl expose deployment hello-minikube --type=NodePort
--port=8080
service/hello-minikube exposed
```

Let's describe the service:

```
$ kubectl describe services/hello-minikube
Name:                     hello-minikube
Namespace:                default
Labels:                   app=hello-minikube
Annotations:              <none>
Selector:                 app=hello-minikube
Type:                     NodePort
IP:                       10.97.95.146
Port:                     <unset>  8080/TCP
TargetPort:               8080/TCP
NodePort:                 <unset>  31286/TCP
Endpoints:                172.17.0.5:8080
Session Affinity:         None
External Traffic Policy:  Cluster
```

From the preceding output, you can see that a Kubernetes service named `hello-minikube` has been created and is accessible on port `31286`, also called `NodePort`. We also see that there is an `Endpoints` object with the `172.17.0.5:8080` value. Soon, we will see the connection between `NodePort` and `Endpoints`.

Let's dig deeper and look at what is happening to iptables. If you would like to see what the preceding service returns, then you can simply type `minikube service`. We are using macOS, where minikube is running itself as a VM. We will need to use `ssh` on minikube to see what's happening with iptables. On Unix host machines, the following steps are not required:

```
$ minikube ssh
```

Let's check the iptables:

```
$ sudo iptables -L KUBE-NODEPORTS -t nat
Chain KUBE-NODEPORTS (1 references)
target     prot opt source               destination
KUBE-MARK-
MASQ  tcp  --  anywhere             anywhere             /*
default/hello-minikube */ tcp dpt:31286
KUBE-SVC-MFJHED5Y2WHWJ6HX
  tcp  --  anywhere             anywhere             /*
default/hello-minikube */ tcp dpt:31286
```

We can see that there are two iptables rules associated with the `hello-minikube` service. Let's look further into these iptables rules:

```
$ sudo iptables -L KUBE-MARK-MASQ -t nat
Chain KUBE-MARK-MASQ (23 references)
target     prot opt source               destination
MARK       all  --  anywhere             anywhere             M
ARK or 0x4000
$ sudo iptables -L KUBE-SVC-MFJHED5Y2WHWJ6HX -t nat
Chain KUBE-SVC-MFJHED5Y2WHWJ6HX (2 references)
target     prot opt source               destination
KUBE-SEP-EVPNTXRIBDBX2HJK
  all  --  anywhere             anywhere             /*
default/hello-minikube */
```

The first rule, `KUBE-MARK-MASQ`, is simply adding an attribute called `packet mark`, with a `0x400` value for all traffic destined for port `31286`.

The second rule, KUBE-SVC-MFJHED5Y2WHWJ6HX, is routing the traffic to another rule, KUBE-SEP-EVPNTXRIBDBX2HJK. Let's look further into it:

```
$ sudo iptables -L KUBE-SEP-EVPNTXRIBDBX2HJK -t nat
Chain KUBE-SEP-EVPNTXRIBDBX2HJK (1 references)
target     prot opt source               destination
KUBE-MARK-
MASQ  all  --  172.17.0.5          anywhere             /*
default/hello-minikube */
DNAT       tcp  --  anywhere             anywhere
/* default/hello-minikube */ tcp to:172.17.0.5:8080
```

Note that this rule has a **destination network address translation** (**DNAT**) to 172.17.0.5:8080, which is the address of the endpoints when we created the service.

Let's scale the number of Pod replicas:

```
$ kubectl scale deployment/hello-minikube --replicas=2
deployment.apps/hello-minikube scaled
```

Describe the service to find any changes:

```
$ kubectl describe services/hello-minikube
Name:                     hello-minikube
Namespace:                default
Labels:                   app=hello-minikube
Annotations:              <none>
Selector:                 app=hello-minikube
Type:                     NodePort
IP:                       10.97.95.146
Port:                     <unset>  8080/TCP
TargetPort:               8080/TCP
NodePort:                 <unset>  31286/TCP
Endpoints:                172.17.0.5:8080,172.17.0.7:8080
Session Affinity:         None
External Traffic Policy:  Cluster
```

Note that the value of the endpoint has changed; let's also describe the `hello-minikube` endpoint:

```
$ kubectl describe endpoints/hello-minikube
Name:         hello-minikube
...
Subsets:
  Addresses:          172.17.0.5,172.17.0.7
  NotReadyAddresses:  <none>
  Ports:
    Name     Port  Protocol
    ----     ----  --------
    <unset>  8080  TCP
```

Note that the endpoint is now also targeting `172.17.0.7` along with `172.17.0.5`. `172.17.0.7`, the new Pod that has been created as a result of increasing the number of replicas to `2`.

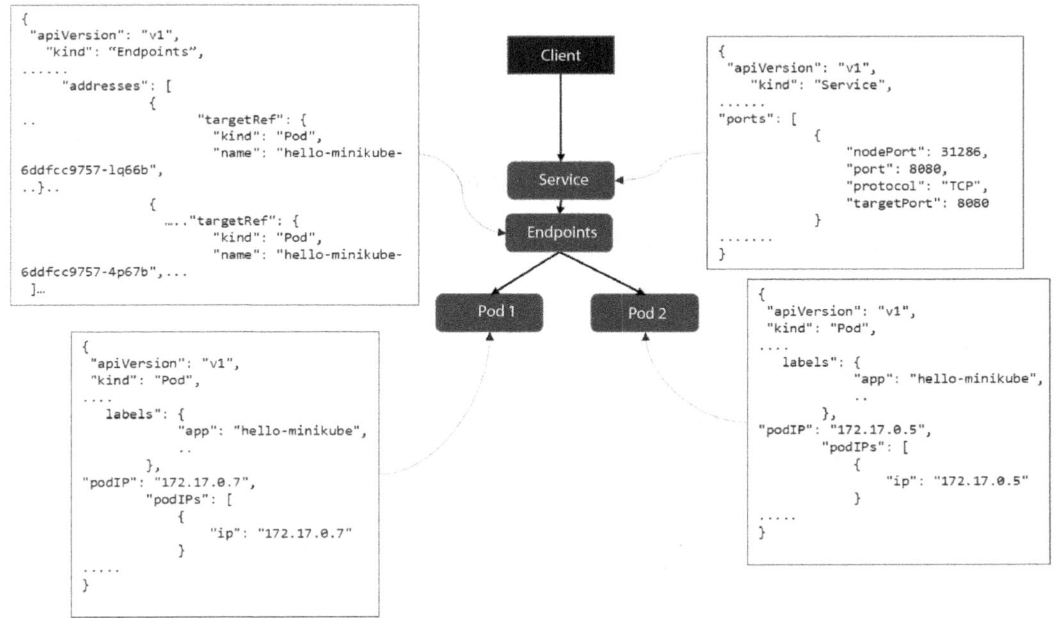

Figure 1.5 – Service, endpoints, and Pods

Let's check the iptables rules now:

```
$ sudo iptables -t nat -L KUBE-SVC-MFJHED5Y2WHWJ6HX
Chain KUBE-SVC-MFJHED5Y2WHWJ6HX (2 references)
target     prot opt source               destination
KUBE-SEP-EVPNTXRIBDBX2HJK  all  --  anywhere
             anywhere            /* default/hello-minikube */
statistic mode random probability 0.50000000000
KUBE-SEP-NXPGMUBGGTRFLABG  all  --  anywhere
             anywhere            /* default/hello-minikube */
```

You will find that an additional rule, `KUBE-SEP-NXPGMUBGGTRFLABG`, has been added, and because of the statistic mode random probability, `0.5`, each packet handled by `KUBE-SVC-MFJHED5Y2WHWJ6HX` is then distributed 50–50 between `KUBE-SEP-EVPNTXRIBDBX2HJK` and `KUBE-SEP-NXPGMUBGGTRFLABG`.

Let's also quickly examine the new chain added after we changed the number of replicas to `2`:

```
$ sudo iptables -t nat -L KUBE-SEP-NXPGMUBGGTRFLABG
Chain KUBE-SEP-NXPGMUBGGTRFLABG (1 references)
target     prot opt source               destination
KUBE-MARK-
MASQ  all  --  172.17.0.7           anywhere             /*
default/hello-minikube */
DNAT       tcp  --  anywhere             anywhere
/* default/hello-minikube */ tcp to:172.17.0.7:8080
```

Note that another `DNAT` entry has been added for `172.17.0.7`. So, essentially, the new chain and the previous one are now routing traffic to corresponding Pods.

So, if we summarize everything, kube-proxy runs on every Kubernetes node and keeps a watch on service and endpoint resources. Based on service and endpoint configurations, kube-proxy then creates iptables rules to take care of routing data packets between the consumer/client and the Pod.

The following diagram depicts the creation of iptables rules via kube-proxy and how consumers connect with Pods.

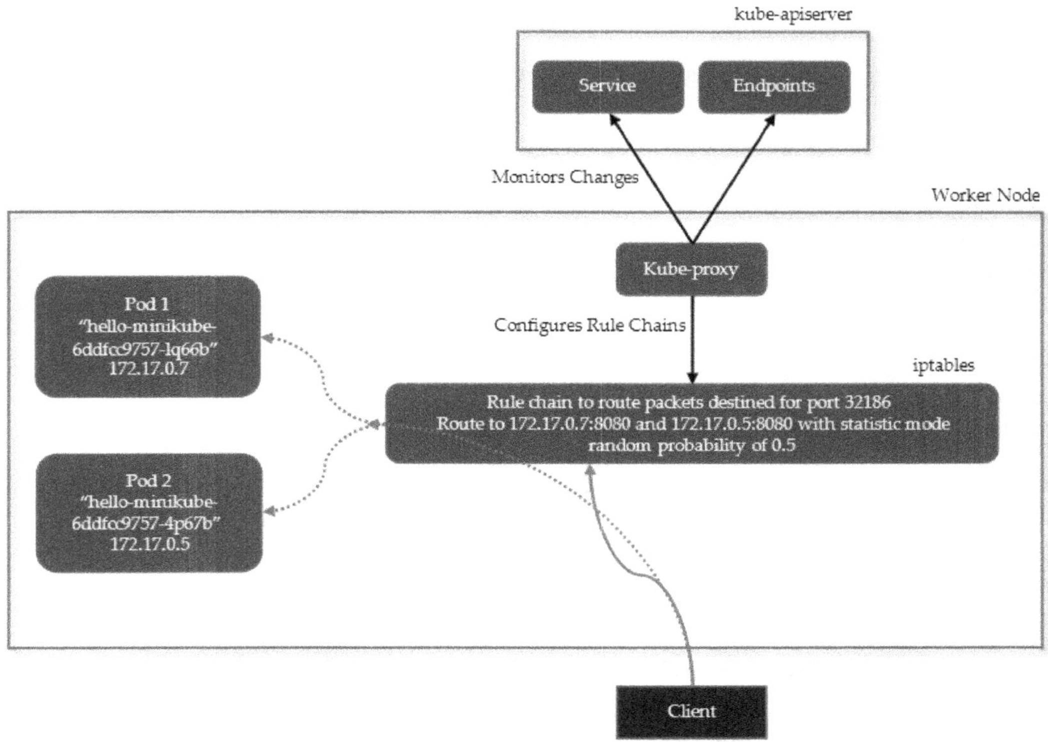

Figure 1.6 – The client connecting to a Pod based on the iptables rule chain

kube-proxy can also run in another mode called **IP Virtual Server** (**IPVS**). For ease of reference, here's how this term is defined on the official Kubernetes website:

> *In IPVS mode, kube-proxy watches Kubernetes Services and Endpoints calls net link interface to create IPVS rules accordingly and synchronizes IPVS rules with Kubernetes Services and Endpoints periodically. This control loop ensures that IPVS status matches the desired state. When accessing a service, IPVS directs traffic to one of the backend Pods.*

> **Tip**
>
> To find out the mode in which kube-proxy is running, you can use `$ curl localhost:10249/proxyMode`. On Linux, you can curl directly, but, on macOS, you need to curl from the minikube VM itself.

So, what is wrong with kube-proxy using iptables or IPVS?

kube-proxy doesn't provide any fine-grained configuration; all settings are applied to all traffic on that node. kube-proxy can only do simple TCP, UDP, and SCTP stream forwarding or round-robin TCP, UDP, and SCTP forwarding across a set of backends. As the number of Kubernetes services grows, so does the number of rulesets in iptables. As the iptables rules are processed sequentially, it causes performance degradation with growth in microservice numbers. Also, iptables only supports the use of simple probability to support traffic distribution, which is very rudimentary. Kubernetes delivers a few other tricks but not enough to cater to resilient communication between microservices. For microservice communication to be resilient, you need more than iptables-based traffic management.

Let's now talk about a couple of capabilities required to have resilient, fault-tolerant communication.

Retry mechanism, circuit breaking, timeouts, and deadlines

If one Pod is not functioning, then the traffic should automatically be sent to another Pod. Also, a retry needs to be done under constraints so as to not make the communication worse. For example, if a call fails, then maybe the system needs to wait before retrying. If a retry is not successful, then maybe it's better to increase the wait time. Even then. If it is not successful, maybe it's worth abandoning retry attempts and breaking the circuit for subsequent connection.

Circuit breaking is a mechanism that usually involves an electric circuit breaker. When there is a fault in a system where it is not safe to operate, the electric circuit breaker automatically trips. Similarly, consider microservices communications where one service is calling another service and the called service is not responding, is responding so slowly that it is detrimental to the calling service, or the occurrence of this behavior has reached a predefined threshold. In such a case, it is better to trip (stop) the circuit (communication) so that when the calling service (downstream) calls the underlying service (upstream), the communication fails straight away. The reason it makes sense to stop the downstream system from calling the upstream system is to stop resources such as network bandwidth, thread, IO, CPU, and memory from being wasted on an activity that has a significantly high probability of failing. Circuit breaking doesn't resolve the communication problem; instead, it stops it from jumping boundaries and impacting other systems. Timeouts are also important during microservices communication so that downstream services wait for a response from the upstream system for a duration in which the response would be valid or worth waiting for. Deadlines build further on timeouts; you can see them as timeouts for the whole request, not just one connection. By specifying a deadline, a downstream system tells the upstream system about the overall maximum time permissible for processing the request, including subsequent calls to other upstream microservices involved in processing the request.

Important note

In a microservices architecture, downstream systems are the ones that rely on the upstream system. If service A calls service B, then service A will be called downstream and service B will be called upstream. When drawing a north–south architecture diagram to show a data flow between A and B, you will usually draw A at the top with an arrow pointing down toward B, which makes it confusing to call A downstream and B upstream. To make it easy to remember, you can draw the analogy that *a downstream system depends on an upstream system*. This way, microservice A depends on microservice B; hence, A is downstream and B is upstream.

Blue/green and canary deployments

Blue/green deployments are scenarios where you would like to deploy a new (green) version of a service side by side with the previous/existing (blue) version of a service. You make stability checks to ensure that the green environment can handle live traffic, and if it can, then you transfer the traffic from a blue to a green environment.

Blue and green can be different versions of a service in a cluster or services in an independent cluster. If something goes wrong with the green environment, you can switch the traffic back to the blue environment. Transfer of traffic from blue to green can also happen gradually (canary deployment) in various ways – for example, at a certain rate, such as 90:10 in the first 10 minutes, 70:30 in the next 10 minutes, 50:50 in the next 20 minutes, and 0:100 after that. Another example can be to apply the previous example to certain traffic, such as transferring the traffic at a previous rate with all traffic with a certain HTTP header value – that is, a certain class of traffic. While in blue/green deployment you deploy like-for-like deployments side by side, in canary deployment you can deploy a subset of what you deploy in green deployment. These features are difficult to achieve in Kubernetes due to it not supporting the fine-grained distribution of traffic.

The following diagram depicts blue/green and canary deployments.

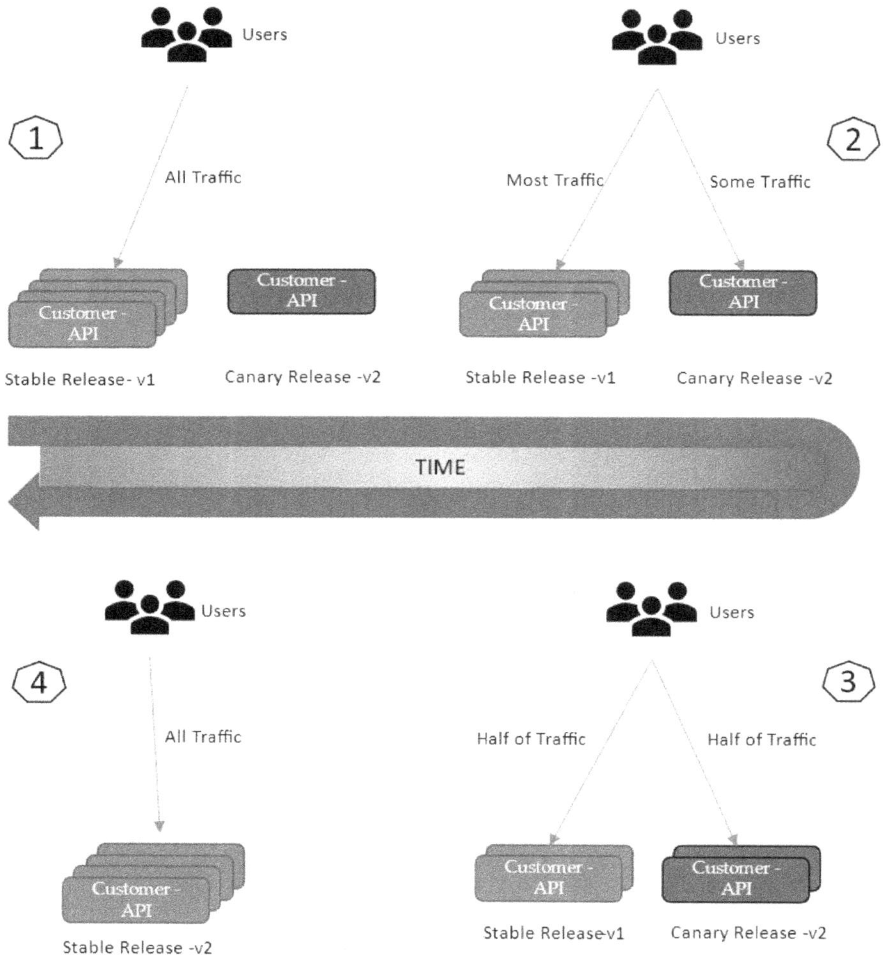

Figure 1.7 – Blue/green deployment

To handle concerns such as blue/green and canary deployments, we need something that can handle the traffic at layer 7 rather than layer 4. There are frameworks such as Netflix **Open Source Software** (**OSS**) and a few others to solve distributed system communication challenges, but in doing so, they shift the responsibility of solving application networking challenges to microservice developers. Solving these concerns in application code is not only expensive and time-consuming but also not conducive to the overall outcome, which is to deliver business outcomes. Frameworks and libraries such as Netflix OSS are written in certain programming languages, which then constrain developers to use only compatible languages for building microservices. These constrain developers to use technologies and programming languages supported by a specific framework, going against the polyglot concept.

What is needed is a kind of proxy that can work alongside an application without requiring the application to have any knowledge of the proxy itself. The proxy should not just proxy the communication but also have intricate knowledge of the services doing the communication, along with the context of the communication. The application/service can then focus on business logic and let the proxy handle all concerns related to communication with other services. `ss` is one such proxy working at layer 7, designed to run alongside microservices. When it does so, it forms a transparent communication mesh with other Envoy proxies running alongside respective microservices. The microservice communicates only with nvoy as localhost, and Envoy takes care of the communication with the rest of the mesh. In this communication model, the microservices don't need to know about the network. Envoy is extensible because it has a pluggable filter chain mechanism for network layers 3, 4, and 7, allowing new filters to be added as needed to perform various functions, such as TLS client certificate authentication and rate limiting.

So, how are Service Meshes related with Envoy? A service Mesh is an infrastructure responsible for application networking. The following diagram depicts the relationship between the Service Mesh control plane, the Kubernetes API server, the Service Mesh sidecar, and other containers in the Pod.

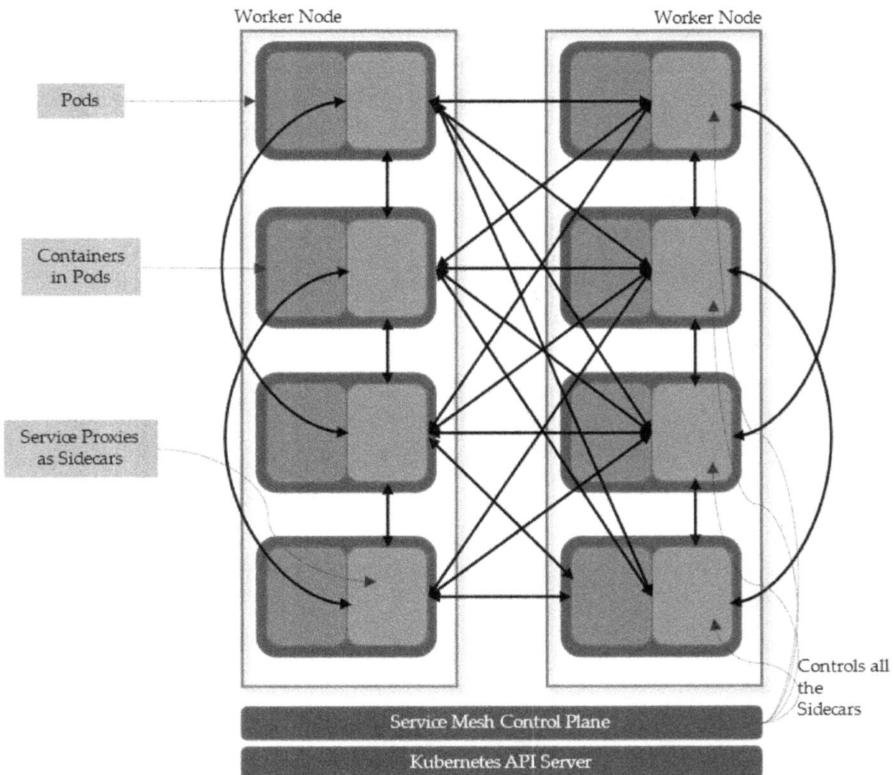

Figure 1.6 – Service Mesh sidecars, data, and the control plane

A Service Mesh provides a data plane, which is basically a collection of application-aware proxies such as Envoy that are then controlled by a set of components called the control plane. In a Kubernetes-based environment, the service proxies are inserted as a sidecar to Pods without needing any modification to existing containers within the Pod. A Service Mesh can be added to Kubernetes and traditional environments, such as virtual machines, as well. Once added to the runtime ecosystem, the Service Mesh takes care of the application networking concerns we discussed earlier, such as load balancing, timeouts, retries, canary and blue-green deployment, security, and observability.

Summary

In this chapter, we started with monolithic architecture and discussed the drag it causes in being able to expand with new capabilities as well as time to market. Monolithic architectures are brittle and expensive to change. We read about how microservices architecture breaks that inertia and provides the momentum required to meet the ever-changing and never-ending appetite of digital consumers. We also saw how microservices architecture is modular, with every module being self-contained, and can be built and deployed independently of each other. Applications built using microservices architecture make use of best-of-breed technologies that are suitable for solving individual problems.

We then discussed the cloud and Kubernetes. The cloud provides utility-style computing with a pay-as-you-go model. Common cloud services include IaaS, PaaS, and SaaS. The cloud provides access to all infrastructure you may need without you needing to worry about the procurement of expensive hardware, data center costs, and so on. The cloud also provides you with software building blocks with which you can reduce your software development cycle. In microservices architecture, containers are the way to package application code. They provide consistency of environments and isolation between services, solving the *noisy neighbor* problem.

Kubernetes, on the other hand, makes the usage of containers easier by providing container life cycle management and solving many of the challenges of running containers in production. As the number of microservices grows, you start facing challenges regarding traffic management between microservices. Kubernetes does provide traffic management based on kube-proxy and iptables-based rules, but it falls short of providing application networking.

We finally discussed Service Mesh, an infrastructure layer on top of Kubernetes that is responsible for application networking. The way it works is by providing a data plane, which is basically a collection of application-aware service proxies, such as Envoy, that are then controlled by a set of components called the control plane.

In the next chapter, we will read about Istio, one of the most popular Service Mesh implementations.

2
Getting Started with Istio

In the previous chapter, we discussed monolithic architecture and its drawbacks. We discussed microservice architecture and how it provides modularity to large complex applications. Microservice architectures are scalable, easier to deploy, resilient, and fault-tolerant via isolation and modularization, leveraging cloud containers and Kubernetes. Containers are the default packaging format for cloud-native applications, and Kubernetes is the de facto platform for container life cycle management and deployment orchestration. The ability of microservices to be distributed, highly scalable, and work in parallel with other microservices amplifies the communication challenges between microservices, and also operational challenges such as visibility in the communication and execution of microservices.

Microservices need to have secure communication with each other to avoid exploitation and attacks such as man-in-the-middle attacks. To solve such challenges in a cost-efficient and performant manner, there is a need for an application networking infrastructure, also called a Service Mesh. Istio is one such implementation of the Service Mesh that has been developed and supported by some great organizations, including Google, Red Hat, VMware, IBM, Lyft, Yahoo, and AT&T.

In this chapter, we will install and run Istio, and while doing that, we will go through its architecture and its various components as well. This chapter will help you understand the difference between Istio and other Service Mesh implementations. By the end, you should be able to configure and set up your environment and then install Istio, after getting a good understanding of how installation works. Once installed, you will then enable Istio sidecar injection to a sample application that comes along with Istio installation. We will take a step-by-step look at pre- and post-enablement of Istio for a sample application and get an idea of how Istio works.

We will be doing this by exploring the following topics:

- Why is Istio the most popular Service Mesh?
- Preparation of your workstation environment to install and run Istio
- Installing Istio
- Installing observability tools
- An introduction to Istio architecture

Why is Istio the most popular Service Mesh?

Istio stands for the Greek word *ιστίο*, pronounced as *Iss-tee-oh*. Istio means *sail*, which is a non-bending, non-compressing structure made of fabric or similar. It propels sailing ships via the lift and drag produced by the wind. What made the initial contributors select Istio as the name probably has something to do with the naming of Kubernetes, which also has a Greek origin, pronounced as *koo-burr-net-eez* and written as *κυβερνήτης*. Kubernetes means *helmsman* – that is, the person standing at the helm of a ship and steering it.

Istio is an open source services mesh distributed under Apache License 2.0. It is platform-independent, meaning it is independent of underlying Kubernetes providers. It also supports not only Kubernetes but also non-Kubernetes environments such as virtual machines. Having said that, Istio development is much more mature for the Kubernetes environment and is adapting and evolving very quickly for other environments. Istio has a very mature development community, a strong user base, and is highly extensible and configurable, providing solid operational control of traffic and security within a Service Mesh. Istio also provides behavioral insights using advanced and fine-grained metrics. It supports WebAssembly, which is very useful for extensibility and tailoring for specific requirements. Istio also offers support and easy configuration for multi-cluster and multi-network environments.

Exploring alternatives to Istio

There are various other alternatives to Istio, all with their own pros and cons. Here, we will list a few of the other Service Mesh implementations available.

Kuma

At the time of writing (2022), **Kuma** is a **Cloud Native Computing Foundation** (**CNCF**) sandbox project and was originally created by Kong Inc., the company that also provides the Kong API management gateway both in open source and commercial variants. Kuma is advertised by Kong Inc. as a modern distributed control plane with bundled Envoy proxy integration. It supports multi-cloud and multi-zone connectivity for highly distributed applications. The Kuma data plane is composed of Envoy proxies, which are then managed by Kuma control planes, and it supports workloads deployed on not only Kubernetes but also virtual machines and bare-metal environments. Kong Inc also provides an enterprise Service Mesh offering called Kong Mesh, which extends CNCF's Kuma and Envoy.

Linkerd

Linkerd was originally created by Buoyant, Inc. but was later made open source, and it is now licensed under Apache V2. Buoyant, Inc. also provides a managed cloud offering of Linkerd, as well as an enterprise support offering for customers who want to run Linkerd themselves but need enterprise support. Linkerd makes running services easier and safer by providing runtime debugging, observability, reliability, and security. Like Istio, you don't need to change your application source code; instead, you install a set of ultralight transparent Linkerd2-proxy next to every service.

The Linkerd2-proxy is a micro-proxy written in Rust and deployed as a sidecar in the Pod along with the application. Linkerd proxies have been written specifically for Service Mesh use cases and are arguably faster than Envoy, which is used as a sidecar in Istio and many other Service Mesh implementations like Kuma. Envoy is a great proxy but designed for multiple use cases – for example, Istio uses Envoy as an Ingress and also Egress gateway, as well as a sidecar simultaneously to go along with applications. Many Linkerd implementations use Linkerd as a Service Mesh and Envoy-based Ingress controllers.

Consul

Consul is a Service Mesh solution from Hashicorp; it is open source but also comes with a cloud and enterprise support offering from Hashicorp. Consul can be deployed on Kubernetes as well as VM-based environments. On top of the Service Mesh, Consul also provides all functionality for service catalogs, TLS certificates, and service-to-service authorizations. The data plane of Consul provides two options; the user can either choose an Envoy-based sidecar model similar to Istio, or native integration via Consul Connect SDKs, which takes away the need to inject a sidecar and provides better performance than Envoy proxies. Another difference is that you need to run a consul agent as a daemon on every worker node in the Kubernetes cluster and every node in non-Kubernetes environments.

AWS App Mesh

App Mesh is a Service Mesh offering from AWS and, of course, is available for workloads deployed in AWS on **Elastic Container Service (ECS)**, Elastic Container Service for Kubernetes, or self-managed Kubernetes clusters running in AWS. Like Istio, App Mesh also uses Envoy as a sidecar proxy in the Pod, while the control plane is provided as a managed service by AWS, similar to EKS. App Mesh provides integration with various other AWS services such as Amazon Cloudwatch and AWS X-Ray.

OpenShift Service Mesh

Red Hat OpenShift Service Mesh is based on Istio; in fact, Red Hat is also a contributor to Istio open source projects. The offering is bundled with Jaeger for distributed tracing and Kiali for visualizing the mesh, viewing configuration, and traffic monitoring. As with other products from Red Hat, you can buy enterprise support for OpenShift Service Mesh.

F5 NGINX Service Mesh

NGINX is part of F5, and hence, its Service Mesh offering is called **F5 NGINX Service Mesh**. It uses NGINX Ingress controller with NGINX App Protect to secure the traffic at the edge and then route to the mesh using Ingress controllers. NGINX Plus is used as a sidecar to the application, providing seamless and transparent load balancing, reverse proxy, traffic routing, and encryption. Metrics collection and analysis are performed using OpenTracing and Prometheus, while inbuilt Grafana dashboards are provided for the visualization of Prometheus metrics.

This briefly covers Service Mesh implementation, and we will cover some of them in greater depth in *Appendix A*. For now, let's return our focus to Istio. We will read more about the benefits of Istio in the upcoming sections and the rest of the book, but let's first get things going by installing Istio and enabling it for an application that is packaged along with Istio.

Preparing your workstation for Istio installation

We will be using minikube for installing and playing with Istio in the first few chapters. In later chapters, we will install Istio on AWS EKS to mimic real-life scenarios. First, let's prepare your laptop/desktop with minikube. If you already have minikube installed in your environment, it is strongly recommended to upgrade to the latest version.

If you don't have minikube installed, then follow the instructions to install minikube. minikube is a local Kubernetes installed on your workstation that makes it easy for you to learn and play with Kubernetes and Istio, without needing a contingent of computers to install a Kubernetes cluster.

System specifications

You will need Linux or macOS or Windows. This book will primarily follow macOS as the target operating system. Where there is a big difference in commands between Linux and macOS, you will find corresponding steps/commands in the form of little notes. You will need at least two CPUs, 2 GB of available RAM, and either Docker Desktop (if macOS or Windows) or Docker Engine for Linux. If you don't have Docker installed, then just follow the instructions at `https://docs.docker.com/` to install Docker on your computer, based on the respective operating system.

Installing minikube and the Kubernetes command-line tool

We will be using **Homebrew** to install minikube. However, if you don't have Homebrew installed, you can install Homebrew using the following command:

```
$/bin/bash -c "$(curl -fsSL https://raw.githubusercontent.com/
Homebrew/install/HEAD/install.sh)"
```

Let's get started:

1. Install minikube using `brew install minikube`:

```
$ brew install minikube
Running `brew update --preinstall`...
..
==> minikube cask is installed, skipping link.
==> Caveats
```

```
Bash completion has been installed to:
  /usr/local/etc/bash_completion.d
==> Summary
  /usr/local/Cellar/minikube/1.25.1: 9 files, 70.3MB
==> Running `brew cleanup minikube`...
```

Once installed, create a symlink to the newly installed binary in the Homebrew `Cellar` folder:

```
$ brew link minikube
Linking /usr/local/Cellar/minikube/1.25.1... 4 symlinks
created.
$ which minikube
/usr/local/bin/minikube
$ ls -la /usr/local/bin/minikube
lrwxr-xr-x  1 arai  admin  38 22 Feb 22:12 /usr/local/
bin/minikube -> ../Cellar/minikube/1.25.1/bin/minikube
```

To test the installation, use the following command to find the minikube version:

```
$ minikube version
minikube version: v1.25.1
commit: 3e64b11ed75e56e4898ea85f96b2e4af0301f43d
```

Attention, Linux users!

If you are installing on Linux, you can use the following commands to install minikube:

```
$ curl -LO https://storage.googleapis.com/minikube/releases/
latest/minikube-linux-amd64
$ sudo install minikube-linux-amd64 /usr/local/bin/minikube.
```

2. The next step is to install kubectl if you do not have it already installed on your machine.

   kubectl is a short form of the Kubernetes command-line tool and is pronounced as *kube-control*. kubectl allows you to run commands against Kubernetes clusters. You can install kubectl on Linux, Windows, or macOS. The following steps install kubectl on macOS using Brew:

   ```
   $ brew install kubectl
   ```

   You can use the following steps to install kubectl on Debian-based machines:

   I. `sudo apt-get update`

   II. `sudo apt-get install -y apt-transport-https ca-certificates curl`

III. `sudo curl -fsSLo /usr/share/keyrings/kubernetes-archive-keyring.gpg https://packages.cloud.google.com/apt/doc/apt-key.gpg`

IV. `echo "deb [signed-by=/usr/share/keyrings/kubernetes-archive-keyring.gpg] https://apt.kubernetes.io/ kubernetes-xenial main" | sudo tee /etc/apt/sources.list.d/kubernetes.list`

V. `echo "deb [signed-by=/usr/share/keyrings/kubernetes-archive-keyring.gpg] https://apt.kubernetes.io/ kubernetes-xenial main" | sudo tee /etc/apt/sources.list.d/kubernetes.list`

VI. `sudo apt-get update`

VII. `sudo apt-get install -y kubectl`

The following steps can be used to install kubectl on Red Hat machines:

I. `cat <<EOF | sudo tee /etc/yum.repos.d/kubernetes.repo`

II. `[kubernetes]`

III. `name=Kubernetes`

IV. `baseurl=https://packages.cloud.google.com/yum/repos/kubernetes-el7-x86_64`

V. `enabled=1`

VI. `gpgcheck=1`

VII. `repo_gpgcheck=1`

VIII. `gpgkey=https://packages.cloud.google.com/yum/doc/yum-key.gpg https://packages.cloud.google.com/yum/doc/rpm-package-key.gpg`

IX. `ckages.cloud.google.com/yum/doc/rpm-package-key.gpg`

X. `EOF`

XI. `sudo yum install -y kubectl`

You have now all that you need to run Kubernetes locally, so go ahead and type the following command. Make sure you are logged in as a user with administrative access.

You can use `minikube start` with the Kubernetes version as follows:

```
$ minikube start --kubernetes-version=v1.23.1
  minikube v1.25.1 on Darwin 11.5.2
  Automatically selected the hyperkit driver
..
```

```
  Done! kubectl is now configured to use the "minikube" cluster
and "default" namespace by default
```

You can see in the console output that minikube is using the HyperKit driver. **HyperKit** is an open source hypervisor used on macOS. We could have also explicitly specified to minikube to use the Hyperkit driver by passing –`driver=hyperkit`.

For Linux users

For Linux, you can use `minikube start --driver=docker`. In this case, minikube will run as a Docker container. For Windows, you can use `minikube start -driver=virtualbox`. To avoid typing `--driver` during every minikube start, you can configure the default driver by using `minikube config set driver DRIVERNAME`, where `DRIVERNAME` can be either Hyperkit, Docker, or VirtualBox.

You can verify that kubectl is working properly and that minikube has also started properly by using the following:

```
$ kubectl cluster-info
Kubernetes control plane is running at
https://192.168.64.6:8443
CoreDNS is running at https://192.168.64.6:8443/api/v1/
namespaces/kube-system/services/kube-dns:dns/proxy
```

In the output, you can see that both the Kubernetes control plane and the DNS servers are running. This concludes the installation of minikube and `kubernetes-cli`. You now have a locally running Kubernetes cluster and a means to communicate with it via kubectl.

Installing Istio

This section is the one you must have been eagerly waiting to read. The wait is over, and you are all set to install Istio. Just follow the instructions provided.

The first step is to download Istio from `https://github.com/istio/istio/releases`. You can download using `curl` as well with the following command. It is a good idea to make a directory where you want to download the binaries and run the following command from within that directory. Let's name that directory `ISTIO_DOWNLOAD`, from which we can run following commands:

```
$ curl -L https://istio.io/downloadIstio | sh -
Downloading istio-1.13.1 from https://github.com/istio/istio/
releases/download/1.13.1/istio-1.13.1-osx.tar.gz ...
Istio 1.13.1 Download Complete!
```

The preceding command downloads the latest version of Istio into the `ISTIO_DOWNLOAD` location. If we dissect this command, it has two parts:

```
$ curl -L https://istio.io/downloadIstio
```

The first part of the command downloads a script from `https://raw.githubusercontent.com/istio/istio/master/release/downloadIstioCandidate.sh` (the location might change), and the second part of the script is then fed to `sh` for execution. The scripts analyze the processor architecture and operating system and, based on that, decide what are the appropriate values of the Istio version (`ISTIO_VERSION`), the operating system (`OSEXT`), and the processor architecture (`ISTIO_ARCH`). The script then populates these values into the following URL, `https://github.com/istio/istio/releases/download/${ISTIO_VERSION}/istio-${ISTIO_VERSION}-${OSEXT}-${ISTIO_ARCH}.tar.gz`, and then downloads the `gz` file and decompresses it.

Let's investigate what has been downloaded into the `ISTIO_DOWNLOAD` location:

```
$ ls
istio-1.13.1
$ ls istio-1.13.1/
LICENSE  README.md bin  manifest.yaml manifests samples  tools
```

The following is a brief description of the folders:

- `bin` contains `istioctl`, also called `Istio-control`, which is the Istio command line to debug and diagnose Istio, as well as creating, listing, modifying, and deleting configuration resources.
- `samples` contains a sample application that we will be using for learning.
- `manifest` has Helm charts, which you don't need to worry about for now. They have relevance when we want the installation process to pick up the charts from `manifest` rather than the default ones.

Since we will be making use of `istioctl` to perform the installation, let's add it to the executable path:

```
$ pwd
/Users/arai/istio/istio-1.13.1
$ export PATH=$PWD/bin:$PATH
$ istioctl version
no running Istio pods in "istio-system"
1.13.1
```

We are one command away from installing Istio. Go ahead and type in the following command to complete the installation:

```
$ istioctl install --set profile=demo
This will install the Istio 1.13.1 demo profile with ["Istio
core" "Istiod" "Ingress gateways" "Egress gateways"] components
into the cluster. Proceed? (y/N) y
✓ Istio core installed
✓ Istiod installed
✓ Egress gateways installed
✓ Ingress gateways installed
✓ Installation complete
Making this installation the default for injection and
validation.
Thank you for installing Istio 1.13.
```

Tip

You can pass `-y` to avoid the (Y/N) question. Just use `istioctl install --set profile=demo -y`.

Viola! You have successfully completed the installation of Istio, including platform setup, in eight commands. If you have been using minikube and kubectl, then hopefully you should have been able to install in three commands. If you have installed this on an existing minikube setup, then it is advisable at this stage to install Istio on a new cluster, rather than an existing one with your other applications.

Let's look at what has been installed. We'll start first by analyzing the namespaces:

```
$ kubectl get ns
NAME              STATUS   AGE
default           Active   19h
istio-system      Active   88m
kube-node-lease   Active   19h
kube-public       Active   19h
kube-system       Active   19h
```

We can see that the installation has created a new namespace called `istio-system`.

Let's check what Pods and Services are in the `istio-system` namespace:

```
$ kubectl get pods -n istio-system
NAME                               READY   STATUS    RESTARTS   AGE
pod/istio-egressgateway-76c96658fd-
pgfbn    1/1     Running   0          88m
pod/istio-ingressgateway-569d7bfb4-
8bzww    1/1     Running   0          88m
pod/istiod-74c64d89cb-
m44ks                 1/1     Running   0          89m
```

While the preceding part of the output shows various Pods running under the `istio-system` namespace, the following will show Services in the `istio-system` namespace:

```
$ kubectl get svc -n istio-system
NAME              TYPE         CLUSTER-IP       EXTERNAL-IP
   PORT(S)            AGE
service/istio-egressgate way    ClusterIP      10.97.150.168
    <none>        80/TCP,443/TCP               88m
service/istio-ingressgateway   LoadBalancer   10.100.113.119
   <pending>     15021:31391/TCP,80:32295/TCP,443:31860/
TCP,31400:31503/TCP,15443:31574/TCP   88m
service/istiod            ClusterIP      10.110.59.167
    <none>        15010/TCP,15012/TCP,443/TCP,15014/TCP    89m
```

You can check all resources by using the following command:

```
$ kubectl get all -n istio-system
```

In the `istio-system` namespace, Istio installs the `istiod` component, which is the control plane of Istio. There are various other custom configs such as Kubernetes Custom Resource Definitions, ConfigMaps, Admission Webhooks, Service Accounts, Role Bindings, as well as Secrets installed.

We will look into `istiod` and other control plane components in more detail in the next chapter. For now, let's enable Istio for a sample application that is packaged with it.

Enabling Istio for a sample application

To keep our work in a sample application segregated from other resources, we will first create a Kubernetes namespace called `bookinfons`. After creating the namespace, we will deploy the sample application in the `bookinfons` namespace.

You need to run the second command from within the Istio installation directory – that is, `$ISTIO_DOWNLOAD/istio-1.13.1`:

```
$ kubectl create ns bookinfons
namespace/bookinfons created
$ kubectl apply -f samples/bookinfo/platform/kube/bookinfo.yaml
-n bookinfons
```

All the created resources are defined in `samples/bookinfo/platform/kube/bookinfo.yaml`.

Check what Pods and Services have been created using the following commands:

```
$ kubectl get po -n bookinfons
$ kubectl get svc -n bookinfons
```

Note that there is one Pod each for `details`, `productpage`, and `ratings`, and three Pods for various versions of `review`. There is one service for each microservice. All of them are similar, except for the `kubectl review` service, which has three endpoints. Using the following command, let's check how the `review` service definition is different from other Services:

```
$ kubectl describe svc/reviews -n bookinfons
...
Endpoints:
         172.17.0.10:9080,172.17.0.8:9080,172.17.0.9:9080
...
$ kubectl get endpoints -n bookinfons
NAME          ENDPOINTS
                                                    AGE
details     172.17.0.6:9080
                                              18h
productpage   172.17.0.11:9080
                                             18h
ratings   172.17.0.7:9080
                                              18h
reviews
      172.17.17.0.10:9080,172.17.0.8:9080,172.17.0.9:9080   18h
```

Now that the bookinfo application has successfully deployed, let's access the product page of the bookinfo application using the following commands:

```
$ kubectl port-forward svc/productpage 9080:9080 -n bookinfons
Forwarding from 127.0.0.1:9080 -> 9080
Forwarding from [::1]:9080 -> 9080
Handling connection for 9080
```

Go ahead and type in `http://localhost:9080/productpage` in your internet browser. If you don't have one, you can do it via `curl`:

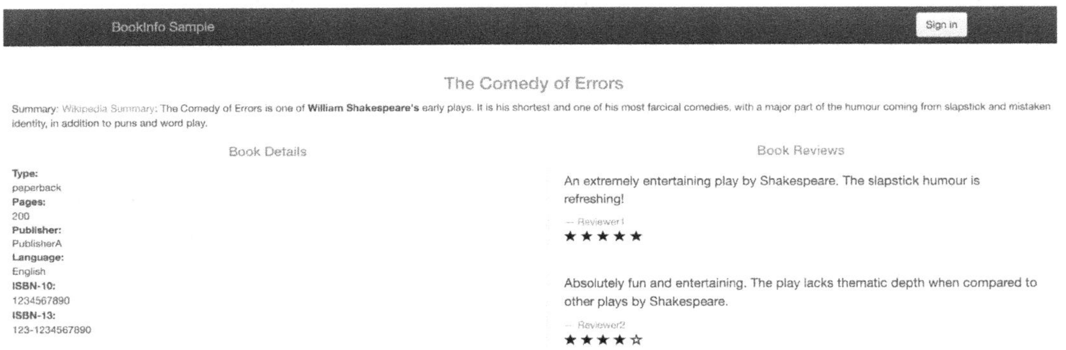

Figure 2.1 – The product page of the BookInfo app

If you can see `productpage`, then you have successfully deployed the sample application.

> **What if I do not have a browser?**
>
> If you don't have a browser, you can use this:
>
> `curl -sS localhost:9080/productpage`

So, now that we have successfully deployed the sample application that comes along with Istio, let's move on to enabling Istio for it.

Sidecar injection

Sidecar injection is the means through which `istio-proxy` is injected into the Kubernetes Pod as a sidecar. Sidecars are additional containers that run alongside the main container in a Kubernetes Pod. By running alongside the main container, the sidecars can share the network interfaces with other containers in the Pod; this flexibility is leveraged by the `istio-proxy` container to mediate and control all communication to and from the main container. We will read more about sidecars in the *Chapter 3*. For now, we will keep the ball rolling by enabling Istio for the sample application.

Let's check out some interesting details before and after we enable Istio for this application:

```
$ kubectl get ns bookinfons -show-labels
NAME          STATUS   AGE    LABELS
bookinfons    Active   114m   kubernetes.io/metadata.
name=bookinfons
```

Let's look at one of the Pods, `productpage`:

```
$ kubectl describe pod/productpage-v1-65b75f6885-8pt66 -n
bookinfons
```

Copy the output to a safe place. We will use this information to compare the findings once you have enabled Istio for the `bookinfo` application.

We will need to delete what we have deployed:

```
$ kubectl delete -f samples/bookinfo/platform/kube/bookinfo.
yaml -n bookinfons
```

Wait for a few seconds and check that all the resources in the `bookinfons` namespace have been terminated. After that, enable `istio-injection` for `bookinfons`:

```
$ kubectl label namespace bookinfons istio-injection=enabled
namespace/bookinfons labeled
$ kubectl get ns bookinfons -show-labels
NAME          STATUS   AGE   LABELS
bookinfons    Active   21h   istio-injection=enabled,kubernetes.
io/metadata.name=bookinfons
```

Manual injection of sidecars

The other option is to manually inject the sidecar by making use of `istioctl kube-inject` to augment the deployment descriptor file and then applying it using kubectl:

```
$ istioctl kube-inject -f deployment.yaml -o deployment-injected.
yaml | kubectl apply -f -
```

Go ahead and deploy the bookinfo application:

```
$ kubectl apply -f samples/bookinfo/platform/kube/bookinfo.yaml
-n bookinfons
```

Let's check what has been created:

```
$ kubectl get po -n bookinfons
```

We can see that the number of containers in Pods is now not one but instead two. Before we enabled istio-injection, the number of containers in Pods was one. We will discuss shortly what the additional container is. Let's also check for any change in the number of Services:

```
$ kubectl get svc -n bookinfons
```

Alright, so there is a change in Pod behavior but no noticeable change in service behavior. Let's look deeper into one of the Pods:

```
$ kubectl describe po/productpage-v1-65b75f6885-57vnb -n
bookinfons
```

The complete output of this command can be found in `Output references/Chapter 2/productpage pod.docx` on the GitHub repository of this chapter.

Note that the Pod description of the productpage Pod as well as every other Pod in the bookinfons have another container named istio-proxy and an init container named istio-init. They were absent when we initially created them but got added after we applied the istio-injection=enabled label, using the following command:

```
kubectl label namespace bookinfons istio-injection=enabled
```

The sidecars can be injected either manually or automatically. Automatic is the easier way to inject sidecars. However, once we have familiarized ourselves with Istio, we will look at injecting sidecars manually by modifying application resource descriptor files in *Part 2* of the book. For now, let's briefly look at how automatic sidecar injection works.

Istio makes use of Kubernetes **admission controllers**. Kubernetes admission controllers are responsible for intercepting a request to the Kubernetes API server. Interception happens post-authentication and authorization but pre-modification/creation/deletion of objects. You can find these admission controllers using the following:

```
$ kubectl describe po/kube-apiserver-minikube -n kube-system |
grep enable-admission-plugins
--enable admission plugins=NamespaceLifecycle, LimitRanger,
ServiceAccount,DefaultStorageClass, DefaultTolerationSeconds,
```

```
NodeRestriction, MutatingAdmissionWebhook,
ValidatingAdmissionWebhook, ResourceQuota
```

Istio makes use of mutating admission webhooks for automatic sidecar injection. Let's find out what mutating admission webhooks are configured in our cluster:

```
$ kubectl get --raw /apis/admissionregistration.k8s.io/v1/
mutatingwebhookconfigurations | jq '.items[].metadata.name'
"istio-revision-tag-default"
"istio-sidecar-injector"
```

The following diagram describes the role of admission controllers during API calls to the Kubernetes API server. The mutating admission Webhook controllers are responsible for the injection of the sidecar.

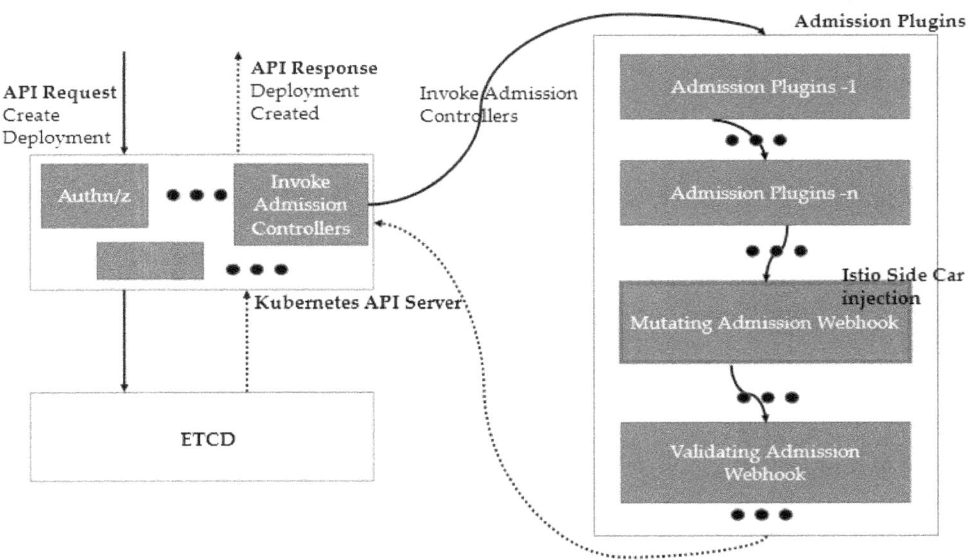

Figure 2.2 – Admission controllers in Kubernetes

We will cover sidecar injection in more detail in *Chapter 3*. For now, let's switch our focus back to what has changed in Pod descriptors due to `istio-injection`.

`istio-iptables` was mentioned in the `istio-init` configuration of the product page Pod description using the following command:

```
kubectl describe po/productpage-v1-65b75f6885-57vnb -n
bookinfons
```

The following is a snippet from the Pod descriptor:

```
istio-iptables -p 15001 -z 15006 -u 1337 -m REDIRECT -I '*' -x
"" -b '*' -d 15090,15021,15020
```

`istio-iptables` is an initialization script responsible for setting up port forwarding via `iptables` for the Istio sidecar proxy. The following are various arguments that are passed during the execution of the script:

- **-p** specifies the Envoy port to which all TCP traffic will be redirected
- **-z** specifies the port to which all inbound traffic to the Pod should be redirected
- **-u** is the UID of the user for which the redirection is not to be applied
- **-m** is the mode to be used for redirecting inbound connections
- **-I** is a list of IP ranges in CIDR block destinations of outbound connections that need to be redirected to Envoy
- **-x** is a list of CIDR block destinations of outbound connections that need to be exempted from being redirected to Envoy
- **-b** is a list of inbound ports for which traffic needs to be redirected to Envoy
- **-d** is a list of inbound ports that need to be excluded from being redirected to Envoy

To summarize the preceding argument in the `istio-init` container, the container is executing a script, `istio-iptables`, which is basically creating `iptables` rules at the Pod level – that is, applied to all containers within the Pod. The script configures an `iptables` rule that applies the following:

- All traffic should be redirected to port `15001`
- Any traffic to the Pod should be redirected to port `15006`
- This rule doesn't apply to UID `1337`
- The mode for redirection to be used is REDIRECT
- All outbound connections to any destination (`*`) should be redirected to `15001`
- No outbound destination is exempt from this rule
- The redirection needs to happen for all inbound connections coming from any IP address, except when the destination ports are `15090`, `15021`, or `15020`

We will dig deeper into this in *Chapter 3*, but for now, remember that the init container basically sets up an `iptables` rule at the Pod level, which will redirect all traffic coming to the product page container at port `9080` to `15006`, while all traffic going out from product page container will be redirected to port `15001`. Both ports `15001` and `15006` are exposed by the `istio-proxy` container, which is created from `docker.io/istio/proxyv2:1.13.1`. The istio-proxy container runs alongside the product page container. Along with `15001` and `15006`, it also has ports `15090`, `15021`, and `15020`.

`Istio-iptables.sh` can be found here: `https://github.com/istio/cni/blob/master/tools/packaging/common/istio-iptables.sh`.

You will also notice that both `istio-init` and `istio-proxy` are spun from the same Docker image, `docker.io/istio/proxyv2:1.13.1`. Inspect the Docker file here: `https://hub.docker.com/layers/proxyv2/istio/proxyv2/1.13.4/images/sha256-1245211d2fdc0f86cc374449e8be25166b9d06f1d0e4315deaaca4d81520215e?context=explore`. The dockerfile gives more insight into how the image is constructed:

```
# BASE_DISTRIBUTION is used to switch between the old base
distribution and distroless base images
..
ENTRYPOINT ["/usr/local/bin/pilot-agent"]
```

The entry point is an Istio command/utility called `pilot-agent` that bootstraps Envoy to run as a sidecar when the *proxy sidecar* argument is passed in the `istio-proxy` container. `pilot-agent` also sets `iptables` during initialization when the *istio-iptables* argument is passed during initialization in the `istio-init` container.

More information on pilot-agent

You can find more details about the pilot agent by executing `pilot-agent` from outside the container, picking any Pod that has the `istio-proxy` sidecar injected. In the following command, we have to use the Ingress gateway Pod in the `istio-system` namespace:

```
$ kubectl exec -it  po/istio-ingressgateway-569d7bfb4-8bzww
-n istio-system -c istio-proxy   -- /usr/local/bin/pilot-agent
proxy router --help
```

Like in the earlier section, you can still access the product page from your browser using `kubectl port-forward`:

```
$ kubectl port-forward svc/productpage 9080:9080 -n bookinfons
Forwarding from 127.0.0.1:9080 -> 9080
Forwarding from [::1]:9080 -> 9080
Handling connection for 9080
```

So far, we have looked at sidecar injection and what effects it has on Kubernetes resource deployments. In the following section, we will read about how Istio manages the Ingress and Egress of traffic.

Istio gateways

Rather than using `port-forward`, we can also make use of the Istio Ingress gateway to expose the application. Gateways are used to manage inbound and outbound traffic to and from the mesh. Gateways provide control over inbound and outbound traffic. Go ahead and try the following command again to list the Pods in the `istiod` namespace and to discover about gateways already installed during the Istio installation:

```
$ kubectl get pod -n istio-system
NAME                                    READY   STATUS    RESTARTS   AGE
istio-egressgateway-76c96658fd-
pgfbn    1/1      Running    0          5d18h
istio-ingressgateway-569d7bfb4-
8bzww    1/1      Running    0          5d18h
istiod-74c64d89cb-
m44ks                   1/1     Running    0          5d18h

$ kubectl get po/istio-ingressgateway-569d7bfb4-8bzww -n istio-
system -o json  | jq '.spec.containers[].image'
"docker.io/istio/proxyv2:1.13.1"
$ kubectl get po/istio-egressgateway-76c96658fd-pgfbn -n istio-
system -o json  | jq '.spec.containers[].image'
"docker.io/istio/proxyv2:1.13.1"
```

You can see that the gateways are also another set of Envoy proxies that are running in the mesh. They are similar to Envoy proxies deployed as a sidecar in the Pods, but in the gateway, they run as standalone containers in the Pod deployed via `pilot-agent`, with *proxy router* arguments. Let's investigate the Kubernetes descriptors of the Egress gateway:

```
$ kubectl get po/istio-egressgateway-76c96658fd-pgfbn -n istio-
system -o json  | jq '.spec.containers[].args'
[
  "proxy",
  "router",
  "--domain",
  "$(POD_NAMESPACE).svc.cluster.local",
  "--proxyLogLevel=warning",
```

```
"--proxyComponentLogLevel=misc:error",
"--log_output_level=default:info"
]
```

Let's look at the Gateway Services next:

```
$ kubectl get svc -n istio-system
NAME                    TYPE          CLUSTER-
IP            EXTERNAL-IP
   PORT(S)                                                        AGE
istio-egressgateway    ClusterIP      10.97.150.168     <none>
        80/TCP,443/TCP                                              5d18h
istio-ingressgateway   LoadBalancer
   10.100.113.119    <pending>     15021:31391/TCP,80:32295/
TCP,443:31860/TCP,31400:31503/TCP,15443:31574/TCP    5d18h
istiod   ClusterIP      10.110.59.167     <none>        15010/
TCP,15012/TCP,443/TCP,15014/
TCP                                              5d18h
```

Now, let's try to make sense of ports for the Ingress gateway using the following command:

```
$ kubectl get svc/istio-ingressgateway -n istio-system -o json
| jq '.spec.ports'
[
…
  {
    "name": "http2",
    "nodePort": 32295,
    "port": 80,
    "protocol": "TCP",
    "targetPort": 8080
  },
  {
    "name": "https",
    "nodePort": 31860,
    "port": 443,
    "protocol": "TCP",
    "targetPort": 8443
  },
  ….
```

You can see that the Ingress gateway service takes `http2` and `https` traffic at ports `32295` and `31860` from outside the cluster. From inside the cluster, the traffic is handled at ports `80` and `443`. The `http2` and `https` traffic is then forwarded to ports `8080` and `8443` to underlying Ingress Pods.

Let's enable the Ingress gateway for the `bookinfo` service:

```
$ kubectl apply -f samples/bookinfo/networking/bookinfo-
gateway.yaml -n bookinfons
gateway.networking.istio.io/bookinfo-gateway created
virtualservice.networking.istio.io/bookinfo created
```

Let's look at the `bookinfo` virtual service definition:

```
$ kubectl describe virtualservice/bookinfo -n bookinfons
Name:         bookinfo
..
API Version:  networking.istio.io/v1beta1
Kind:         VirtualService
...
Spec:
  Gateways:
    bookinfo-gateway
  Hosts:
    *
  Http:
    Match:
      Uri:
        Exact:  /productpage
      Uri:
        Prefix:  /static
      Uri:
        Exact:  /login
      Uri:
        Exact:  /logout
      Uri:
        Prefix:  /api/v1/products
    Route:
      Destination:
        Host:  productpage
```

```
      Port:
        Number:  9080
```

The virtual service is not restricted to any particular hostname. It routes /productpage, login, and /logout, and any other URI with the /api/v1/products or /static prefix to the productpage service at port 9080. If you remember, 9080 was also the port exposed by the productpage service. The spec.gateways annotation implies that this virtual service config should be applied to bookinfo-gateway, which we will investigate next:

```
$ kubectl describe gateway/bookinfo-gateway -n bookinfons
Name:         bookinfo-gateway
..
API Version:  networking.istio.io/v1beta1
Kind:         Gateway
..
Spec:
  Selector:
    Istio:  ingressgateway
  Servers:
    Hosts:
      *
    Port:
      Name:      http
      Number:    80
      Protocol:  HTTP
..
```

The gateway resource describes a load balancer receiving incoming and outgoing connections to and from the mesh. The preceding example first defines that the configuration should be applied to the Pod with the Istio: ingressgateway labels (Ingress gateway Pods in the istiod namespace). The config is not bound to any hostnames, and it takes connection at port 80 for HTTP traffic.

So, to summarize, you have a load balancer configuration defined in the form of a gateway along with routing configuration to backend in the form of virtual services. These configs are applied to a proxy Pod, which in this case is istio-ingressgateway-569d7bfb4-8bzww.

Go ahead and check the logs of the proxy Pod while opening the product page in the browser.

First, find the IP and the port (the HTTP2 port in the Ingress gateway service):

```
$ echo $(minikube ip)
192.168.64.6
$ echo $(kubectl -n istio-system get service istio-
ingressgateway -o jsonpath='{.spec.ports[?(@.name=="http2")].
nodePort}')
32295
```

Fetch the products via following URL: `http://192.168.64.6:32295/api/v1/products`. You can do this either in the browser or through `curl`.

Stream the log of the `istio-ingressgateway` Pod to `stdout`:

```
$ kubectl logs -f pod/istio-ingressgateway-569d7bfb4-8bzww -n
istio-system
"GET /api/v1/products HTTP/1.1" 200 - via_upstream - "-"
0 395 18 16 "172.17.0.1" "Mozilla/5.0 (Macintosh; Intel
Mac OS X 10_15_7) AppleWebKit/537.36 (KHTML, like Gecko)
Chrome/98.0.4758.102 Safari/537.36" "cfc414b7-10c8-9ff9-
afa4-a360b5ad53b8" "192.168.64.6:32295" "172.17.0.10:9080"
outbound|9080||productpage.bookinfons.svc.cluster.local
172.17.0.5:56948 172.17.0.5:8080 172.17.0.1:15370 - -
```

From the logs, you can infer that an inbound request `GET /api/v1/products HTTP/1.1` arrived at `192.168.64.6:32295`, which was then routed to `172.17.0.10:9080`. This is the endpoint – that is, the IP address of the `productpage` Pod.

The following diagram illustrates the composition of the `bookinfo` Pods with injected `istio-proxy` sidecars and the Istio Ingress gateway.

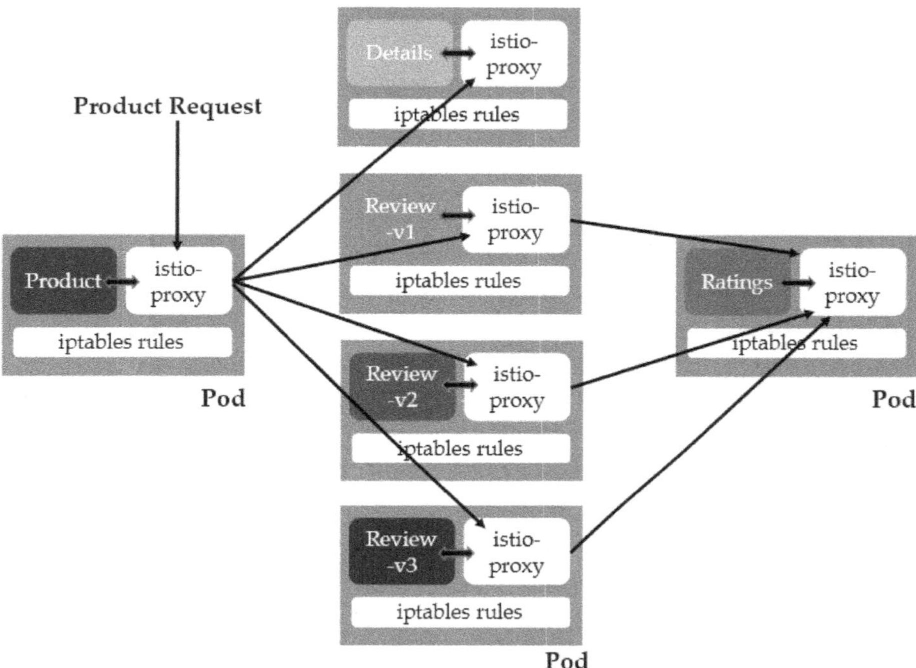

Figure 2.3 – The BookInfo app with sidecar injection and the Istio Ingress gateway for traffic Ingress

> **Tip**
>
> If you are getting TLS errors such as certificate expired or any other OpenSSL error, then just try restarting the BookInfo application and Istio components using the following command:
>
> ```
> $ kubectl rollout restart deployment --namespace bookinfons
> $ kubectl rollout restart deployment --namespace istio-system.
> ```

I hope by now you are familiarized with the basic concepts of Istio and its installation on your workstations. In the next section, we will continue with the installation of add-on components in Istio.

Observability tools

Istio produces various metrics that can then be fed into various telemetry applications. The out-of-the-box installation is shipped with add-ons that include **Kiali**, **Jaeger**, **Prometheus**, and **Grafana**. Let's take a look at them in the following sections.

Kiali

The first component to install will be Kiali, the default management UI for Istio. We'll start by enabling the telemetry tools by running the following command:

```
$ kubectl apply -f samples/addons
serviceaccount/grafana created
……. 
$ kubectl rollout status deployment/kiali -n istio-system
Waiting for deployment "kiali" rollout to finish: 0 of 1
updated replicas are available...
deployment "kiali" successfully rolled out
```

Once all the resources have been created and Kiali has successfully deployed, you can then open the dashboard of Kiali by using the following command:

```
$ istioctl dashboard kiali
http://localhost:20001/kiali
```

Kiali is very handy when you want to visualize or troubleshoot the mesh topology as well as underlying mesh traffic. Let's take a quick look at some of the visualizations.

The **Overview** page provides an overview of all the namespaces in the cluster.

Figure 2.4 – The Kiali dashboard Overview section

You can click on the three dots in the top-right corner to dive further into that namespace and also to change the configuration for it.

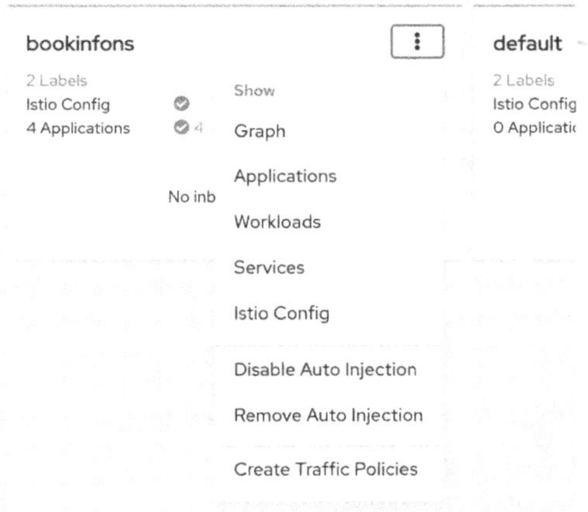

Figure 2.5 – Istio configuration for a namespace on the Kiali dashboard

You can also check out individual applications, Pods, Services, and so on. One of the most interesting visualizations is **Graph**, which represents the flow of traffic in the mesh for a specified period.

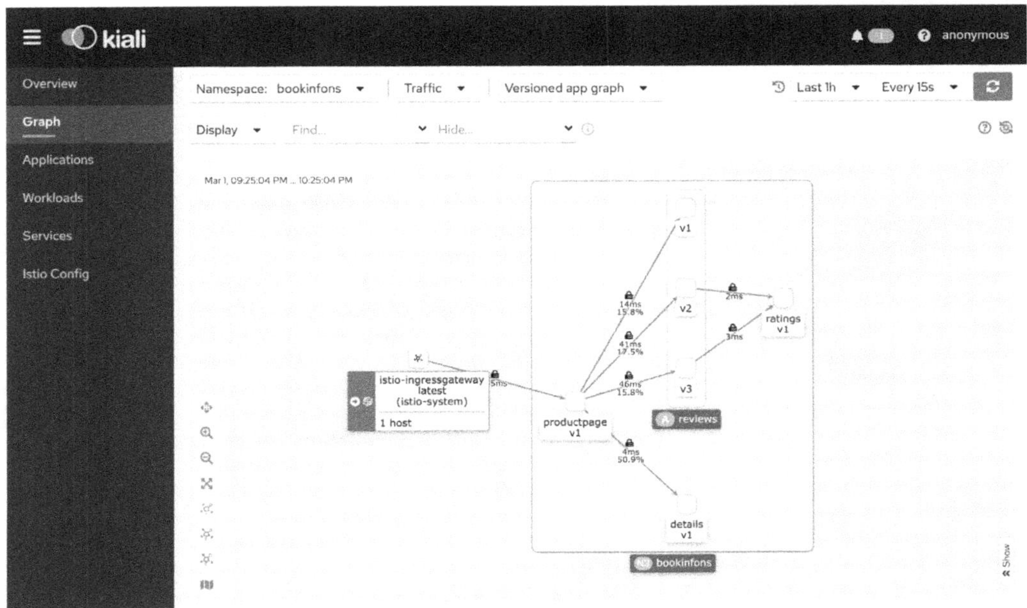

Figure 2.6 – A versioned app graph on the Kiali dashboard

The preceding screenshot is of a versioned app graph, where multiple versions of an application are grouped together; in this case, it is a reviews app. We will look into this in much more detail in *Chapter 8*.

Jaeger

Another add-on is Jaeger. You can open the Jaeger dashboard type with the following command:

```
$ istioctl dashboard jaeger
http://localhost:16686
```

The preceding command should open your browser on the Jaeger dashboard. Jaeger is an open source, end-to-end, distributed transaction monitoring software. The need for such a tool will become explicit when we build and deploy a hands-on application in *Chapter 4*.

In the Jaeger dashboard under **Search**, select any service for which you are interested to look at traffic. Once you select the service and click on **Find Traces**, you should be able to see all traces involving the Details app in the `bookinfons` namespace.

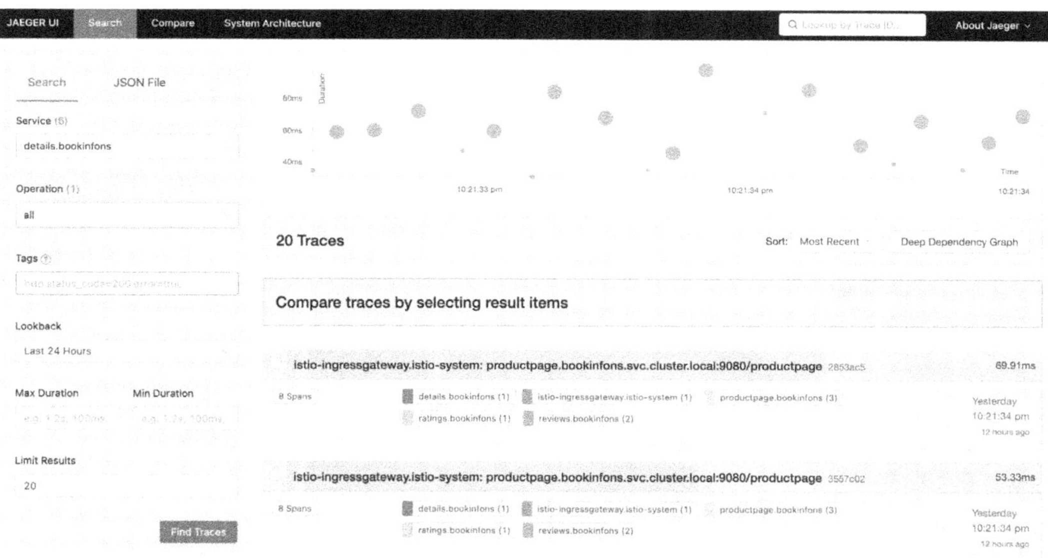

Figure 2.7 – The Jaeger dashboard Search section

You can then click on any of the entries for further details:

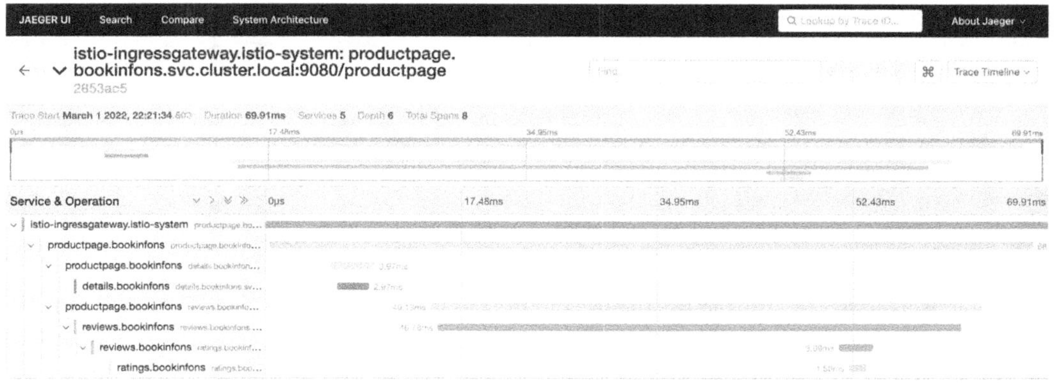

Figure 2.8 – The Jaeger dashboard details section

You can see that the overall invocation took 69.91 ms. The details were called by `productpage`, and it took 2.97 ms for them to return the response. You can then click further on any of the services to see a detailed trace.

Prometheus

Next, we will look into Prometheus, which is also an open source monitoring system and time series database. Prometheus is used to capture all metrics against time to track the health of the mesh and its constituents.

To start the Prometheus dashboard, use the following command:

```
$ istioctl dashboard prometheus
http://localhost:9090
```

This should open the Prometheus dashboard in your browser. With our installation, Prometheus is configured to collect metrics from `istiod`, the Ingress and Egress gateways, and the istio-proxy.

In the following example, we are checking the total requests handled by Istio for the `productpage` application.

Figure 2.9 – The Istio total request on the Prometheus dashboard

Another add-on to look at is Grafana, which, like Kiali, is another visualization tool.

Grafana

To start the Grafana dashboard, use the following command:

```
$ istioctl dashboard grafana
http://localhost:3000
```

The following is a visualization of the total requests handled by Istio for `productpage`:

Figure 2.10 – The Grafana dashboard Explore section

The following is another visualization of the Istio performance metrics.

Figure 2.11 – The Grafana Istio Performance Dashboard

Note that by just applying a label, `istio-injection: enabled`, we enabled the Service Mesh for the BookInfo application. Sidecars were injected automatically and mTLS was enabled by default for communication between different microservices of the application. Moreover, a plethora of monitoring tools provide information about the BookInfo application and its underlying microservices.

Istio architecture

Now that we have installed Istio, enabled it for the BookInfo application, and also analyzed it's operations, it is time to simplify what we have seen so far with a diagram. The following figure is a representation of Istio architecture.

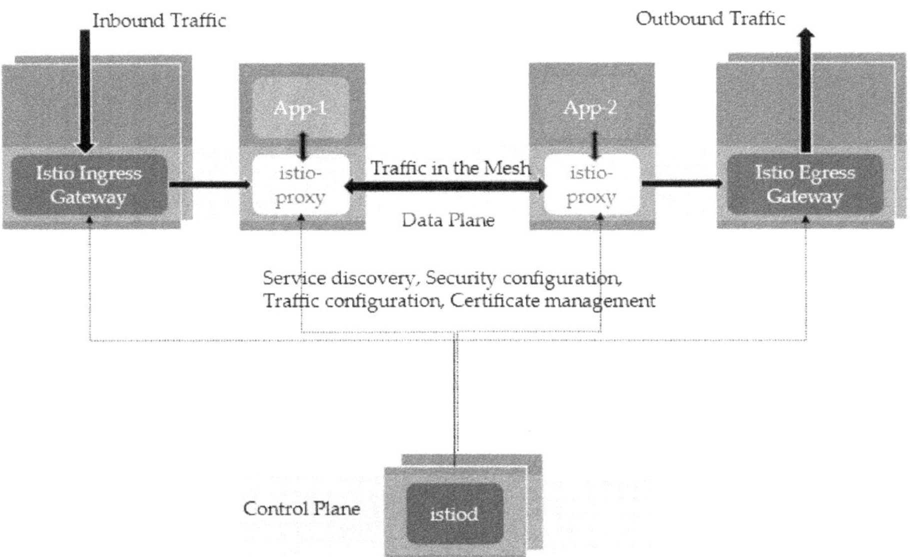

Figure 2.12 – Istio architecture

The Istio Service Mesh comprises a data plane and a control plane. The example we followed in this chapter installs both of them on one node. In a production or non-production environment, the Istio control plane will be installed on its own separate set of nodes. The **control plane** comprises istiod components as well as a few other Kubernetes configs, which, altogether, are responsible for managing and providing service discovery to the data plane, propagation of configuration related to security and traffic management, as well as providing and managing identity and certificates to data plane components.

The **data plane** is another part of the Service Mesh that consists of Istio proxies deployed alongside the application container in the Pod. Istio proxies are basically Envoy. Envoy is an application-aware service proxy that mediates all network traffic between microservices, based on instructions from the control plane. Envoy also collects various metrics and reports back telemetry to various add-on tools.

Subsequent chapters will be dedicated to the control plane and data plane, in which we will dive deeper into understanding their functions and behavior.

Summary

In this chapter, we prepared a local environment to install Istio using `istioctl` which is the Istio command-line utility. We then enabled sidecar injection by applying a label called `istio-injection: enabled` to the namespace that hosts the microservices.

We briefly looked at Kubernetes admission controllers and how mutating admission webhooks inject sidecars to the deployment API calls to the Kubernetes API server. We also read about gateways and looked at the sample Ingress and Egress gateways that are installed with Istio. The gateway is a standalone istio-proxy, aka an Envoy proxy, and is used to manage Ingress and Egress traffic to and from the mesh. Following this, we looked at how various ports are configured to be exposed on the Ingress gateway and how traffic is routed to upstream services.

Istio provides integration with various telemetry and observability tools. The first tool we looked at was Kiali, the visualization tool providing insight into traffic flows. It is also the management console for the Istio Service Mesh. Using Kiali, you can also perform Istio management functions such as checking/modifying various configurations and checking infrastructure status. After Kiali, we looked at Jaeger, Prometheus, and Grafana, all of which are open source and can be integrated easily with Istio.

The content of this chapter sets the foundations for and prepares you to deep dive into Istio in the upcoming chapters. In the next chapter, we will be reading about Istio's control and data planes, taking a deep dive into their various components.

3
Understanding Istio Control and Data Planes

The previous chapter gave you an overview of Istio, what a simple installation looks like, and how to apply a Service Mesh to a sample application. In this chapter, we will dive deeper into Istio's control plane and data plane. We will understand the role of each of these components by going through the following topics:

- Components of Istio control plane
- Deployment models for Istio control plane
- Envoy, the Istio data plane

This chapter will help you understand the Istio control plane so you can plan the installation of control planes in a production environment. After reading this chapter, you should be able to identify the various components of the Istio control plane including istiod, along with the functionality they each deliver in the overall working of Istio.

Exploring the components of a control plane

The following diagram summarizes the Istio architecture along with the interaction between various components. We used the Ingress gateway and istio-proxy in the previous chapter so we will not go into further details on those here. We will, however, unravel some of the other components of the Istio control plane not directly depicted in the following illustration.

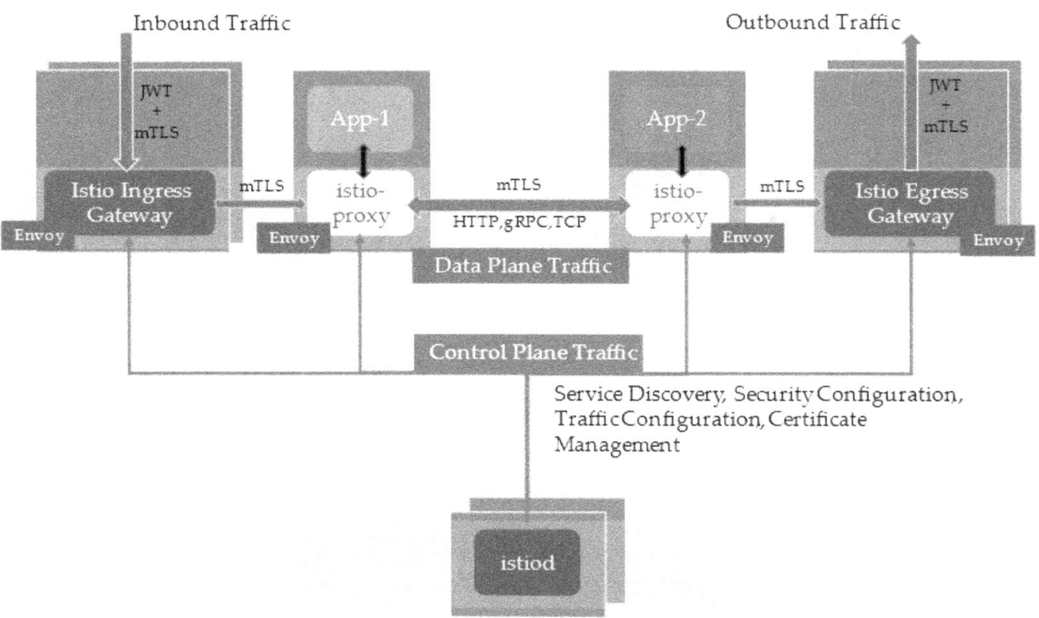

Figure 3.1 – Istio control plane

Before we delve into the components of the control plane, let's first get the definition of the term out of the way – the **control plane** is a set of Istio services that are responsible for the operations of the Istio data plane. There is no single component that constitutes the control plane – rather, there are several.

Let's look at the first component of the Istio control plane called istiod.

istiod

istiod is one of the Istio control plane components, providing service discovery, configuration, and certificate management. In prior versions of Istio, the control plane was made up of components called Galley, Pilot, Mixer, Citadel, WebHook Injector, and so on. istiod unifies the functionality of these components (Pilot, Galley, and Citadel) into a single binary to provide simplified installation, operation, and monitoring, as well as seamless upgrades between various Istio versions.

Let's look at the `istiod` Pod running in the `istio-system` namespace:

```
$ kubectl get po -n istio-system
NAME                                    READY   STATUS    RESTARTS   AGE
istio-egressgateway-84f95886c7-
5gxps     1/1     Running   0          10d
istio-ingressgateway-6cb4bb68df-
qbjmq    1/1     Running   0          10d
istiod-65fc7cdd7-
r95jk                1/1     Running   0          10d
$ kubectl exec -it pod/istiod-65fc7cdd7-r95jk -n istio-system
-- /bin/sh -c «ps -ef"
UID          PID    PPID  C STIME TTY          TIME CMD
istio-p+       1       0  0 Mar14 ?        00:08:26 /usr/
local/bin/pilot-discovery discovery --monitoringAddr=:15014
--log_output_level=default:info --domain cluster.local
--keepaliveMaxServerConnectionAge 30m
```

You must have noticed that the Pod itself is running `pilot-discovery` and is based on the following image:

```
$ kubectl get pod/istiod-65fc7cdd7-r95jk -n istio-system -o
json | jq '.spec.containers[].image'
"docker.io/istio/pilot:1.13.1"
```

You must have also noticed that the image for the `istiod` Pod is different to the istio-proxy image inserted as a sidecar. The istiod image is based on `pilot-discovery`, whereas the sidecar is based on `proxyv2`.

The following command shows that the sidecar container is created from `proxyv2`:

```
$ kubectl get pod/details-v1-7d79f8b95d-5f4td -n bookinfons -o
json|jq '.spec.containers[].image'
"docker.io/istio/examples-bookinfo-details-v1:1.16.2"
"docker.io/istio/proxyv2:1.13.1"
```

Now that we know that the `istiod` Pod is based on `pilot-discovery`, let's look at some of the functions performed by istiod.

Configuration watch

istiod watches Istio **Custom Resource Definitions** (**CRDs**) and any other Istio-related configuration being sent to the Kubernetes API server. Any such configuration is then processed and distributed internally to various subcomponents of istiod. You interact with Istio Service Mesh via the Kubernetes API server, but all interactions with Kubernetes API server are not necessarily destined for the Service Mesh.

istiod keeps an eye on various config resources, typically Kubernetes API server resources identified by certain characteristics such as labels, namespaces, annotations, and so on. These configuration updates are then intercepted, collected, and transformed into Istio-specific formats and distributed via the **Mesh Configuration Protocol** (**MCP**) to other components of istiod. istiod also implements configuration forwarding, which we will be looking at in later chapters when we do a multi-cluster installation of Istio. For now, let's just say that istiod can also pass configurations to another istiod instance over MCP in both pull and push modes.

API validation

istiod also adds an admission controller to enforce the validation of Istio resources before they are accepted by the Kubernetes API server. In the previous chapter, we saw two admission controllers: the **mutating webhook** and the **validation webhook**.

The mutating webhook is responsible for augmenting the API calls for resources such as deployments by adding configuration for Istio sidecar injection. Similarly, the validation webhook auto registers itself with the Kubernetes API server to be called for each incoming call for Istio CRDs. When such calls to add/update/delete Istio CRDs arrive at the Kubernetes API server, they are passed to the validation webhook, which then validates the incoming request and, based on the outcome of the validation, the API calls are accepted or rejected.

Istio Certificate Authority

Istio provides comprehensive security for all communication in the mesh. All Pods are assigned an identity through the Istio PKI with x.509 key/cert in **Spifee Verifiable Identity Document** (**SVID**) format. The Istio **Certificate Authority** (**CA**) is responsible for signing requests from node agents deployed along with istio-proxy. The Istio CA is built on top of Citadel and is responsible for approving and signing the **Certificate signature requests(CSRs)** sent by Istio node agents. The Istio CA also performs the rotation and revocation of certificates and keys. It offers pluggability of different CAs as well as the flexibility to use the Kubernetes CA.

Some of the other functions and components of the Istio control plane are as follows:

- **Sidecar injection**: The Istio control plane also manages sidecar injection via mutating webhooks.
- **Istio node agent**: Node agents are deployed along with Envoy and take care of communication with the Istio CA, providing the cert and keys to Envoy.

- **Identity directory and registry**: The Istio control plane manages a directory of identities for various types of workloads that will be used by the Istio CA to issue key/certs for requested identities.
- **End-user context propagation**: Istio provides a secure mechanism to perform end user authentication on Ingress and then propagate the user context to other services and apps within the Service Mesh. The user context is propagated in JWT format, which helps to pass on user information to services within the mesh without needing to pass the end user credentials.

istiod is a key control plane component performing many key functions of the control plane, but is not the only control plane component worth remembering. In the next section, we will examine other components that are not part of istiod but still important components of the Istio control plane.

The Istio operator and istioctl

The Istio operator and istioctl are both control plane components and are optional to install. Both provide administrative functions to install and configure components of the control and data planes. You have used istioctl quite a lot in the previous chapter as a command-line tool to interact with the Istio control plane to pass on instructions. The instructions can be to fetch information and create, update, or delete a configuration related to the workings of the Istio data plane. The Istio operator and istioctl essentially perform the same functions with the exception that istioctl is explicitly invoked to make a change, whereas the Istio operator functions per the *operator* framework/pattern of Kubernetes.

We will not be using the Istio operator, but if you want, you can install it using the following command:

```
$ istioctl operator init
Installing operator controller in namespace: istio-operator
using image: docker.io/istio/operator:1.13.1
Operator controller will watch namespaces: istio-system
✓ Istio operator installed
✓ Installation complete
```

The two main components of the Istio operator are the customer resource called **IstioOperator**, represented by high-level APIs, and a controller that has logic to transform the high-level API into low-level Kubernetes actions. The IstioOperator CRD wraps a second component called **IstioOperatorSpec**, a status field, and some additional metadata.

You can use the following command to find details of the IstioOperator **Custom Resource** (**CR**):

```
$ kubectl get istiooperators.install.istio.io -n istio-system
-o json
```

You can find the output of the command here: `https://github.com/PacktPublishing/Bootstrap-Service-Mesh-Implementations-with-Istio/blob/main/Output%20references/Chapter%203/IstioOperator%20CR.docx`

As you can see in the output, the API is structured in line with the control plane components around the base Kubernetes resources, pilot, Ingress and Egress gateways, and finally, the optional third-party add-ons.

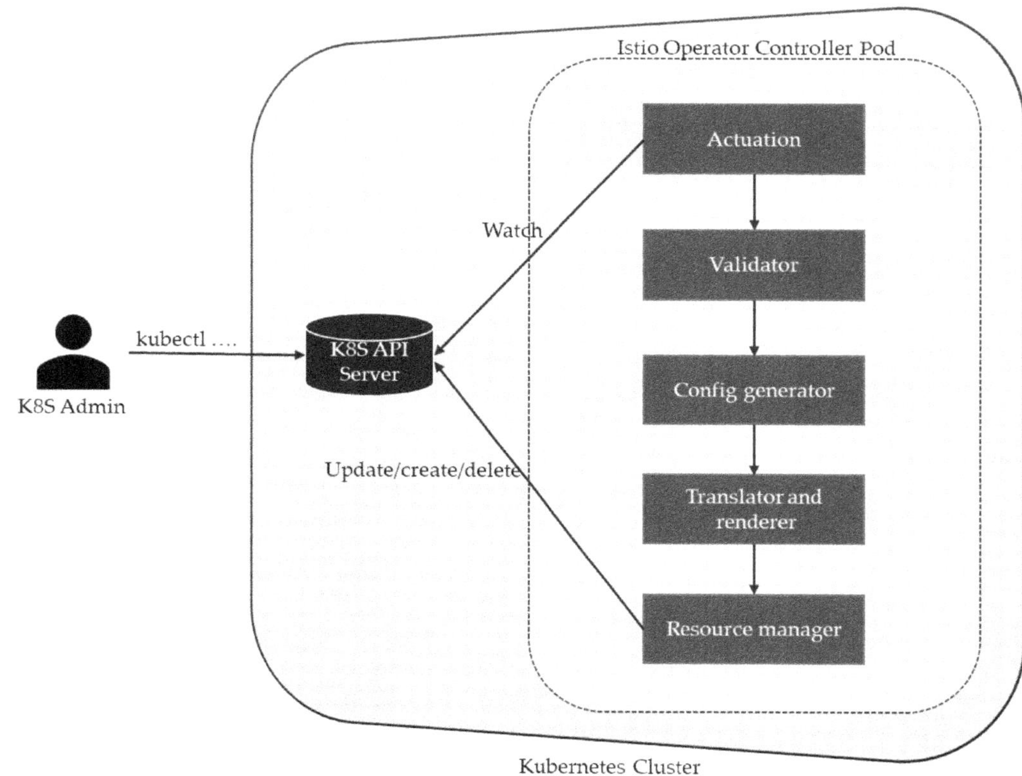

Figure 3.2 – The Istio operator

The preceding diagram describes the operations of the IstioOperator, while the following describes the operations of istioctl:

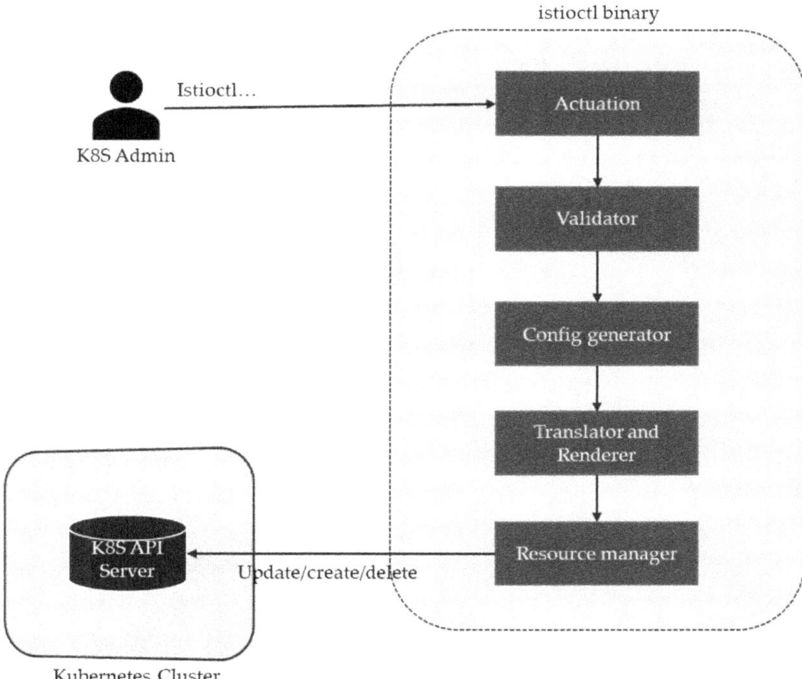

Figure 3.3 – istioctl

istioctl and the operator are very similar to each other except when in the *Actuation* phase. istioctl is a user-run command that takes an IstioOperator CR as input, while the controller runs whenever the in-cluster IstioOperator CR changes, but the remaining components are similar, if not the same.

The following is a brief summary of the various components of the Istio operator and istioctl:

- **Actuation**: Triggers the validator component in response to an event, for example, a request for a CR update. For istioctl, the actuation logic is triggered by the operator invoking the istioctl CLI, which is written in Go using Cobra, a library for creating powerful CLI applications.
- **Validator**: Verifies the input (the IstioOperator CR) against the original schema of the CR.

- **Config generator**: In this phase, a full-blown configuration is created. The configuration includes parameters and values provided in the original event, as well as parameters that were omitted in the original event. The configuration contains the omitted parameters, along with their respective default values.
- **Translator and renderer**: The translator maps IstioOperator's Kubernetes resources specs to Kubernetes resources, while the renderer produces the output manifest after applying all configurations.
- **Resource Manager**: This is responsible for managing the resources in the cluster. It caches the recent state of resources in a built-in cache, which is then compared with the output manifests, and every time there is a deviation or inconsistency between the state of Kubernetes objects (namespaces, CRDs, ServiceAccounts, ClusterRoles, ClusterRoleBindings, MutatingWebhookConfigurations, ValidatingWebhookConfigurations, Services, Deployments, or ConfigMaps) and the output manifest, then the Resource Manager updates them as per the manifest.

Steps to uninstall IstioOperator

As we will not be using IstioOperator for the rest of book, I suggest uninstalling it using the following commands:

```
$ istioctl operator remove
  Removing Istio operator...
  Removed Deployment:istio-operator:istio-operator.
  Removed Service:istio-operator:istio-operator.
  Removed ServiceAccount:istio-operator:istio-operator.
  Removed ClusterRole::istio-operator.
  Removed ClusterRoleBinding::istio-operator.
✓ Removal complete
$ kubectl delete ns istio-operator
namespace "istio-operator" deleted
```

We briefly looked at the istio-proxy in the previous chapter. In the next section, we will examine the Istio agent, which is one of the containers deployed in the istio-proxy.

Istio agent

The Istio agent (also called `pilot-agent`) is part of the control plane deployed in every istio-proxy to help connect to the mesh by securely passing configuration and secrets to the Envoy proxies. Let's look at the istio-agent in one of the microservices of `bookinfo` by listing all running process in the istio-proxy sidecar of `details-v1`:

```
$ kubectl exec -it details-v1-7d79f8b95d-5f4td -c istio-proxy
-n bookinfons --/bin/sh -c "ps -ef"
UID          PID    PPID  C STIME TTY          TIME CMD
istio-p+       1       0  0 Mar14 ?        00:02:02 /usr/local/
bin/pilot-agent p
istio-p+      15       1  0 Mar14 ?        00:08:17 /usr/local/
bin/Envoy -c etc/
```

You must have noticed that pilot-agent is also running within the sidecar. pilot-agent not only bootstraps the Envoy proxy but also generates key and certificate pairs for Envoy proxies to establish the identity of Envoy proxies during mesh communication.

Before we talk about the role of the Istio agent in certificate generation, let's just briefly talk about the Istio **Secret Discovery Service** (**SDS**). The SDS simplifies certificate management and was originally created by the Envoy project to provide a flexible API to deliver secrets/certificates to the Envoy proxy. The components needing the certificates are called SDS clients, and the component generating the certificates is called the SDS server. In the Istio data plane, the Envoy proxy acts as an SDS client and the Istio agent acts as the SDS server. The communication between the SDS client and SDS server happens using the SDS API specifications, mostly implemented over gRPC.

The following steps, also illustrated in *Figure 3.4*, are performed between the Istio agent, Envoy, and istiod to generate the certificate:

1. During sidecar injection, istiod passes information about the SDS, including the location of the SDS server to the Envoy proxy.
2. Envoy sends a request to pilot-agent (SDS server) for certificate generation over a **Unix domain socket** (**UDS**) via SDS protocols. pilot-agent generates a certificate signing request.
3. pilot-agent then communicates with istiod and provides its identity along with the certificate signing request.
4. istiod authenticates pilot-agent and if all is OK, signs the certificate.
5. pilot-agent passes the certificate and keys to the Envoy proxy over UDS.

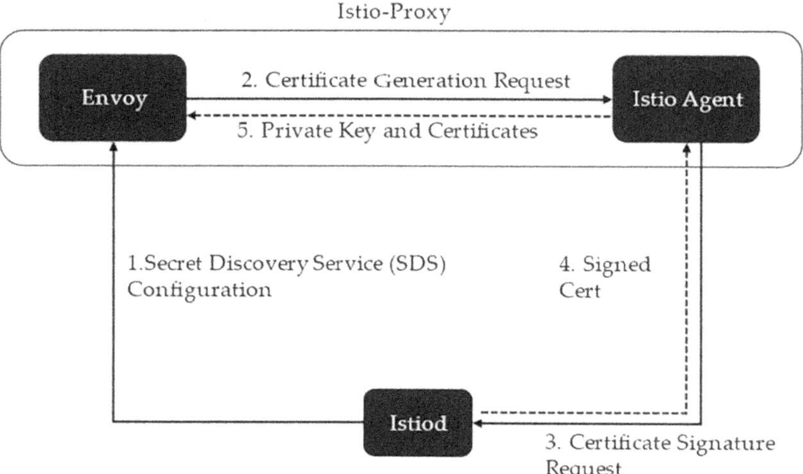

Figure 3.4 – Certificate generation for Envoy communication

In this and prior sections, we covered the Istio control plane. Now it's time to go through various options to deploy the Istio control plane.

Deployment models for the Istio control plane

In the previous chapters, we installed Istio on minikube, which is one local cluster meant for development purposes on your local workstation. When deploying Istio in enterprise environments, the deployment will be not on minikube but rather on an enterprise-grade Kubernetes cluster. The Service Mesh might run on one Kubernetes cluster or be spread across multiple Kubernetes clusters. It might also be the case that all services will be on one network or may be on different networks with no direct connectivity between them. Every organization will have a different network and infrastructure disposition, and the deployment model for Istio will change accordingly.

> **What is a cluster?**
>
> There are many definitions of a cluster depending on what context they are being referred to. In this section, when we say cluster, we are basically referring to a set of compute nodes hosting containerized applications interconnected with each other. You can also think of a cluster as a Kubernetes cluster.

We will be discussing various architecture options for Istio in *Chapter 8*, but for now, let's just briefly go through various deployment models for the control plane.

Single cluster with a local control plane

All sidecar proxies across all namespaces in the cluster connect to the control plane deployed in the same cluster. Similarly, the control plane is watching, observing, and communicating with the Kubernetes API server and sidecars within the same cluster where it is deployed.

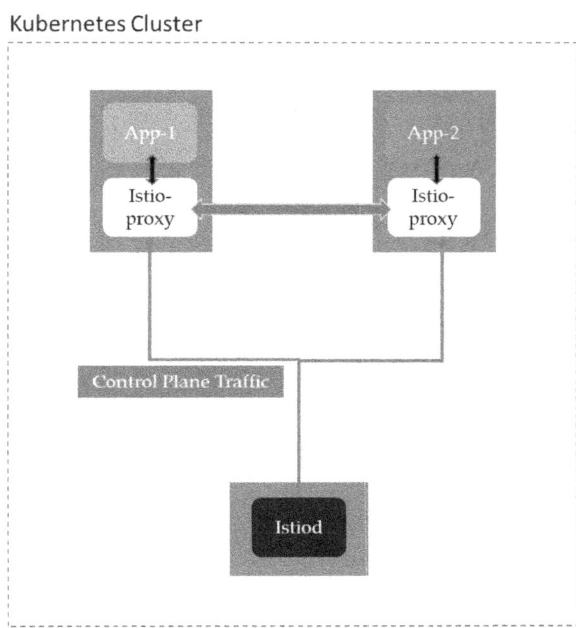

Figure 3.5 – The data plane and control plane residing in the same Kubernetes cluster

The preceding illustration describes the deployment model we used in the previous chapter to deploy Istio. From the diagram, you can see that the Istio control plane and data plane both reside in the same Kubernetes cluster; in our case, it was minikube. istiod is installed in the `istio-system` namespace or any other namespace of your choosing. The data plane comprises various namespaces where applications are deployed along with istio-proxy sidecars.

Primary and remote cluster with a single control plane

A Service Mesh cluster, where the data plane and control plane are deployed in the same Kubernetes cluster, is also called a *primary cluster*. A cluster where the control plane is not collocated with the data plane is called a *remote cluster*.

In this architecture, there is a primary cluster and a remote cluster both sharing a common control plane. With this model, additional configuration is required to provide interconnectivity between the control plane in the primary cluster and the data plane in the remote cluster. The connectivity between the remote cluster and primary cluster control plane can be achieved by adding an Ingress gateway to protect and route communication to the primary control plane. This is shown in the following diagram:

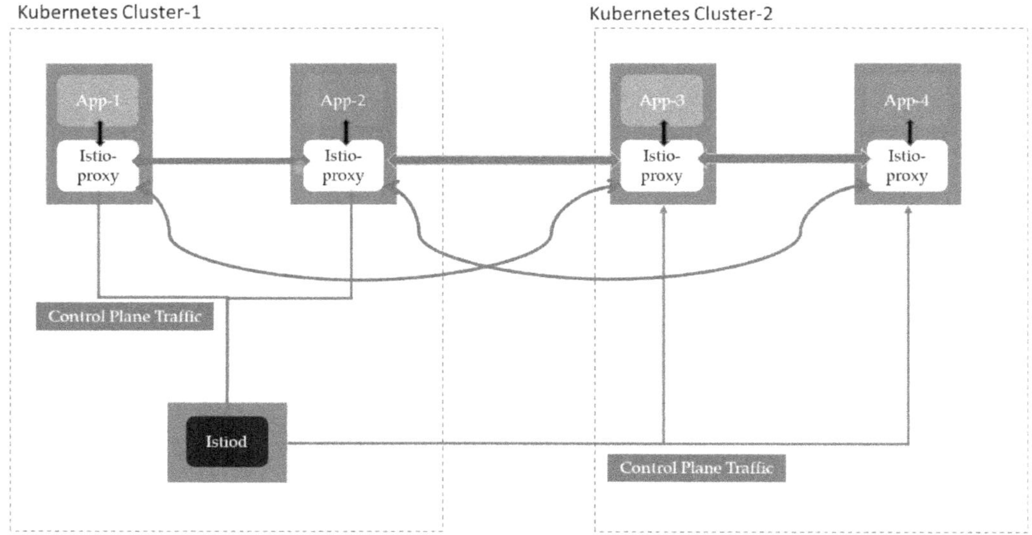

Figure 3.6 – Uni-cluster control plane with data plane spread across multiple Kubernetes clusters

The Istio control plane also needs to be configured to establish the following communications:

- Communication with the remote plane Kubernetes API server
- Patch mutating webhooks into the remote plane to watch the namespaces configured for automated injection of istio-proxy
- Provide endpoints for CSR requests from Istio agents in the remote plane

Single cluster with an external control plane

In this configuration, instead of running a primary cluster with the control and data planes collocated on the same Kubernetes cluster, you can separate them from each other. This is done by deploying the control plane remotely on one Kubernetes cluster and having the data plane deployed on its own dedicated Kubernetes cluster. This deployment can be seen in the following diagram:

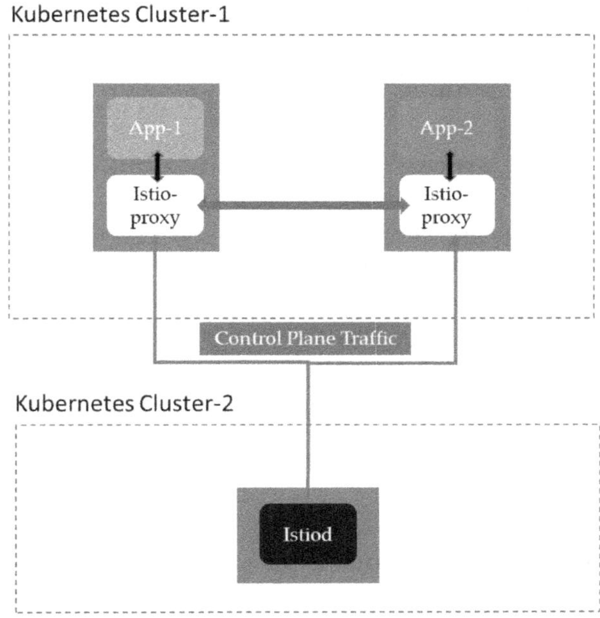

Figure 3.7 – The control plane and data plane residing in separate Kubernetes clusters

For security, separation of concerns, and compliance requirements such as **Federal Information Processing Standards** (**FIPS**), we can be required to deploy the Istio control plane separately from the data plane. Separating the control plane from the data plane allows the enforcement of strict traffic and access policies for control plane traffic without impacting the traffic flow of the data plane. Also, in an enterprise environment, where you have teams who can provide control planes as a managed service to project teams, then this model of deploying the control plane is highly suitable.

So far, the deployment models we have discussed reside over one or many Kubernetes clusters within a shared network. Where the network is not shared, the deployment model becomes more complex. We will go through some of those deployment models, along with the ones we covered in this chapter, with some hands-on exercises in *Chapter 10*.

In the next section, we will look at the Istio data plane and we will do that by understanding Envoy.

Exploring Envoy, the Istio data plane

Envoy is the key component of the Istio data plane. To understand the Istio data plane, it is important to understand and know about Envoy.

Envoy is an open source project and CNCF graduate. You can find more details about Envoy as a CNCF project at `https://www.cncf.io/projects/Envoy/`. In this section, we will learn about Envoy and why it was selected as the service proxy for the Istio data plane.

What is Envoy?

Envoy is a lightweight, highly performant Layer 7 and Layer 4 proxy with an easy-to-use configuration system, making it highly configurable and suitable for serving as a standalone edge-proxy in the API gateway architecture pattern, as well as running as a sidecar in the Service Mesh architecture pattern. In both architecture patterns, Envoy runs in its own single process alongside the applications/services, which makes it easier to upgrade and manage and also allows Envoy to be deployed and upgraded transparently across the entire infrastructure.

To understand Envoy, let's look at the following three distinctive features that make Envoy different from other proxies available today.

Threading model

One of the highlights of the Envoy architecture is its unique threading model. In Envoy, the majority of the threads run asynchronously without blocking each other. Instead of having one thread per connection, multiple connections share the same worker thread running in non-blocking order. The threading model helps to process requests asynchronously but in a non-blocking manner, resulting in very high throughput.

Broadly, Envoy has three types of threads:

- **Main thread**: This owns the startup and shutdown of Envoy and xDS (more on xDS in the next section), API handling, runtime, and general process management. The main thread coordinates all management functionality in general, which does not require too much CPU. Therefore, the Envoy logic related to general management is single-threaded, making the code base simpler to write and manage.
- **Worker thread**: Generally, you run a worker thread per CPU core or per hardware thread if the CPU is hyper-threaded. The worker threads open one or more network locations (ports, sockets, etc) to which downstream systems can connect; this function of Envoy is called *listening*. Each worker thread runs a non-blocking event loop to perform listening, filtering, and forwarding.
- **File flusher thread**: This thread takes care of writing to files in a non-blocking fashion.

Architecture

Another highlight of the Envoy architecture is its filter architecture. Envoy is also an L3/L4 network proxy. It features a pluggable filter chain to write filters to perform different TCP/UDP tasks. A **filter chain** is basically a set of steps where the output from one step is fed into the input of the second step, and so on, just as with pipes in Linux. You can construct logic and behavior by stacking your desired filters to form a filter chain. There are many filters available out of the box to support tasks, such as raw TCP proxy, UDP proxy, HTTP proxy, and TLS client cert authentication. Envoy also supports an additional HTTP L7 filter layer. Through filters, we can perform different tasks, such as buffering, rate limiting, routing, forwarding, and so on.

Envoy supports both HTTP 1.1 and HTTP 2 and can operate as a transparent proxy in both HTTP protocols. This is particularly useful when you have legacy applications that support HTTP 1.1, but when you deploy them alongside Envoy proxy, you can bridge the transformation – meaning the application can communicate over HTTP 1.1 with Envoy, which then uses HTTP 2 to communicate with others. Envoy supports a comprehensive routing subsystem that allows a very flexible routing and redirection functionality, making it suitable for building Ingress/Egress API gateways as well as being deployed as a proxy in the sidecar pattern.

Envoy also supports modern protocols such as gRPC. **gRPC** is an open source remote procedure call framework that can run anywhere. It is widely used for service-to-service communication and is very performant and easy to use.

Configuration

The other highlight of Envoy is how it can be configured. We can configure Envoy using static configuration files that describe the services and how to communicate with them. For advanced scenarios where statically configuring Envoy would be impractical, Envoy supports dynamic configuration and can automatically reload configuration at runtime without needing a restart. A set of discovery services called xDS can be used to dynamically configure Envoy through the network and provide Envoy information about hosts, clusters HTTP routing, listening sockets, and cryptographic material. This makes it possible to write different kind of control planes for Envoy. The control plane basically implements the specification of xDS API and keeps up-to-date information of various resources and information required to be fetched dynamically by Envoy via xDS APIs. There are many open source control plane implementations for Envoy; a couple are linked as follows:

- `https://github.com/envoyproxy/go-control-plane`
- `https://github.com/envoyproxy/java-control-plane`

Various Service Mesh implementations such as Istio, Kuma, Gloo, and so on., which use Envoy as a sidecar, implement xDS APIs to provide configuration information to Envoy.

Envoy also supports the following:

- **Automatic retries**: Envoy supports the retrying of requests any number of times or under a retry budget. The request can be configured to be retried for certain retry conditions depending on application requirement. If you want to read further about retry, head to `https://www.abhinavpandey.dev/blog/retry-pattern`.

- **Circuit breaking**: Circuit breaking is important for microservices architecture. Envoy provides circuit breaking at network level, so as to protect upstream systems across all HTTP request executions. Envoy provides various circuit breaking limits based on configurations such as maximum number of connections, maximum number of pending requests, maximum request, maximum active retries, and maximum concurrent connection pools supported by upstream systems. More details about circuit breaker pattern are available at `https://microservices.io/patterns/reliability/circuit-breaker.html`.
- **Global rate limiting**: Envoy supports global rate limiting to control downstream systems from overwhelming upstream systems. The rate limiting can be performed at the network level as well at HTTP request level.
- **Traffic mirroring**: Envoy supports the shadowing of traffic from one cluster to another. This is very useful for testing as well as a myriad of other use cases, such as machine learning. An example of traffic mirroring at network level is AWS VPC, which provides options to mirror all traffic to VPC. You can read about AWS traffic mirroring at `https://docs.aws.amazon.com/vpc/latest/mirroring/what-is-traffic-mirroring.html`.
- **Outlier detection**: Envoy supports dynamically determining unhealthy upstream systems and removing them from the healthy load-balancing set.
- **Request hedging**: Envoy supports request hedging to deal with tail latency by issuing requests to multiple upstream systems and returning the most appropriate response to the downstream system. You can read more about request hedging at `https://medium.com/star-gazers/improving-tail-latency-with-request-hedging-700c77cabeda`.

We discussed earlier how filter chain-based architecture is one of the differentiating features of Envoy. Now let's read about those filters that make up the filter chain.

HTTP filters

HTTP is one of the most common application protocols, and it's not unusual for the majority of a given workload to operate over HTTP. To support HTTP, Envoy ships with various HTTP-level filters.

When configuring Envoy, you will have to deal primarily with the following configurations:

- **Envoy listeners**: These are the ports, sockets, and any other named network locations that downstream systems connect to
- **Envoy routes**: These are Envoy configurations describing how the traffic should be routed to upstream systems
- **Envoy clusters**: These are logical services formed of a group of similar upstream systems to which envoy routes or forwards the requests
- **Envoy endpoints**: These are individual upstream systems that serve requests

> **Important note**
>
> We will now be using Docker to play with Envoy. If you are running minikube, it will be a good idea to stop minikube now. If you don't have Docker, you can install it by following the instructions at `https://docs.docker.com/get-docker/`.

Armed with the knowledge we've obtained so far, let's go and create some Envoy listeners.

Download the `envoy` Docker image:

```
$ docker pull envoyproxy/envoy:v1.22.2
```

Once you have pulled the Docker image, go ahead and run the following from the Git repository of this chapter:

```
docker run -rm -it -v $(pwd)/envoy-config-1.yaml:/envoy-custom.
yaml -p 9901:9901 -p 10000:10000 envoyproxy/envoy:v1.22.2 -c /
envoy-custom.yaml
```

In the preceding command, we are mounting the `envoy-config-1.yaml` file as a volume and passing it to the Envoy container with the `-c` option. We are also exposing `10000` to the localhost, which is mapped to port `10000` of the Envoy container.

Let's now check the contents of `envoy-config-1.yaml`.The root of Envoy configuration is called bootstrap configuration. The first line describes whether it is in static or dynamic configuration. In this instance, we are proving a static configuration by specifying `static_resources`:

```
Static_resources:
  listeners:
  - name: listener_http
```

In this instance, the configuration is very straightforward. We have defined a listener called `listener_http`, which is listening on `0.0.0.0` and port `10000` for incoming requests:

```
Listeners:
  - name: listener_http
    address:
      socket_address:
        address: 0.0.0.0
        port_value: 10000
```

We have not applied any filter specific to the listener, but we have applied a network filter called `HTTPConnectionManager`, or HCM:

```
Filter_chains:
    - filters:
      - name: envoy.filters.network.http_connection_manager
        typed_config:
          "@type": type.googleapis.com/envoy.
extensions.filters.network.http_connection_manager.
v3.HttpConnectionManager
          stat_prefix: chapter3-1_service
```

The HCM filter is capable of translating raw bytes into HTTP-level messages. It can handle access logging, generate request IDs, manipulate headers, manage route tables, and collect statistics. Envoy also supports defining multiple HTTP-level filters within the HCM filter. We can define these HTTP filters under the `http_filters` field.

In the following configuration, we have applied an HTTP router filter:

```
http_filters:
          - name: envoy.filters.http.router
            typed_config:
              "@type": type.googleapis.com/envoy.extensions.
filters.http.router.v3.Router
          route_config:
            name: my_first_route_to_nowhere
            virtual_hosts:
            - name: dummy
              domains: ["*"]
              routes:
              - match:
                  prefix: "/"
                direct_response:
                  status: 200
                  body:
                    inline_string: "Bootstrap Service Mesh
Implementations with Istio"
```

The router filter is responsible for performing routing tasks and is also the last filter to be applied in the HTTP filter chain. The router filter defines the routes under the `route_config` field. Within the route configuration, we can match the incoming requests by looking at metadata such as the URI, headers, and so on., and based on that, we define where the traffic should be routed or processed.

A top-level element in routing configuration is a virtual host. Each virtual host has a name that's used when emitting statistics (not used for routing) and a set of domains that get routed to it. In `envoy-config-1.yaml`, for all requests, irrespective of the host header, a hardcoded response is returned.

To check the output of `envoy-config1.yaml`, you can use `curl` to test the response:

```
$ curl localhost:10000
Bootstrap Service Mesh Implementations with Istio
```

Let's manipulate the virtual host definition in `route_config` of `envoy-config1.yaml` with the following:

```
            route_config:
              name: my_first_route_to_nowhere
              virtual_hosts:
              - name: acme
                domains: ["acme.com"]
                routes:
                - match:
                    prefix: "/"
                  direct_response:
                    status: 200
                    body:
                      inline_string: "Bootstrap Service Mesh
Implementations with Istio And Acme.com"
              - name: ace
                domains: ["acme.co"]
                routes:
                - match:
                    prefix: "/"
                  direct_response:
                    status: 200
                    body:
                      inline_string: "Bootstrap Service Mesh
Implementations with Istio And acme.co"
```

Here, we have defined two entries under `virtual_hosts`. If an incoming request's host header is `acme.com`, then the routes defined in the `acme` virtual host will get processed. If the incoming request is destined for `acme.co`, then the routes defined under the `ace` virtual host will get processed.

Stop the Envoy container and restart it using the following commands:

```
docker run -rm -it -v $(pwd)/envoy-config-1.yaml:/envoy-custom.
yaml -p 9901:9901 -p 10000:10000 envoyproxy/envoy:v1.22.2 -c /
envoy-custom.yaml
```

Check the output by passing different host headers to `curl`:

```
$ curl -H host:acme.com localhost:10000
Bootstrap Service Mesh Implementations with Istio And Acme.com
$ curl -H host:acme.co localhost:10000
Bootstrap Service Mesh Implementations with Istio And acme.co
```

In most cases, you will not send a hardcoded response to HTTP requests. Realistically, you will want to route requests to real upstream services. To demonstrate this scenario, we will be making use of nginx to mock a dummy upstream service.

Run the `nginx` Docker container using the following command:

```
docker run -p 8080:80 nginxdemos/hello:plain-text
```

Check the output from another terminal using `curl`:

```
$ curl localhost:8080
Server address: 172.17.0.3:80
Server name: a7f20daf0d78
Date: 12/Jul/2022:12:14:23 +0000
URI: /
Request ID: 1f14eb809462eca57cc998426e73292c
```

We will route the request being processed by Envoy to nginx by making use of cluster subsystem configurations. Whereas the `Listener` subsystem configurations handle downstream request processing and managing the downstream request life cycle, the cluster subsystem is responsible for selecting and connecting the upstream connection to an endpoint. In the cluster configuration, we define clusters and endpoints.

Let's edit `envoy-config-2.yaml` and modify the virtual host for `acme.co` with the following:

```
          - name: ace
            domains: ["acme.co"]
```

```
            routes:
            - match:
                prefix: "/"
              route:
                cluster: nginx_service
  clusters:
  - name: nginx_service
    connect_timeout: 5s
    load_assignment:
      cluster_name: nginx_service
      endpoints:
      - lb_endpoints:
        - endpoint:
            address:
              socket_address:
                address: 172.17.0.2
                port_value: 80
```

We have removed the `direct_response` attribute and replaced it with the following:

```
route:
                cluster: nginx_service
```

We have added cluster to the definition, which sits at the same level as the listener configuration. In the cluster definition, we defined the endpoints. In this case, the endpoint is the `nginx` Docker container running on port `80`. Please note that we are assuming that both Envoy and nginx are running on the same Docker network.

You can find the IP of the `nginx` container by inspecting the container. The config is saved in `envoy-config-3.yaml`. Please update the `address` value with the correct IP address of the nginx container and run the Envoy container with the updated `envoy-config-3.yaml`:

```
$ docker run -rm -it -v $(pwd)/envoy-config-3.yaml:/
envoy-custom.yaml -p 9901:9901 -p 10000:10000 envoyproxy/
envoy:v1.22.2 -c /envoy-custom.yaml
```

Perform the `curl` test and you will notice the response for the request destined for `acme.co` is coming from the nginx container:

```
$ curl -H host:acme.com localhost:10000
Bootstrap Service Mesh Implementations with Istio And Acme.com
$ curl -H host:acme.co localhost:10000
Server address: 172.17.0.2:80
Server name: bfe8edbee142
Date: 12/Jul/2022:13:05:50 +0000
URI: /
Request ID: 06bbecd3bc9901d50d16b07135fbcfed
```

Envoy provides several built-in HTTP filters. You can find the complete list of HTTP filters here: `https://www.envoyproxy.io/docs/envoy/latest/configuration/http/http_filters/http_filters#config-http-filters`.

Listener filters

We read previously that the listener subsystem handles the processing of incoming requests and the response to and from downstream systems. In addition to defining which addresses and ports Envoy *listens* on for incoming requests, we can optionally configure each listener with **listener filters**. The listener filters operate on newly accepted sockets and can stop or subsequently continue execution to further filters.

The order of the listener filters matters, as Envoy processes them sequentially right after the listener accepts a socket and before the connection is created. We use results from the listener filters to do filter matching to select appropriate network filter chains. For example, using a listener filter, we can determine the protocol type, and based on that, we might run specific network filters related to that protocol.

Let's look at a simple example of listener filters in `envoy-config-4.yaml` under `listener_filters`. You will notice that we are using `envoy.filters.listener.http_inspector` of the following type: `type.googleapis.com/envoy.extensions.filters.listener.http_inspector.v3.HttpInspector`.

The `HTTPInspector` listener filter can detect the underlying application protocol and whether it is `HTTP/1.1` or `HTTP/2`. You can read more about the `HTTPInspector` listener filter here: `https://www.envoyproxy.io/docs/envoy/latest/configuration/listeners/listener_filters/http_inspector`.

In this example, we are using the listener filter to find the application protocol via the filter chain. Depending on which HTTP protocol is used by the downstream system, we then apply a variety of HTTP filters, as discussed in previous sections.

You can find this example in the `envoy-config-4.yaml` file. Go ahead and apply the configuration to Envoy, but do also remember to close down the Docker containers you created for previous examples:

```
$ docker run -rm -it -v $(pwd)/envoy-config-4.yaml:/
envoy-custom.yaml -p 9901:9901 -p 10000:10000 envoyproxy/
envoy:v1.22.2 -c /envoy-custom.yaml
```

Perform `curl` with the `HTTP 1.1` and `HTTP 2` protocols, and you will see that Envoy is able to figure out the application protocol and route the request to the correct destination:

```
$ curl localhost:10000 -http1.1
HTTP1.1
$ curl localhost:10000 -http2-prior-knowledge
HTTP2
```

As I mentioned earlier when introducing Envoy, it is highly configurable and can be configured dynamically. I believe the dynamic configurability of Envoy is what makes it so popular and makes it standout from the other proxies available today. Let's look more into this next!

Dynamic configuration via xDS APIs

So far, in our previous examples, we have been using static configuration by specifying `static_resources` at the beginning of the config file. Every time we wanted to change the config, we had to restart the Envoy container. To avoid this, we can make use of **dynamic configuration**, where Envoy dynamically reloads the configuration either by reading it from disk or over the network.

For dynamic configuration where Envoy fetches the configuration over the network, we need to make use of xDS APIs, which are basically a collection of various service discovery APIs related to various Envoy configurations. To make use of xDS APIs, you need to implement an application that can fetch the latest values of various Envoy configurations and then present them via gRPC # as per the xDS *protobuf* specifications (also called *protocol buffers*; you can find details about protocol buffers at `https://developers.google.com/protocol-buffers`, and more on gRPC at `https://grpc.io/`). This application is commonly referred to as the control plane. The following diagram describes this concept.

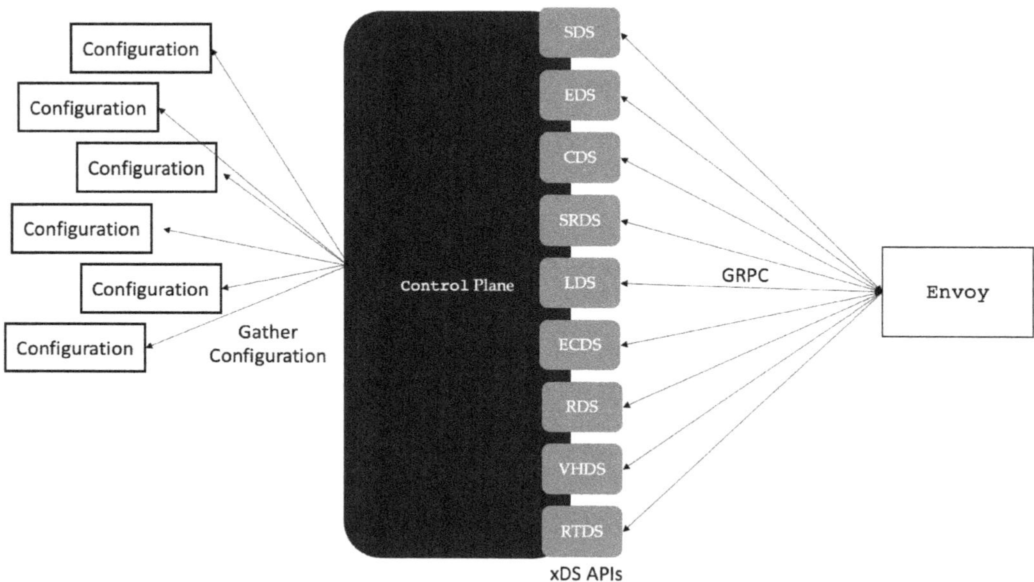

Figure 3.8 – Control plane implementation of the xDS API

Let's see what the service discovery APIs provide:

- **Secret Discovery Service (SDS)**: Provides secrets, such as certificates and private keys. This is required for MTLS, TLS, and so on.
- **Endpoint Discovery Service (EDS)**: Provides details of members of the cluster.
- **Cluster Discovery Service (CDS)**: Provides cluster-related information including references to endpoints.
- **Scope Route Discovery Service (SRDS)**: Provides route information in chunks when the route confirmation is large.
- **Listener Discovery Service (LDS)**: Provides details of listeners including the ports, addresses, and all associated filters.
- **Extension Config Discovery Service (ECDS)**: Provides extension configuration, such as HTTP filters, and so on. This API helps to fetch information independently from the listener.
- **Route Discovery Service (RDS)**: Provides route information including a reference to the cluster.
- **Virtual Host Discovery Service (VHDS)**: Provides information about the virtual hosts .

- **Runtime Discovery Service** (**RTDS**): This service provides information about the runtime. The runtime configuration specifies a virtual filesystem tree that contains reloadable configuration elements. This virtual filesystem can be realized via a series of local filesystems, static bootstrap configuration, RTDS, and admin console-derived overlays.
- **Aggregated Discovery Service** (**ADS**): ADS allows all APIs and their resources to be delivered via a single API interface. Through ADS APIs, you can sequence changes related to various resource types, including listeners, routes, and clusters, and deliver them via a single stream.
- **Delta Aggregated Discovery Service** (**DxDS**): With other APIs, every time there is a resource update, the API needs to include all resources in the API response. For example, every RDS update must contain every route. If we don't include a route, Envoy will consider the route to have been deleted. Doing updates this way results in high bandwidth usage and computational costs, especially when a lot of resources are being sent over the network. Envoy supports a delta variant of xDS where we can include only resources we want to add/remove/update to improve on this scenario.

We covered Envoy filters in the previous section, but note that you are not limited to the built-in filters – you can easily build new filters, as we'll see in the next section.

Extensibility

The filter architecture of Envoy makes it highly extensible; you can make use of various filters from the filter library as part of the filter chain. When you need some functionality not available in the filter library, then Envoy also provides the flexibility to write your own custom filter, which is then dynamically loaded by Envoy and can be used like any other filter. By default, Envoy filters are written in C++, but they can also be written using Lua script or any other programming language compiled into **WebAssembly** (**Wasm**).

The following is a brief description of all three options currently available for writing Envoy filters:

- **Native C++ API**: The most favorable option is to write native C++ filters and then package them with Envoy. But this option requires recompiling Envoy, which might not be ideal if you are not a big enterprise that wants to maintain its own version of Envoy.
- **Lua filter**: Envoy provides a built-in HTTP Lua filter named `envoy.filters.http.lua` that allows you to define a Lua script either inline or as an external file, and execute it during both the request and response flows. Lua is a free, fast, portable, and powerful scripting language that runs over LuaJIT, which is a just-in-time compiler for Lua. At runtime, Envoy creates a Lua environment for each worker thread and runs Lua scripts as coroutines. As the HTTP Lua filters are executed during request and response flows, you can do the following:
    - Inspect and modify headers and trailers during request/response flows
    - Inspect, block, or buffer the body during request/response flows
    - Invoke upstream systems asynchronously

- **Wasm filter**: Last but not least is Wasm-based filters. We write these filters using our preferred programming language and then compile the code into a low-level assembly-like programming language called Wasm, which is then loaded by Envoy dynamically at runtime. Wasm is widely used in the open web, where it executes inside JavaScript virtual machines within web browsers. Envoy embeds a subset of the V8 VM (`https://v8.dev/`) with every worker thread to execute Wasm modules. We will read more about Wasm and do hands-on exercises in *Chapter 9*.

The ability to write custom filters makes Envoy extensible enough to implement for custom use cases. Support for Wasm-based filters brings down the learning curve of writing new filters as you can use a language you are most comfortable with. We hope that with the growing adoption of Envoy, there will be more tooling available to the developer to easily extend it using custom filters.

Summary

This chapter provided you with details about Istio control plane components including istiod and its architecture. We also read about the Istio operator, the CLI, and how certificate distribution works. The Istio control plane can be deployed in various architecture patterns, and we had an overview of some of these deployment patterns as well.

After covering the Istio control plane, we read about Envoy, a lightweight, highly performant `l3/l4/l7` proxy. It provides a range of configurations via the listener and cluster subsystems to control request processing. The filter-based architecture is easy to use, as well as extensible, as new filters can be written in Lua, Wasm, or C++ and can easily be plugged into Envoy. Last but not least is the ability of Envoy to support dynamic configuration via xDS APIs. Envoy is the best choice for the Istio data plane because of its flexibility and performance when serving as a proxy, as well as its easy configurability via xDS APIs, which are implemented by the Istio control plane. The istio-proxy, as discussed in the previous chapter, is made up of Envoy and the Istio agent.

In the next chapter, we will put Istio to one side and instead immerse ourselves in experiencing a real-life application. We will take the application to a production-like environment and then discuss the problems that engineers would face in building and operating such an application. In *Part 2* and *Part 3* of this book, we will make use of this application in hands-on exercises. So, sit tight and brace yourselves for the next chapter.

Part 2: Istio in Practice

This part describes the application of Istio and how it is used to manage application traffic, provide resiliency to applications, and secure communication between microservices. With the help of numerous hands-on examples, you will learn about various Istio traffic management concepts and use them to perform application networking. The part concludes with a chapter on observability, in which you will read about how to observe the Service Mesh and use it to understand system behavior and the underlying causes behind faults so that you can confidently troubleshoot issues and analyze the effects of potential fixes.

This part contains the following chapters:

- *Chapter 4, Managing Application Traffic*
- *Chapter 5, Managing Application Resiliency*
- *Chapter 6, Securing Microservices Communication*
- *Chapter 7, Service Mesh Observability*

4
Managing Application Traffic

Microservices architecture creates a sprawl of loosely coupled applications deployed as containers on platforms such as Kubernetes. With the loose coupling of applications, inter-service traffic management becomes complex. If exposed insecurely to external systems, it can cause exposure of sensitive data, making your system vulnerable to external threats. Istio provides various mechanisms to secure and govern the following kinds of application traffic:

- Ingress traffic coming to your application from outside
- Inter-mesh traffic generated between various components of the application
- Egress traffic going out from your application to other applications outside the mesh

In this chapter, we will read about and practice managing application traffic by going through the following topics in detail.

- Managing Ingress traffic using the Kubernetes Ingress resource and an Istio Gateway
- Traffic routing and canary release
- Traffic mirroring
- Routing traffic to services outside the mesh
- Exposing Ingress over HTTPS
- Managing Egress

It will be a good idea to delete Istio and install it again to get a clean slate, as well as practice what you learned in *Chapter 2*. Istio releases a minor version at a 3-month cadence, as described at `https://istio.io/latest/docs/releases/supported-releases/`; therefore, it is recommended to keep your Istio version up to date using the documentation on the Istio website (`https://istio.io/latest/docs/setup/getting-started/#download`) and the concepts you learned about in *Chapter 2*.

Technical requirements

In this section, we will create an AWS cloud setup, which will be used to perform hands-on exercises in this and subsequent chapters. You can use any cloud provider of your choice, but to introduce some variety in this book, I have selected AWS for *Part 2* and Google Cloud for *Part 3*. You can also use minikube for the exercises, but you will need at least a quad-core processor and 16 GB or more of RAM assigned to minikube for smooth, lag-free operations.

Setting up the environment

Let's go!

1. Create an AWS account. If you don't already have an AWS account, then it's time to sign up with AWS using `https://portal.aws.amazon.com/billing/signup#/start/email`.
2. Set up the AWS CLI:
    I. Install the AWS CLI using the steps provided at `https://docs.aws.amazon.com/cli/latest/userguide/getting-started-install.html`.
    II. Configure the AWS CLI using the steps provided at `https://docs.aws.amazon.com/cli/latest/userguide/cli-chap-configure.html`.
3. Install the AWS IAM authenticator using the steps provided at `https://docs.aws.amazon.com/eks/latest/userguide/install-aws-iam-authenticator.html`.
4. Install Terraform, which is infrastructure-as-code software that automates the provisioning of infrastructure. This helps you to create an infrastructure that is consistent with the infrastructure used for the exercises in this book. I hope this provides a hassle-free experience where you can spend more time learning about Istio rather than troubleshooting infrastructure issues. Follow the steps provided at `https://learn.hashicorp.com/tutorials/terraform/install-cli?in=terraform/aws-get-started`.

Creating an EKS cluster

Next, you will need to create an EKS cluster. **EKS** stands for **Elastic Kubernetes Service**, which is a managed service offering from AWS. EKS provides a hassle-free and easy-to-use Kubernetes cluster where you don't need to worry about the setup and operation of the Kubernetes control plane. We will make use of Terraform to set up an EKS cluster; the Terraform code and config are available at `sockshop/devops/deploy/terraform` in the source code repo of this book.

Perform the following steps from the `sockshop/devops/deploy/terraform/src` folder:

1. Initialize Terraform. Prepare the working directory so that Terraform can run the configurations:

   ```
   % terraform init
   Initializing the backend...
   Initializing provider plugins...
   - Reusing previous version of hashicorp/aws from the
   dependency lock file
   - Using previously-installed hashicorp/aws v4.26.0
   Terraform has been successfully initialized!
   ```

   You can read about `init` at `https://developer.hashicorp.com/terraform/tutorials/cli/init`.

2. Configure Terraform variables by modifying `sockshop/devops/deploy/terraform/src/variables.tf`. Default values will work, but you can also modify them to suit your requirements.

3. Plan the deployment. In this step, Terraform creates an execution plan that you can inspect to find any discrepancies and get a preview of the infrastructure, although it is not provisioned yet.

   The output in the following code snippet is shortened to save space:

   ```
   % terraform plan
   .........
   ~ cluster_endpoint          =
   "https://647937631DD1A55F1FDDAB99E08DEE0C.gr7.us-east-1.
   eks.amazonaws.com" -> (known after apply)
   ```

   You can read more about `plan` at `https://developer.hashicorp.com/terraform/tutorials/cli/plan`.

4. Provision the infrastructure. In this step, Terraform will create the infrastructure as per the execution plan created in previous steps:

   ```
   % terraform apply
   ```

   Once the infrastructure is provisioned, Terraform will also set a variable as defined in `sockshop/devops/deploy/terraform/src/outputs.tf`.

   You can read more about `apply` at `https://developer.hashicorp.com/terraform/tutorials/cli/apply`.

Setting up kubeconfig and kubectl

Next, we will configure kubectl to be able to connect the newly created EKS cluster using Terraform. Use the following aws cli command to update kubeconfig with cluster details:

```
% aws eks --region $(terraform output -raw region) update-
kubeconfig --name $(terraform output -raw cluster_name)
```

Next, check that kubectl is using the correct context:

```
% kubectl config current-context
```

If the value is not as expected, you can perform the following:

```
% kubectl config view -o json | jq '.contexts[].name'
"arn:aws:eks:us-east-1:803831378417:cluster/MultiClusterDemo-
Cluster1-cluster"
"minikube"
```

Find the correct name of the cluster and then use the following command to set the kubectl context:

```
% kubectl config use-context "arn:aws:eks:us-east-
1:803831378417:cluster/MultiClusterDemo-Cluster1-cluster"
Switched to context "arn:aws:eks:us-east-
1:803831378417:cluster/MultiClusterDemo-Cluster1-cluster".
```

There will be instances where you will need to use minikube. In such a case, simply use the following command to switch context, and vice versa for switching back to EKS:

```
% kubectl config use-context minikube
Switched to context "minikube".
```

Deploying the Sockshop application

Finally, to add some variety to hands-on exercises, we will make use of a demo application called Sockshop available at `https://github.com/microservices-demo/microservices-demo`. You can find the deployment files in `sockshop/devops/deploy/kubernetes/manifests`:

```
% kubectl create -f sockshop/devops/deploy/kubernetes/
manifests/00-sock-shop-ns.yaml
% kubectl create -f  sockshop/devops/deploy/kubernetes/
manifests/* -n sock-shop
```

This will deploy the Sockshop application, and you are all set with the environment. The next step for you is to install the latest version of Istio using the instructions at `https://istio.io/latest/docs/setup/install/istioctl/` and the concepts you learned about in *Chapter 2*.

In the rest of the chapter and the book, we will be making use of the Sockshop application to demonstrate various Service Mesh concepts. Feel free to use the sample `BookInfo` application that is provided with Istio, or any other application you may like for performing hands-on exercises. Let's get started!

Managing Ingress traffic using the Kubernetes Ingress resource

When building applications that need to be consumed by other applications from outside the network boundary in which the application is deployed, you will need to build an Ingress point using which the consumers can reach the application. In Kubernetes, a Service is an abstraction through which a set of Pods are exposed as a network service. When these services need to be consumed by other applications, they need to be made externally accessible. Kubernetes supports `ClusterIP` for consuming services internally from within the cluster, `NodePort` for consuming the service outside the cluster but within the network, `LoadBalancer` for consuming the services externally via the cloud load balancer, and there are also options for exposing an internal-facing load balancer for internal traffic outside of the Kubernetes cluster. In this section, we will read about how we can configure Istio to expose a service using the Kubernetes Ingress resource.

In the previous chapter, we exposed the frontend service as a `NodePort` type and accessed it via minikube tunnel as well as AWS `Loadbalancer`. This approach takes away any control we might need on how the traffic to the frontend service should be managed.

So, instead of making use of `Loadbalancer` service types to expose the frontend service, let's make the frontend service internal facing and rather make use of the Kubernetes Ingress resource. Update the frontend service Kubernetes configuration by removing `NodePort` (if using minikube) or `LoadBalancer` (if deploying on AWS) by removing the following lines from the YAML file:

```
type: NodePort
…..
    nodePort: 30001
```

The preceding changes make the service type take the default value of `ClusterIp`.

The updated file is also available in the `Chapter4` folder under the name `10-1-front-end-svc.yaml`.

Go ahead and change the frontend service type to `ClusterIP` using the following command:

```
$ kubectl apply -f Chapter4/ClusterIp-front-end-svc.yaml
```

After the changes, you will notice that the Sockshop website is not accessible from the browser due to obvious reasons.

Now, we will make use of the Kubernetes Ingress resource to provide access to the Sockshop frontend service. Kubernetes Ingress is a way to provide access to `ClusterIP` services in the cluster. Ingress defines the addressed host accepted by Ingress, along with a list of URIs and the services to which the request needs to be routed. The following is an illustration highlighting this concept:

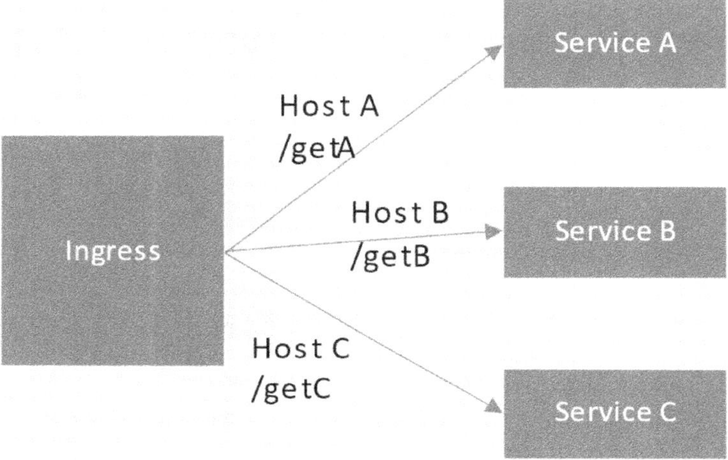

Figure 4.1 – Kubernetes Ingress resource

Along with defining Ingress, we also need to define Ingress controllers, which are another Kubernetes resource that is responsible for handling the traffic as per the specification defined in the Ingress resource.

The following illustrates the relationship between Ingress, Ingress controllers, and Services. Please note that Ingress is a logical construct – that is, a set of rules enforced by the Ingress controller.

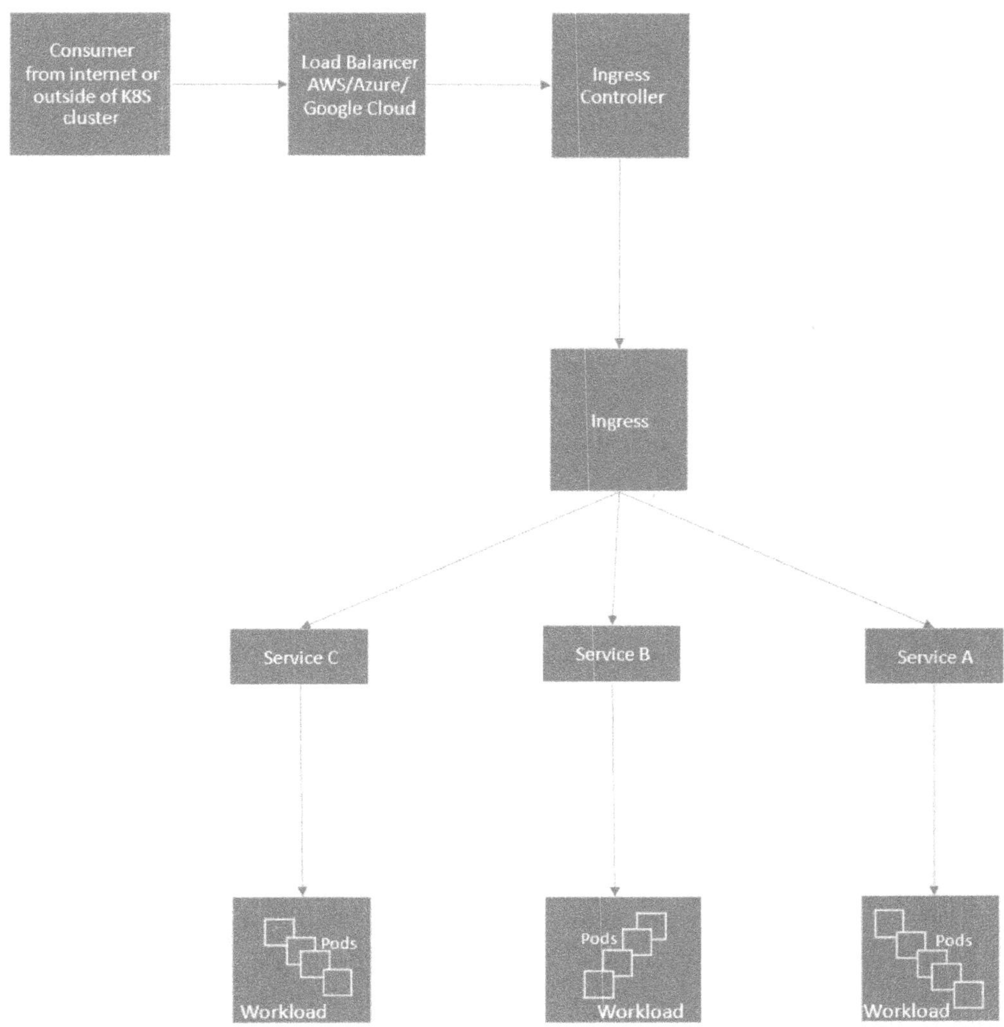

Figure 4.2 – Ingress controllers

Next, we will be making use of the Istio Gateway controller to handle the Ingress; we read about the Istio Gateway in *Chapter 3*.

We will need to provide the following configuration, also defined in `Chapter4/1-istio-ingress.yaml`, to make the changes:

```
apiVersion: networking.k8s.io/v1
kind: Ingress
```

```
metadata:
  annotations:
    kubernetes.io/ingress.class: istio
  name: sockshop-istio-ingress
  namespace: sock-shop
spec:
  rules:
  - host: "sockshop.com"
    http:
      paths:
      - path: /
        pathType: Prefix
        backend:
          service:
            name: front-end
            port:
              number: 80
```

In the preceding configuration, we are doing the following:

- Creating an Ingress resource with an annotation of `kubernetes.io/ingress.class: istio`, which, via admission controllers, as we discussed in *Chapter 3*, tells Istio that this Ingress is to be handled by the Istio Gateway.
- The Ingress resource is defined in the `sock-shop` namespace, as that's where our Sockshop frontend service exists.
- A rule that says that any request specified by `path` of the / value and destined for the `"sockshop.com"` host (specified by `host` and the `sockshop.com` value) should be handled by this Ingress.
- Within the `path` configuration, we are configuring `pathType` of `Prefix`, which basically means that any request of the `hostname/` format will be matched. Other values for `pathType` are as follows:
  - `Exact`: the path is matching exactly as specified in `path`
  - `ImplementationSpecific`: the matching of `path` is decided by the underlying implementation of the Ingress controller

Apply the rule using the following command:

```
$ kubectl create -f Chapter4/1-istio-ingress.yaml
```

If you are using minikube for this exercise, then run `minikube tunnel` in a separate terminal and get the external IP from the output. Find the port at which the service is exposed by the Istio Ingress gateway using the following command:

```
$ kubectl get svc istio-ingressgateway -n istio-system -o wide
NAME                    TYPE            CLUSTER-IP
EXTERNAL-IP     PORT(S)
                                          AGE    SELECTOR
istio-ingressgateway    LoadBalancer    10.97.245.106
10.97.245.106    15021:32098/TCP,80:31120/TCP,443:30149/
TCP,31400:30616/TCP,15443:32339/TCP    6h9m    app=istio-
ingressgateway,istio=ingressgateway
```

In this instance, the Ingress gateway is exposing traffic from port `80` to Ingress port `31120`, and `443` to `30149`, but it may be different for your setup.

If you followed the instruction in *Chapter 4* to use AWS EKS, then the IP and ports will be different; the following is an equivalent of minikube for AWS EKS:

```
$ kubectl get svc istio-ingressgateway -n istio-system
NAME                    TYPE            CLUSTER-IP
 EXTERNAL-IP
                          PORT(S)
                                                 AGE
istio-ingressgateway    LoadBalancer    172.20.143.136
 a816bb2638a5e4a8c990ce790b47d429-1565783620.us-east-1.elb.
amazonaws.com    15021:30695/TCP,80:30613/TCP,443:30166/
TCP,31400:30402/TCP,15443:31548/TCP    29h
```

In this example, the Ingress gateway is exposed via an AWS classic load balancer at the following:

- `http://a816bb2638a5e4a8c990ce790b47d429-1565783620.us-east-1.elb.amazonaws.com:80` for HTTP traffic
- `https://a816bb2638a5e4a8c990ce790b47d429-1565783620.us-east-1.elb.amazonaws.com:443` for HTTPS traffic

Going forward, please use appropriate IPs and ports depending on your choice of environment. The examples in the rest of the chapters are deployed on the AWS EKS cluster, but they will also work for any other Kubernetes provider.

Go ahead and test the Ingress to frontend service via `curl`:

```
curl -HHost:sockshop.com http://
a816bb2638a5e4a8c990ce790b47d429-1565783620.us-east-1.elb.
amazonaws.com/
```

Or if using the Chrome browser, then use extensions such as ModHeader, available at `http://modheader.com/`. In either case, you will need to provide the `host` header with the value of `sockshop.com`.

So, we saw how the Istio Ingress gateway can be configured to handle Kubernetes Ingress.

Let's add another Ingress rule to see how the Istio Ingress controller can handle multiple Ingress rules. We will be making use of the `envoy` configuration we did in *Chapter 3*, where we used a router filter to return a dummy string:

1. In the following command, we are creating a `chapter4` namespace so that we can be organized, and it will be easier to clean up:

   ```
   $ kubectl create ns chapter4
   ```

2. At this stage, we don't need automatic sidecar injection, but from a visibility point of view and for getting meaningful information from Kiali, it will be a good idea to enable Istio sidecar injection using the following command we discussed in *Chapter 2*:

   ```
   $ kubectl label namespace chapter4 istio-
   injection=enabled --overwrite
   ```

3. We will then go ahead with the creation of `configmap` to load `envoy` config (also discussed in *Chapter 3*), which will be required by the Pods we will be creating in the next step:

   ```
   $ kubectl create configmap envoy-dummy --from-
   file=Chapter3/envoy-config-1.yaml -n chapter4
   ```

4. Next, we are creating the service and deployment to run `envoy` to return a dummy response for all HTTP requests:

   ```
   $ kubectl apply -f Chapter4/01-envoy-proxy.yaml
   ```

5. And finally, we create an Ingress rule to route all traffic destined for `mockshop.com` to the `envoy` service we created in the previous step:

   ```
   $ kubectl apply -f Chapter4/2-istio-ingress.yaml
   ```

6. Go ahead and test using the `sockshop.com` and `mockshop.com` hosts headers; the Istio Ingress controller will manage the routing to the appropriate destination as per defined Ingress rules.

The following illustration describes what we have configured so far. Note how the Ingress rules define the routing of traffic to Service A and B based on hostnames:

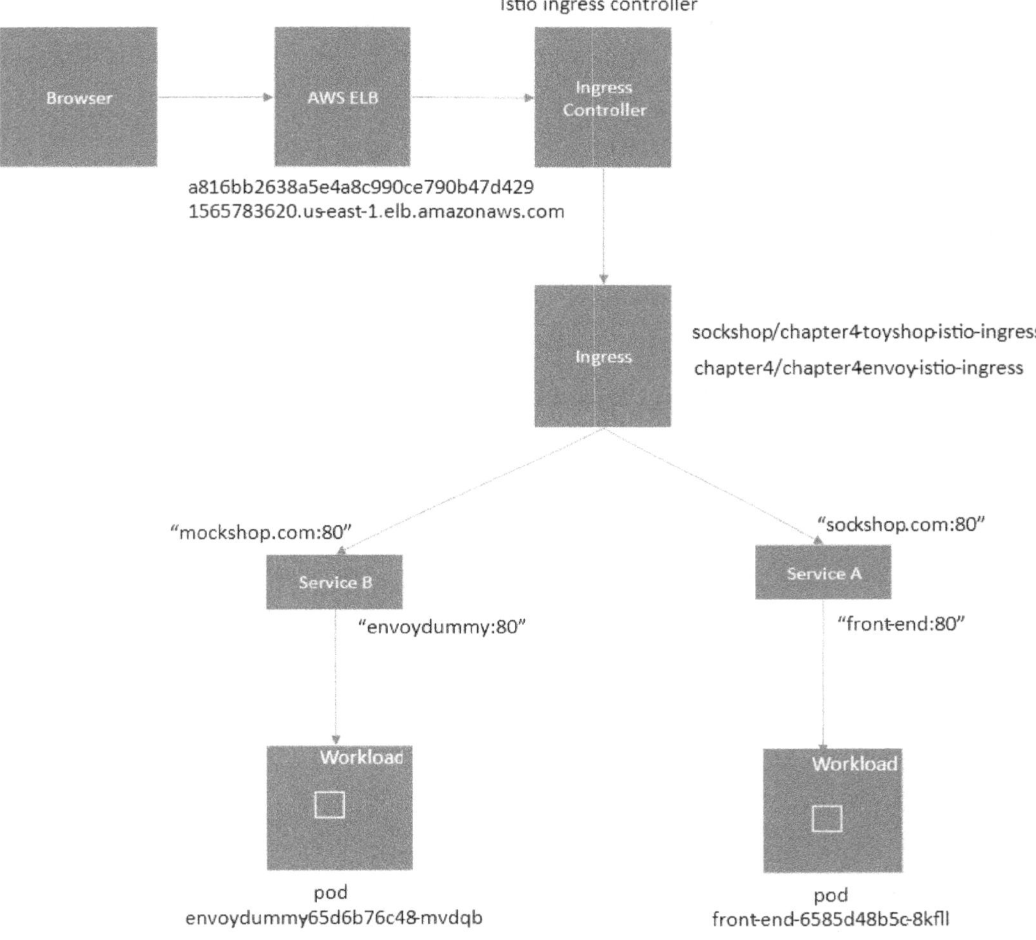

Figure 4.3 – Snapshot of Ingress configuration

In this section, we discussed how to expose services outside of the Kubernetes cluster using the Kubernetes Ingress resource and Istio Ingress controller. In this kind of Ingress configuration, although we are using Istio to manage the Ingress, we are limited by the spec of Kubernetes Ingress, which allows Ingress controllers to perform limited functions such as load balancing, SSL termination, and name-based virtual hosting. When using the Kubernetes Ingress resource type, we are not leveraging a wide range of functionality provided by Istio to manage Ingress. When using Istio, it is recommended to use the Istio Gateway CRD to manage Ingress; we will be discussing that in the next section.

Before moving on, let's do some technical cleanup of your environment so that it doesn't conflict with upcoming exercises:

```
$ kubectl delete -f Chapter4/2-istio-ingress.yaml
$ kubectl delete -f Chapter4/1-istio-ingress.yaml
```

Important note

Throughout this book, we will be leaving you reminders to reverse or clean up the configurations. You can use the preceding commands to execute the cleanup.

Managing Ingress using the Istio Gateway

When managing Ingress, it is recommended to make use of Istio Gateway over the Kubernetes Ingress resource. Istio Gateway is like a load balancer running at the edge of the mesh receiving incoming HTTP and TCP connections.

When configuring Ingress via Istio Gateway, you need to perform the following tasks.

Creating the gateway

The following code block creates an Istio Gateway resource:

```
apiVersion: networking.istio.io/v1alpha3
kind: Gateway
metadata:
  name: chapter4-gateway
  namespace: chapter4
spec:
  selector:
    istio: ingressgateway
  servers:
  - port:
      number: 80
      name: http
      protocol: HTTP
    hosts:
    - "sockshop.com"
    - "mockshop.com"
```

Here, we are declaring a Kubernetes resource named `chapter4-gateway` of the `gateway.networking.istio.io` type custom resource definition in the `chapter4` namespace. This is also equivalent to defining a load balancer.

In the `servers` property, we are defining the following:

- `hosts`: These are one or more DNS names exposed by the gateway. In the preceding example, we are defining two hosts: `sockshop.com` and `mockshop.com`. Any other hosts apart from these two will be rejected by the Ingress gateway.
- `port`: In the port configuration, we define port numbers and the protocols, which can be either `HTTP`, `HTTPS`, `gRPC`, `TCP`, `TLS`, or `Mongo`. The name of the port can be anything you like to use. In this example, we are exposing port `80` over the `HTTP` protocol.

To summarize, the gateway will accept any HTTP request over port `80` with the host header of `sockshop.com` or `mockshop.com`.

Creating virtual services

A virtual service is another set of abstractions between the Ingress gateway and destination services. Using virtual services, you declare how the traffic for a single host (such as `sockshop.com`) or multiple hosts (such as `mockshop.com` and `sockshop.com`) should be routed to its destination. For example, you can define the following in a virtual service for all traffic addressed to `sockshop.com`:

- Request with the `/path1` URI should go to service 1, and `/path2` should go to service 2
- Route request based on the value of header or query parameters
- Weight-based routing or traffic splitting – for example, 60% of the traffic goes to version 1 of the service and 40% goes to another version of the traffic
- Define timeouts – that is, if a response is not received from the upstream service in *X* seconds, then the request should time out
- Retry – that is, how many times a request should be attempted if the upstream system is not responding or is too slow to respond

All these routing features are implemented via virtual services, and we will read more about them in this chapter as well as the next chapter.

In the following configuration, we are defining two virtual services that contain rules regarding traffic matching and to what destination it should be routed:

```
---
apiVersion: networking.istio.io/v1alpha3
kind: VirtualService
```

```
metadata:
  name: sockshop
  namespace: chapter4
spec:
  hosts:
  - "sockshop.com"
  gateways:
  - chapter4-gateway
  http:
  - match:
    - uri:
        prefix: /
    route:
    - destination:
        port:
          number: 80
        host: front-end.sock-shop.svc.cluster.local
```

In the preceding configuration, we have defined a virtual service named `sockshop`. In the `spec` property, we are defining the following:

- `hosts`: The rules in this virtual service will be applied to traffic destined for `sockshop.com` as defined under `host`.
- `gateway`: This virtual service is associated with `chapter4-gateway` we created in step 1 (*Creating the gateway*); this enforces that any other traffic not associated with the mentioned gateway will not be processed by this virtual service configuration.
- `http`: Here, we will define rules and routing information for HTTP traffic. There is also an option for defining `tls` and `tcp` routes; `tls` is used for passthrough TLS or HTTPS traffic, whereas `tcp` is used for opaque TCP traffic.
- `match`: These contain the matching criteria and can be based on path, headers, and so on. In this example, we are instructing that all traffic will be routed as per the instruction in this section.
- `route`: If the traffic is matched, the traffic is routed as per the information provided here. In this example, we are routing traffic to `front-end.sock-shop.svc.cluster.local` on port `80`.

You can find the declaration of the corresponding virtual service for `envoy-dummy-svc` in `Chapter4/3-istio-gateway.yaml`. The file combines the declaration of the gateway and virtual services.

As a next step, if not already deleted as per the cleanup notes of the previous section, then please delete the Ingress resources you created in the previous section so that they do not conflict with the configuration we will be applying in this section.

Apply the new configuration:

```
$ kubectl apply -f chapter4/3-istio-gateway.yaml
```

Please test that you are able to access `sockshop.com` and `mockshop.com` using your preferred HTTP client, and do not forget to inject the correct `host` header.

If you are finding it difficult to visualize the end-to-end configuration, then take the help of the following illustrations:

- The following diagram summarizes the configuration in this section:

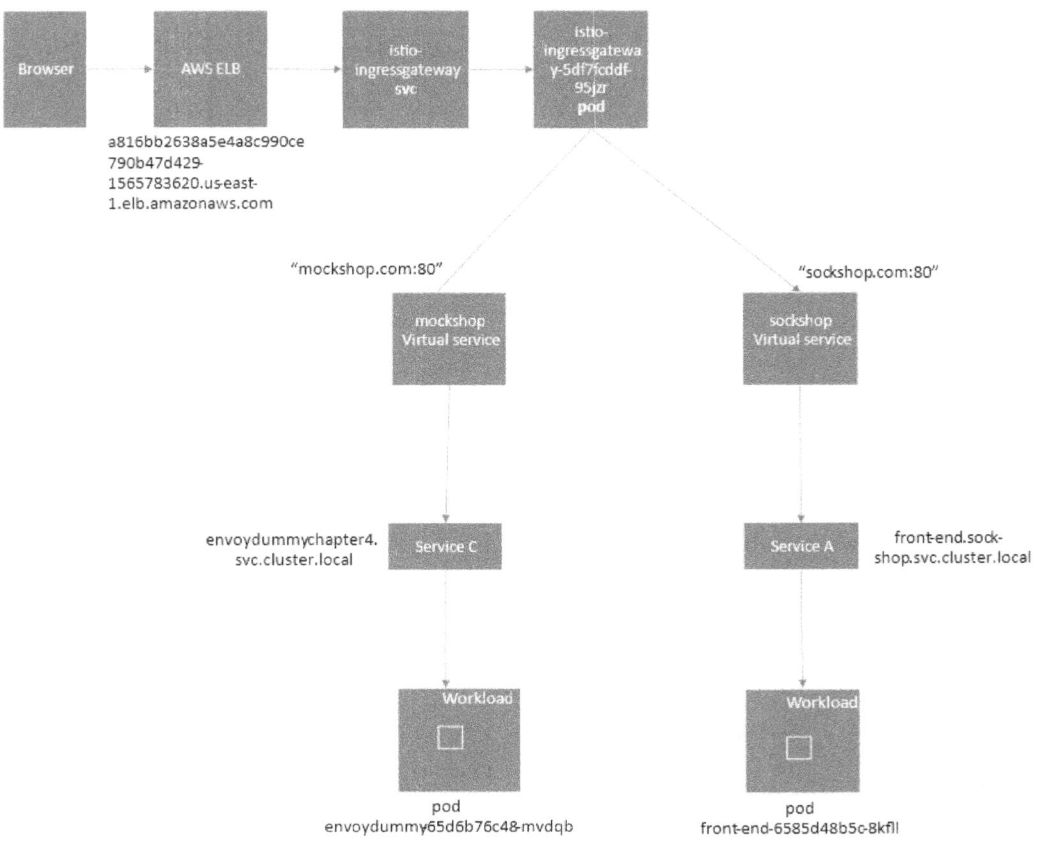

Figure 4.4 – Virtual services

- The following diagram summarizes the association between various Istio CRDs and Kubernetes resources:

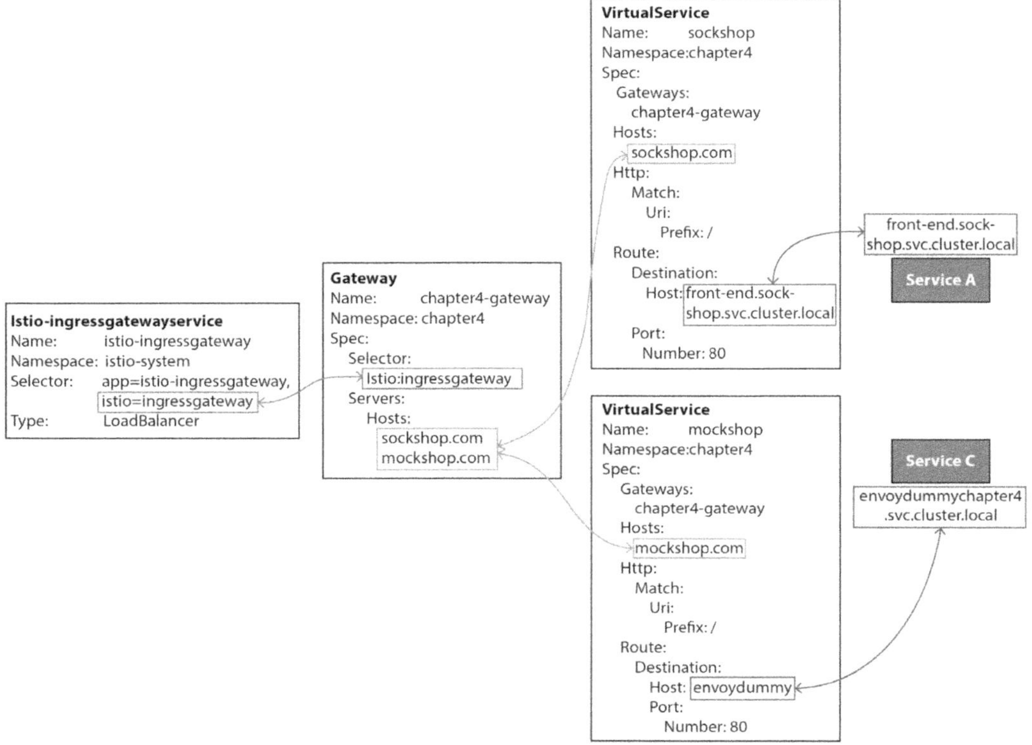

Figure 4.5 – Association between virtual services and other Istio resources

In this section, we learned how to use Istio Gateway and virtual services for managing Ingress.

> **Reminder**
>
> Please clean up `chapter4/3-istio-gateway.yaml` to avoid conflict with the upcoming exercises.

Traffic routing and canary release

In the previous section, we went through some of the functionality of virtual services; in this section, let's go through how you can distribute traffic to multiple destinations.

I'm assuming you have the `envoy-dummy` config map configured and the `envoy` Pod and service running as per the `01-envoy-proxy.yaml` file. If not, follow the instructions in the previous section to get these configured.

In the following exercise, we will be creating another version of the envoydummy Pod called v2, which returns a different response than v1. We will deploy v2 alongside v1 and then configure traffic splitting between the two versions of the envoydummy Pods:

1. Create another version of the envoy mock service but with a different message:

```
direct_response:
                  status: 200
                  body:
                    inline_string: "V2----------Bootstrap
Service Mesh Implementation with Istio----------V2"
```

2. The changes can be found in Chapter4/envoy-config-2.yaml; go ahead and create another config map:

```
$ kubectl create configmap envoy-dummy-2 --from-
file=Chapter4/envoy-config-2.yaml -n chapter4
```

3. Then, create another Deployment, but this time label the Pods as follows:

```
template:
    metadata:
      labels:
        name: envoyproxy
        version: v2
```

4. Apply the changes:

```
$ kubectl apply -f Chapter4/02-envoy-proxy.yaml
```

5. Next, we will be creating another virtual service, but with the following changes:

```
    route:
    - destination:
        port:
          number: 80
        subset: v1
        host: envoy-dummy-svc
      weight: 10
    - destination:
        port:
          number: 80
```

```
          subset: v2
          host: envoy-dummy-svc
        weight: 90
```

You must have noticed that we have two destinations under the same route. `destination` indicates the location of the service to which the requests are eventually routed. Under `destination`, we have the following three fields:

- `host`: This states the service name to which the request should be routed. The service names are resolved against the Kubernetes service registry or hosts registered by Istio service entry. We will read about service entry in the next section.
- `subset`: This is a subset of the service defined by the destination rule, as described next.
- `port`: This is the port on which the service is reachable.

We are also associating weights to the routing rules, specifying that 10% of traffic should be sent to `subset: v1`, whereas 90% should be sent to `subset: v2`.

Following the virtual service definition, we also need to define destination rules. Destination rules are a set of rules applied to the traffic after they have gone through the virtual service routing rules.

In the following configuration, we are defining a destination rule called `envoy-destination`, which will be applied to traffic destined for `envoy-dummy-svc`. It further defines two subsets – `subset: v1` corresponds to the `envoy-dummy-svc` endpoints with the `version = v1` label, while `subset: v2` corresponds to endpoints with the `version = v2` label:

```
apiVersion: networking.istio.io/v1alpha3
kind: DestinationRule
metadata:
  name: envoy-destination
  namespace: chapter4
spec:
  host: envoy-dummy-svc
  subsets:
  - name: v1
    labels:
      version: v1
  - name: v2
    labels:
      version: v2
```

Apply the changes:

```
kubectl apply -f Chapter4/4a-istio-gateway.yaml
```

You will notice that 10% of the request will be returning `Bootstrap Service Mesh Implementation with Istio`, and 90% of the request will be returning the `V2----------Bootstrap Service Mesh Implementation with Istio---------V2` response.

If you are finding it difficult to visualize the end-to-end configuration, then take the help of the following illustration, which summarizes the configuration in this section:

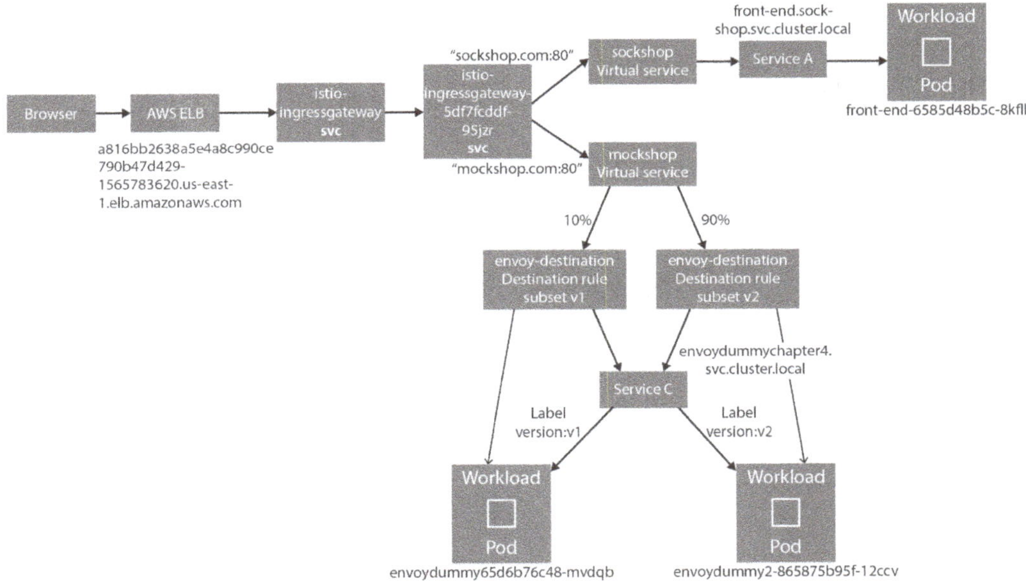

Figure 4.6 – Destination rules

The following diagram summarizes the association between various Istio CRDs and Kubernetes resources:

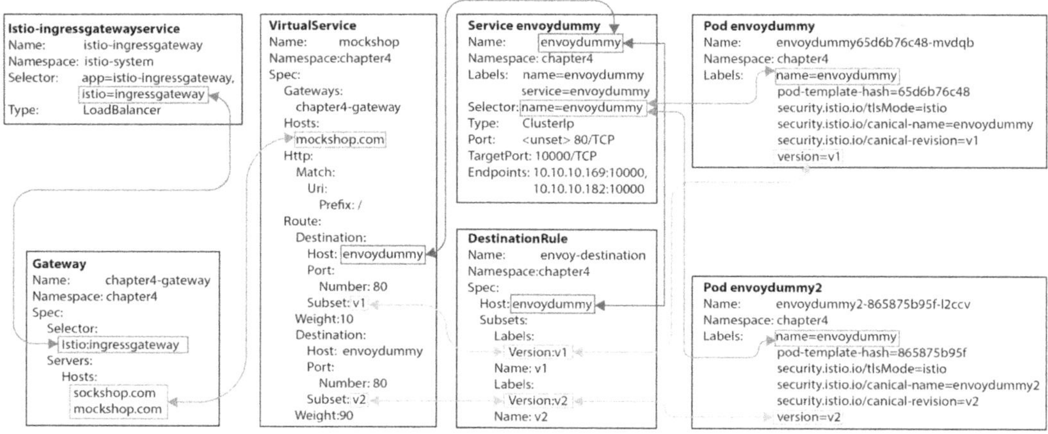

Figure 4.7 – Association between destination rules and other Istio resources

You can also check in the Kiali dashboard that traffic is getting routed in a 1:9 ratio between the two services:

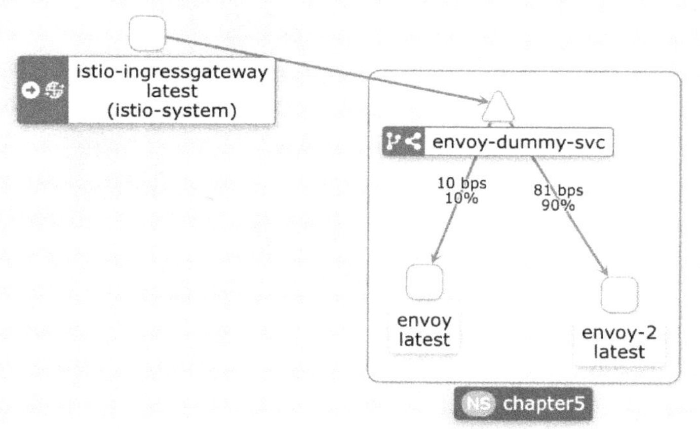

Figure 4.8 – Kiali dashboard showing traffic split

> **Reminder**
>
> Please clean up `Chapter4/4a-istio-gateway.yaml` to avoid conflicts in the upcoming exercises.

In this section, you learned how to route or split traffic between two versions of a service. This is fundamental to various operations related to traffic management, with the canary release being one of them. In the next section, we will read about traffic mirroring, which is also called **traffic shadowing**; it is another example of traffic routing.

Traffic mirroring

Traffic mirroring is another important feature that allows you to asynchronously copy traffic being sent to an upstream to another upstream service as well, also known as **mirrored service**. Traffic mirroring is on a fire-and-forget basis, where the sidecar/gateway will not wait for responses from the mirrored upstream.

The following is an illustration of traffic mirroring:

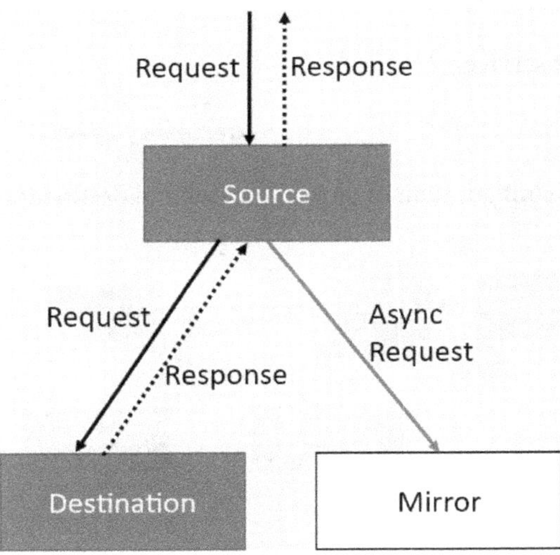

Figure 4.9 – Traffic mirroring

There are very interesting use cases for traffic mirroring, including the following:

- Traffic mirroring to pre-production systems for testing purposes
- Traffic mirroring to sink systems where traffic is recorded for out-of-band analysis

In the following example, in the virtual service definition under route configuration, we are mentioning that 100% of traffic should be mirrored to `subset: v2`:

```
route:
  - destination:
      port:
```

```
          number: 80
        subset: v1
        host: envoydummy
      weight: 100
    mirror:
      host: nginxdummy
      subset: v2
    mirrorPercentage:
      value: 100.0
```

Before applying the preceding changes, first, create an nginx service using the following:

```
kubectl apply -f utilities/nginx.yaml
```

After that, deploy the virtual service:

```
kubectl apply -f chapter4/4b-istio-gateway.yaml
```

The following illustrates the configuration of virtual services and destination rules:

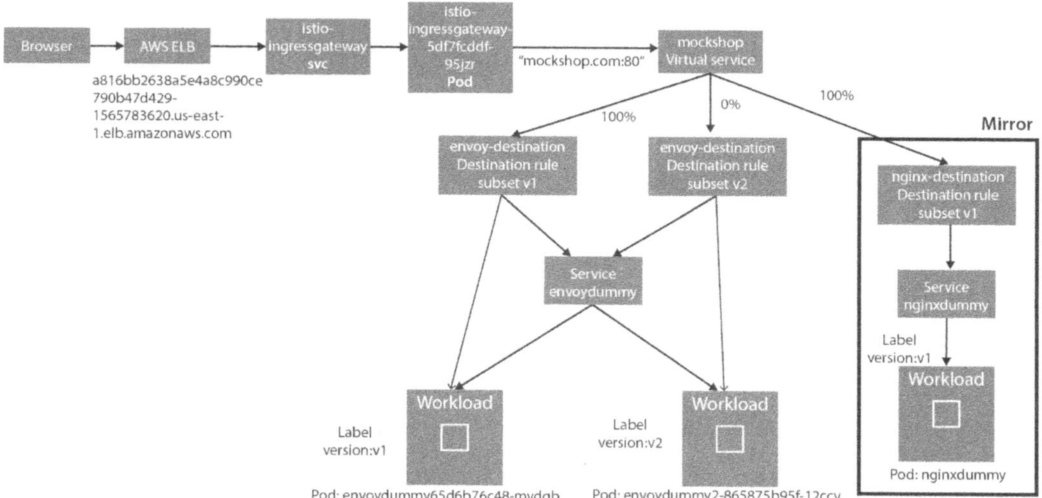

Figure 4.10 – Traffic mirroring via virtual services

When accessing the service using `curl` or the browser with the `mockshop.com` host header, you will notice that you are always receiving a `Bootstrap Service Mesh Implementation with Istio` response.

But if you check the nginx logs using the `kubectl logs nginxdummy -c nginx -n chapter4` command, you will notice that nginx is also receiving the request, indicating that the traffic has been shadowed to nginx.

This completes a short section on traffic mirroring, a simple but powerful feature especially for event-driven architecture, testing, and training models when using machine learning and artificial intelligence.

Reminder

Please clean up `Chapter4/4b-istio-gateway.yaml` to avoid conflict in upcoming exercises.

Routing traffic to services outside of the cluster

In your IT environments, not all services will be deployed within the Kubernetes cluster; there will be services running on traditional VMs or bare metal environments, there will be services that will be provided by SaaS providers as well as your business partners, and there will be services running outside or on a different Kubernetes cluster. In those scenarios, there is a requirement to let services from the mesh reach out to such services. So, as the next steps, let's try building routes to a service outside of the cluster. We will make use of the `httpbin` service, available at `https://httpbin.org/`.

Any request destined for `mockshop.com/get` should be routed to `httpbin`; the rest should be processed by `envoy-dummy-svc, which` we created in the previous section.

In the following virtual service definition, we have defined that any request with `/get` should be routed to `httpbin.org`:

```
- match:
    - uri:
        prefix: /get
    route:
    - destination:
        port:
          number: 80
        host: httpbin.org
```

Next, we will create `ServiceEntry`, which is a way of adding entries to Istio's internal service registry. The Istio control plane manages a registry of all services within the mesh. The registry is populated from two kinds of data sources– one being the Kubernetes API server, which in turn uses etcd for maintaining a registry of all services in the cluster, and the second being a config store that is populated by `ServiceEntry` and `WorkloadEntry`. Now, `ServiceEntry` and `WorkloadEntry` are used to populate details about services that are unknown to the Kubernetes service registry. We will read about `WorkloadEntry` in *Chapter 10*.

The following is the `ServiceEntry` declaration for adding `httpbin.org` to the Istio service registry:

```
apiVersion: networking.istio.io/v1alpha3
kind: ServiceEntry
metadata:
  name: httpbin-svc
  namespace: chapter4
spec:
  hosts:
  - httpbin.org
  location: MESH_EXTERNAL
  ports:
  - number: 80
    name: httpbin
    protocol: http
  resolution: DNS
```

In the `ServiceEntry` declaration, the following configurations are defined:

- `resolution`: Here, we define how the hostname should be resolved; the following are the possible values:
  - `DNS`: Makes use of available DNS to resolve the hostname
  - `DNS_ROUND_ROBBIN`: In this case, the first resolved address is used
  - `NONE`: No DNS resolution is required; the destination is specified in form of an IP address
  - `STATIC`: Uses a static endpoint against the hostnames
- `location`: The service entry location is used to specify whether the requested service is part of the mesh or outside the mesh. Possible values are `MESH_EXTERNAL` and `MESH_INTERNAL`.
- `hosts`: This is the hostname associated with the service being requested; in this example, the host is `httpbin.org`. The host field in `ServiceEntry` is matched with host fields specified in virtual service and destination rules.

Go ahead and apply the changes:

```
$ kubectl apply -f Chapter4/5a-istio-gateway.yaml
```

When executing `curl` to `/get`, you will receive a response from `httpbin.org`, whereas `/ping` should route to the `envoydummy` service.

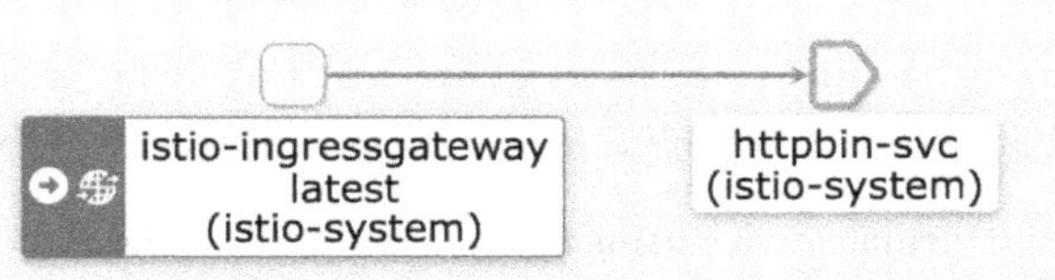

Figure 4.11 – Kiali dashboard showing connection to an external system via ServiceEntry

> **Reminder**
>
> Please clean up `Chapter4/5a-istio-gateway.yaml` to avoid conflict in upcoming exercises.

`ServiceEntry` provides various options to register external services to the Istio registry so that traffic within the mesh can be correctly routed to workloads outside the mesh.

Exposing Ingress over HTTPS

In this section, we will learn how to configure the Istio Gateway to expose the Sockshop frontend application over HTTPs.

Steps 1 and *3* are optional if you already have a **Certificate Authority** (**CA**); usually, for production systems, these steps will be performed by your organization's CA:

1. Create a CA. Here, we are creating a CA with `CN` (**Common Name**) as `sockshop.inc`:

   ```
   $openssl req -x509 -sha256 -nodes -days 365 -newkey
   rsa:2048 -subj '/O=Sockshop Inc./CN=Sockshop.inc' -keyout
   Sockshop.inc.key -out Sockshop.inc.crt
   ```

2. Generate a **Certificate Signing Request** (**CSR**) for the sockshop. Here, we are generating a CSR for `sockshop.com`, which also generates a private key:

   ```
   $openssl req -out sockshop.com.csr -newkey rsa:2048
   -nodes -keyout sockshop.com.key -subj "/CN=sockshop.com/
   O=sockshop.inc"
   ```

3. Sign the CSR using the CA with the following command:

```
$openssl x509 -req -sha256 -days 365 -CA Sockshop.inc.crt
-CAkey Sockshop.inc.key -set_serial 0 -in sockshop.com.
csr -out sockshop.com.crt
```

4. Load the certificate and private key as a Kubernetes Secret:

```
$kubectl create -n istio-system secret tls sockshop-
credential --key=sockshop.com.key --cert=sockshop.com.crt
```

5. Create a gateway and virtual service using the following command:

```
kubectl apply -f Chapter4/6-istio-gateway.yaml
```

This way, we have created a certificate and loaded that along with its private key as a Kubernetes Secret.

Finally, we are configuring Istio Gateway to use the Secret as a credential for TLS communications. In the `Chapter4/6-istio-gateway.yaml` file gateway, we are configuring `IstioGateway` as the Ingress, and listening on port `443` on the `HTTPS` server protocol:

```
servers:
- port:
    number: 443
    name: https
    protocol: HTTPS
  tls:
    mode: SIMPLE
    credentialName: sockshop-credential
  hosts:
  - "sockshop.com"
```

In the gateway configuration, we have changed the protocol version to `HTTPS` from `HTTP`, and we added the following configurations under `servers>tls`:

- `Mode`: Indicates whether this port should be secured using TLS. Possible values for this field are as follows:
  - `SIMPLE`: This is the standard TLS setting that we have selected to expose Sockshop.
  - `MUTUAL`: This is for mutual TLS between the gateway and any system calling the gateway.
  - `PASSTHROUGH`: This is used when a connection needs to be routed to a virtual service with the host value as the **Server Name Indication** (**SNI**) presented during the call.

> **SNI**
>
> SNI is an extension of the TLS protocol, where the hostname or domain name of the destination service is shared at the TLS handshake process rather than Layer 7. SNI is useful where a server is hosting multiple domain names, with each represented by its own HTTPS certificate. By knowing the requested hostname at the Layer 5 handshake, the server is able to present the correct certificate as per the presented SNI during the handshake.

  - `AUTO_PASSTHROUGH`: This is the same as `PASSTHROUGH`, except that there is no need for virtual services. The connection is forwarded to upstream services as per the details in the SNI.
  - `ISTIO_MUTUAL`: This is the same as `MUTUAL`, except that the certificate used for mutual TLS is generated automatically by Istio.

- `Credential name`: This is the Secret that holds the private key and certificate to be used for the server-side connection during TLS. We created the Secret in *step 4*.

Go ahead and access `sockshop.com`; you will have to use `--connect-to` in `curl` to get around the name resolution issue caused by the difference in the replacement name and the actual name of the host:

```
$ curl -v -HHost:sockshop.com --connect-to "sockshop.
com:443:a816bb2638a5e4a8c990ce790b47d429-1565783620.us-east-1.
elb.amazonaws.com" --cacert Sockshop.inc.crt  https://sockshop.
com:443/
```

Please note, `a816bb2638a5e4a8c990ce790b47d429-1565783620.us-east-1.elb.amazonaws.com` is the **fully qualified domain name** (**FQDN**) of the load balancer provided by AWS. If you are using minikube, you can run the command against localhost by using `--resolve` in `curl`, similar to the following:

```
$ curl -v -HHost:sockshop.com --resolve "sockshop.
com:56407:127.0.0.1" http://sockshop.com:56407/
```

In the preceding command, `56407` is the local port on which the Ingress gateway is listening.

During the connection, you will notice in the output that the gateway correctly presented the server-side certificates:

```
* SSL connection using TLSv1.3 / AEAD-CHACHA20-POLY1305-SHA256
* ALPN, server accepted to use h2
* Server certificate:
*  subject: CN=sockshop.com; O=sockshop.inc
*  start date: Aug 12 06:45:27 2022 GMT
```

```
* expire date: Aug 12 06:45:27 2023 GMT
* common name: sockshop.com (matched)
* issuer: O=Sockshop Inc.; CN=Sockshop.inc
* SSL certificate verify ok.
```

One special point to make here is that we exposed `sockshop.com` as an `HTTPS` service without making any changes to the frontend services that host the website.

> **Reminder**
> Please clean up `Chapter4/6-istio-gateway.yaml` to avoid conflict with upcoming exercises.

Enabling HTTP redirection to HTTPS

For downstream systems that are still sending requests to non-HTTPS ports, we can implement HTTP redirection by making the following changes in gateway configuration for non-HTTPS ports:

```
servers:
- port:
    number: 80
    name: http
    protocol: HTTP
  hosts:
  - "sockshop.com"
  tls:
    httpsRedirect: true
```

We have simply added `httpsRedirect: true`, which makes the gateway send a `301` redirect for all non-HTTPS connections. Apply the changes and test the connection:

```
$ kubectl apply -f Chapter4/7-istio-gateway.yaml
$ curl -v -HHost:sockshop.com --connect-to "sockshop.
com:80:a816bb2638a5e4a8c990ce790b47d429-1565783620.us-east-1.
elb.amazonaws.com" --cacert Sockshop.inc.crt  http://sockshop.
com:80/
```

In the output, you will notice the redirection to `sockshop.com`:

```
* Mark bundle as not supporting multiuse
< HTTP/1.1 301 Moved Permanently
< location: https://sockshop.com/
```

> **Reminder**
>
> As usual, please clean up `Chapter4/7-istio-gateway.yaml` to avoid conflict with the next section exercises.

Enabling HTTPS for multiple hosts

In the previous section, we defined the settings for `sockshop.com` on the gateway. We can also apply similar settings for multiple hosts on the gateway. In this section, we will enable TLS on the gateway for `mockshop.com` along with `sockshop.com`:

1. We will make use of the CA we created in the previous section. So, as the next steps, let's generate a CSR for `mockshop.com`:

   ```
   $openssl req -out mockshop.com.csr -newkey rsa:2048
   -nodes -keyout mockshop.com.key -subj "/CN=mockshop.com/
   O=mockshop.inc"
   ```

2. Sign the CSR using the CA:

   ```
   $openssl x509 -req -sha256 -days 365 -CA Sockshop.inc.crt
   -CAkey Sockshop.inc.key -set_serial 0 -in mockshop.com.
   csr -out mockshop.com.crt
   ```

3. Load the certificate and private key as a Kubernetes Secret:

   ```
   $kubectl create -n istio-system secret tls mockshop-
   credential --key=mockshop.com.key --cert=mockshop.com.crt
   ```

4. Add the following configuration for `mockshop.com` under server configuration in the gateway:

   ```
   - port:
         number: 443
         name: https-mockshop
         protocol: HTTPS
   ```

```
    tls:
      mode: SIMPLE
      credentialName: mockshop-credential
    hosts:
    - "mockshop.com"
```

5. Apply the changes:

```
kubectl apply -f Chapter4/8-istio-gateway.yaml
```

After the changes, the gateway will resolve the correct certificates based on the hostname.

6. Let's now access `sockshop.com`:

```
curl -v --head -HHost:sockshop.com --resolve "sockshop.
com:56408:127.0.0.1" --cacert Sockshop.inc.crt https://
sockshop.com:56408/
```

In the response, you can see that the correct certificates have been presented:

```
* SSL connection using TLSv1.3 / AEAD-CHACHA20-POLY1305-
SHA256
* ALPN, server accepted to use h2
* Server certificate:
*  subject: CN=sockshop.com; O=sockshop.inc
*  start date: Aug 12 06:45:27 2022 GMT
*  expire date: Aug 12 06:45:27 2023 GMT
*  common name: sockshop.com (matched)
*  issuer: O=Sockshop Inc.; CN=Sockshop.inc
*  SSL certificate verify ok.
* Using HTTP2, server supports multiplexing
```

7. Similarly, test `mockshop.com`:

```
curl -v -HHost:mockshop.com --connect-to "mockshop.
com:443:a816bb2638a5e4a8c990ce790b47d429-1565783620.
us-east-1.elb.amazonaws.com" --cacert Sockshop.inc.
crt  https://mockshop.com/
```

8. Then, check whether the certificate presented by the gateway belongs to `mockshop.com`:

```
SSL connection using TLSv1.3 / AEAD-CHACHA20-POLY1305-
SHA256
* ALPN, server accepted to use h2
```

```
* Server certificate:
*  subject: CN=mockshop.com; O=mockshop.inc
*  start date: Aug 12 23:47:27 2022 GMT
*  expire date: Aug 12 23:47:27 2023 GMT
*  common name: mockshop.com (matched)
*  issuer: O=Sockshop Inc.; CN=Sockshop.inc
*  SSL certificate verify ok.
* Using HTTP2, server supports multiplexing
```

In this way, we have configured the Istio Ingress gateway to serve multiple TLS certificates depending on hostnames; this is also called SNI. The Istio Ingress gateway can resolve SNI at the TLS Layer 4 level, allowing it to serve multiple domain names over TLS.

Enabling HTTPS for CNAME and wildcard records

The last topic on HTTPS is how to manage certificates for CNAME and wildcard records. Especially for traffic exposed internally, it is important to support wildcard. In this section, we will configure the gateway to support wildcard using SNI support. We will be using the CA we created in previous sections:

1. Create a CSR for `*.sockshop.com` and sign it using the CA certificates, then create the Kubernetes Secret:

```
$openssl req -out sni.sockshop.com.csr -newkey rsa:2048
-nodes -keyout sni.sockshop.com.key -subj "/CN=*.
sockshop.com/O=sockshop.inc"
$openssl x509 -req -sha256 -days 365 -CA Sockshop.inc.crt
-CAkey Sockshop.inc.key -set_serial 0 -in sni.sockshop.
com.csr -out sni.sockshop.com.crt
$kubectl create -n istio-system secret tls sni-sockshop-
credential --key=sni.sockshop.com.key --cert=sni.
sockshop.com.crt
```

2. Then, add the `*.sockshop.com` hostname to the server configuration in the gateway:

```
servers:
  - port:
      number: 443
      name: https-sockshop
      protocol: HTTPS
    tls:
      mode: SIMPLE
```

```
        credentialName: sni-sockshop-credential
      hosts:
      - "*.sockshop.com"
```

3. Also, modify the virtual service with `*.sockshop.com`:

```
kind: VirtualService
metadata:
  name: sockshop
  namespace: chapter4
spec:
  hosts:
  - "*.sockshop.com"
  - "sockshop.com"
```

4. Apply the configuration:

```
$ kubectl apply -f Chapter4/9-istio-gateway.yaml
```

5. You can test `mockshop.com`, `sockshop.com`, or any other CNAME records for `sockshop.com`. The following example is using `my.sockshop.com`:

```
$ curl -v -HHost:my.sockshop.
com --connect-to "my.sockshop.
com:443:a816bb2638a5e4a8c990ce790b47d429-1565783620.
us-east-1.elb.amazonaws.com" --cacert Sockshop.inc.
crt  https://my.sockshop.com/
```

The following is the snippet from the output of *step 5* showing that the correct certificate was presented during the handshake:

```
* SSL connection using TLSv1.3 / AEAD-CHACHA20-POLY1305-
SHA256
* ALPN, server accepted to use h2
* Server certificate:
*  subject: CN=*.sockshop.com; O=sockshop.inc
*  start date: Aug 13 00:27:00 2022 GMT
*  expire date: Aug 13 00:27:00 2023 GMT
*  common name: *.sockshop.com (matched)
*  issuer: O=Sockshop Inc.; CN=Sockshop.inc
```

```
*  SSL certificate verify ok.
* Using HTTP2, server supports multiplexing
```

As you can see, Istio presented the correct wildcard certificate for CNAME. This example demonstrates how Istio gateways can be configured to handle multiple domains and subdomains.

In this and prior sections, we read about the various ways Ingress and the routing of traffic within and outside of the mesh are managed by Istio. It is important that you go through the concepts of gateways, virtual services, destination rules, and service entries and play with the example provided in this chapter, as well as think of other examples and try to implement them. In *Chapter 6*, we will discuss security in more depth and will cover topics such as mTLS, and so on. But for now, we will read about how Istio manages Egress traffic.

> **Reminder**
>
> Don't forget to clean up `Chapter4/8-istio-gateway.yaml` and `Chapter4/9-istio-gateway.yaml`.

Managing Egress traffic using Istio

In the *Routing traffic to services outside of the cluster* section, we discovered how service entries can be used to update the Istio service registry about services external to the mesh and the cluster. Service entries are a way to add additional entries into Istio's internal service registry for virtual services to be able to route to those entries. An Egress gateway, however, is used for controlling how the traffic for external service leaves the mesh.

To get familiar with Egress gateways, we will first deploy a Pod within the mesh from which we can call an external service:

```
$ kubectl apply -f utilities/curl.yaml
```

The command creates a Pod from which you can perform `curl`; this mimics a workload running inside the mesh:

```
$ kubectl exec -it curl sh -n chapter4
```

From the shell, access `httpbin.org` using `curl`:

```
$ curl -v https://httpbin.org/get
```

Now, we will stop all Egress traffic from the mesh using the following command:

```
$ istioctl install -y --set profile=demo --set meshConfig.
outboundTrafficPolicy.mode=REGISTRY_ONLY
```

In the previous command, we are modifying the Istio installation to change the outbound traffic policy from `ALLOW_ANY` to `REGISTRY_ONLY`, which enforces that only hosts defined with `ServiceEntry` resources are part of the mesh service registry.

Go back and try `curl` again; you will see the following output:

```
$ curl -v https://httpbin.org/get
curl: (35) OpenSSL SSL_connect: SSL_ERROR_SYSCALL in connection
to httpbin.org:443
```

Let's now list `httpbin.org` in the Istio service registry by creating a service entry as follows:

```
apiVersion: networking.istio.io/v1alpha3
kind: ServiceEntry
metadata:
  name: httpbin-svc
  namespace: chapter4
spec:
  hosts:
  - httpbin.org
  location: MESH_EXTERNAL
  resolution: DNS
  ports:
  - number: 443
    name: https
    protocol: HTTPS
  - number: 80
    name: http
    protocol: HTTP
```

Now, you may go ahead and apply the configuration:

```
$ kubectl apply -f Chapter4/10-a-istio-egress-gateway.yaml
```

Access `https://httpbin.org/get` from the `curl` Pod; this time, you will succeed.

`ServiceEntry` added `httpbin.org` to the mesh service registry, and hence we were able to access `httpbin.org` from the `curl` Pod.

Though `ServiceEntry` is great for providing external access, it does not provide any control over how the external endpoints should be accessed. For example, you may want only certain workloads or namespaces to be able to send traffic to an external resource. What if there is a need to verify the authenticity of an external resource by verifying its certificates?

> **Reminder**
>
> Don't forget to clean up `Chapter4/10-a-istio-egress-gateway.yaml`.

The Egress gateway, along with a combination of virtual services, destination rules, and service entries, provides flexible options to manage and control traffic egressing out of the mesh. So, let's make configuration changes to route all traffic for `httpbin.org` to the Egress gateway:

1. Configure the Egress gateway, which is very similar to the Ingress gateway configuration. Please note the Egress gateway is attached to `httpbin.org`; you can provide other hosts or `*` to match all hostnames:

```
apiVersion: networking.istio.io/v1alpha3
kind: Gateway
metadata:
  name: istio-egressgateway
  namespace: chapter4
spec:
  selector:
    istio: egressgateway
  servers:
  - port:
      number: 80
      name: http
      protocol: HTTP
    hosts:
    - httpbin.org
```

2. Next, configure the virtual service. Here, we are configuring the virtual service to attach to the Egress gateway as well as to the mesh:

```
spec:
  hosts:
  - httpbin.org
  gateways:
```

```
    - istio-egressgateway
    - mesh
```

In the following part of the virtual service definition, we are configuring that all traffic originating from within the mesh for the `httpbin.org` host will be directed to the Egress gateway:

```
  http:
    - match:
      - gateways:
        - mesh
        port: 80
      route:
      - destination:
          host: istio-egressgateway.istio-system.svc.
cluster.local
          subset: httpbin
          port:
            number: 80
        weight: 100
```

We have configured `subset: httpbin` to apply destination rules; in this example, the destination rules are empty.

Finally, we will add another rule to route traffic from the Egress gateway to `httpbin.org`:

```
    - match:
      - gateways:
        - istio-egressgateway
        port: 80
      route:
      - destination:
          host: httpbin.org
          port:
            number: 80
        weight: 100
```

3. Create a placeholder for any destination rules you might want to implement:

```
apiVersion: networking.istio.io/v1alpha3
kind: DestinationRule
```

```
metadata:
  name: rules-for-httpbin-egress
  namespace: chapter4
spec:
  host: istio-egressgateway.istio-system.svc.cluster.
local
  subsets:
  - name: httpbin
```

4. You also need to add `ServiceEntry` for `httpbin.org`, which we discussed in the previous section.
5. Go ahead and apply the changes:

   ```
   kubectl apply -f Chapter4/10-b-istio-egress-gateway.yaml
   ```

6. Try accessing `httpbin.org` from the `curl` Pod; you will be able to access it now.

Examine the headers in the response, as well as the logs of `istio-egressgateway pods`. You will find information about the Egress gateway under `X-Envoy-Peer-Metadata-Id`. You can also see the request in the Egress gateway logs.

You will notice that you are not able to access `https://httpbin.org/get`, although we have defined `https` in the service entry. Try enabling `https` access to `httpbin.org`; you will find the solution in `Chapter4/10-c-istio-egress-gateway.yaml`.

Egress is important to control traffic leaving the mesh. In *Chapter 6*, we will focus on some of the other security aspects of Egress.

Reminder

Delete `chapter4/10-b-istio-egress-gateway.yaml` along with `chapter4` namespace.

Revert the authorization policy to allow all outgoing traffic from the mesh without needing an Egress gateway using the following command:

```
$ istioctl install -y --set profile=demo --set meshConfig.
outboundTrafficPolicy.mode=ALLOW_ANY
```

Summary

In this chapter, we read about how to manage external traffic coming inside the Service Mesh using the Istio Ingress gateway, as well as how to manage internal traffic leaving the mesh via Istio Egress gateways.

We learned about virtual services and destination rules: how virtual services are used to describe the rules to route traffic to various destinations in the mesh, how destination rules are used to define the end destination, and how the destination processes the traffic routed via rules defined by virtual services. Using virtual services, we can perform weight-based traffic routing, which is also used for canary releases and blue-green deployment.

Additionally, we learned about `ServiceEntry` and how it is used to make Istio aware of external services so that workloads in the mesh can send traffic to services outside the mesh. And finally, we learned how Egress gateways are used to control the Egress to endpoints defined by `ServiceEntry` so that we can access external services securely and reliably from the mesh. This chapter sets you up for the next chapter, where we will discuss how to implement application resiliency using the concepts from this chapter.

5
Managing Application Resiliency

Application resiliency is the ability of software applications to withstand faults and failures without any noticeable degradation of quality and level of service for its consumer. The move from monoliths to microservices has exacerbated the need for application resiliency in software design and architecture. In monolith applications, there is a single code base and single deployment whereas, in microservice-based architecture, there are many independent code bases each with its own deployment. When leveraging Kubernetes and other similar platforms, you also need to cater to deployment flexibility and the fact that multiple application instances are being deployed and scaled elastically; these dynamic instances need to not only coordinate with each other but also coordinate with all other microservices.

In this chapter, we will read about how to make use of Istio to increase the application resiliency of microservices. As we go through each section, we will discuss various aspects of application resiliency and how they are addressed by Istio.

In a nutshell, the following topics will be covered in this chapter:

- Application resiliency using fault injection
- Application resiliency using timeouts and retries
- Building application resiliency using load balancing
- Rate limiting
- Circuit breaker and outlier detection

> **Important note**
> The technical prerequisites for this chapter are the same as the previous chapter.

Application resiliency using fault injection

Fault injections are used for testing the recovery capability of applications in case of any kind of failure. In principle, every microservice should be designed with in-built resiliency for internal and external failures but often, that is not the case. The most complex and difficult work of building resiliency is usually at design and test time.

During design time, you must identify all known and unknown scenarios to which you need to cater. For example, you must address the following:

- What kind of known and unknown errors might happen inside and outside of the microservice?
- How should each of those errors be handled by the application code?

During test time, you should be able to simulate these scenarios to validate the contingencies built into the application code:

- Mimic in real-time different failure scenarios in the behavior of other upstream services to test the overall application behavior
- Mimic not just application failures but also network failures such as delays, outages, etc., as well as infrastructure failure

Chaos engineering is the software engineering term used to refer to the discipline of testing software systems by introducing adverse conditions into software systems and their execution and communication environments and ecosystems. You can read more about Chaos engineering at `https://hub.packtpub.com/chaos-engineering-managing-complexity-by-breaking-things/`. Various tools are available for generating chaos – as in, failures – mostly at the infrastructure level. One such popular tool is Chaos Monkey. It is available at `https://netflix.github.io/chaosmonkey/`. AWS also provides AWS Fault Injection Simulator for running fault injection simulations. You can read about AWS Fault Injection Simulator at `https://aws.amazon.com/fis/`. Another popular open source software is Litmus, which is a chaos engineering platform to identify weaknesses and potential outages in infrastructures by inducing chaos tests in a controlled way. You can read more about it `https://litmuschaos.io/`.

Istio provides fine-grained control for injecting failures because of its access and knowledge of application traffic. With Istio fault injection, you can use application-layer fault injection; this, combined with infrastructure-level fault injectors such as Chaos Monkey and AWS Fault Simulator, provides a very robust capability for testing application resiliency.

Istio supports the following types of fault injections:

- HTTP delay
- HTTP abort

In the following sections, we will discuss both fault injection types in more detail.

What is HTTP delay?

Delays are timing failures. They mimic increased request turnaround times caused either by network latency or an overloaded upstream service. These delays are injected via Istio VirtualServices.

In our sock shop example, let's inject an HTTP delay between the frontend service and catalog service and test how the frontend service behaves when it cannot get images from the catalog service. We will do this specifically for one image rather than all images.

You will find the VirtualService definition in `Chapter5/02-faultinjection_delay.yaml` on GitHub.

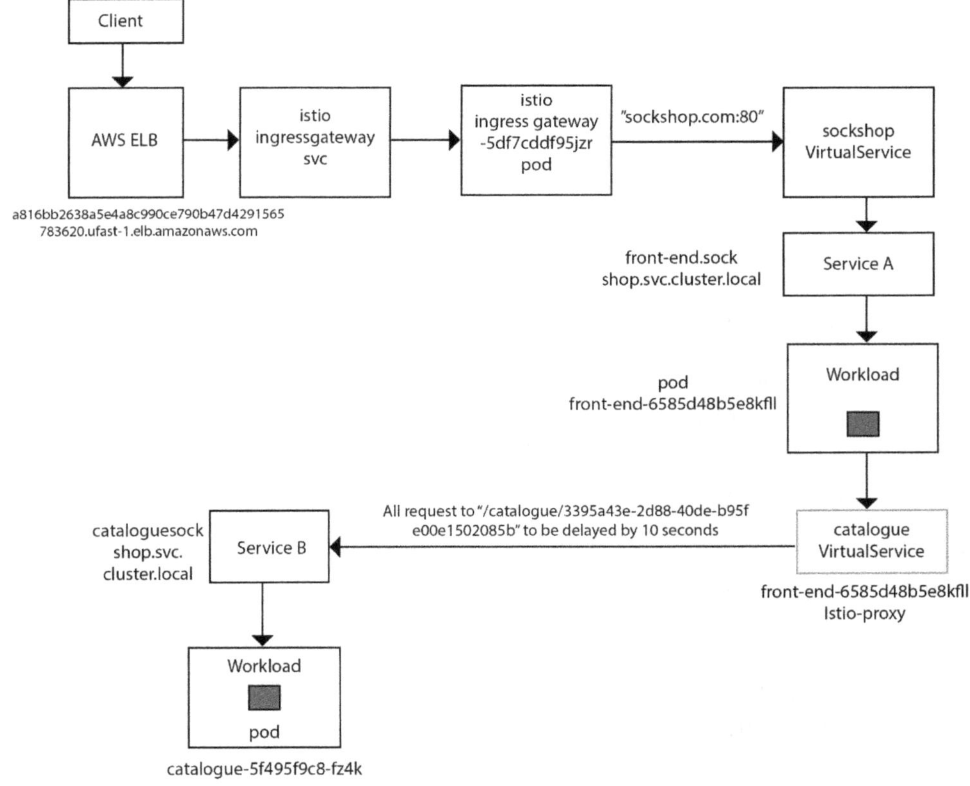

Figure 5.1 – HTTP delay injection in catalogue VirtualService

The following is the snippet from the VirtualService definition:

```
spec:
  hosts:
```

```
  - "catalogue.sock-shop.svc.cluster.local"
  gateways:
  - mesh
  http:
  - match:
    - uri:
        prefix: "/catalogue/3395a43e-2d88-40de-b95f-
e00e1502085b"
      ignoreUriCase: true
    fault:
      delay:
        percentage:
          value: 100.0
        fixedDelay: 10s
    route:
    - destination:
        host: catalogue.sock-shop.svc.cluster.local
        port:
          number: 80
```

The first thing to note is the `fault` definition, which is used to inject a delay and/or abort the fault before forwarding the request to the destination specified in the route. In this case, we are injecting a `delay` type of fault, which is used to emulate slow response times.

In the `delay` configuration, the following fields are defined:

- `fixedDelay`: Specifies the delay duration, the value can be in hours, minutes, seconds, or milliseconds, specified by the `h`, `m`, `s`, or `ms` suffix, respectively
- `percentage`: Specifies the percentage of requests for which the delay will be injected

The other thing to note is that the VirtualService is associated with the `mesh` gateway; you may have noticed that we did not define an Ingress or Egress gateway with `mesh`. So, you must be wondering where this came from, `mesh` is a reserved word used to refer all the sidecars in the mesh. This is also the default value for the gateway configuration, so if you don't provide a value for the gateway, then the VirtualService by default associates itself with all the sidecars in the mesh.

So, let's summarize what we have configured.

The `sock-shop` VirtualService is associated with all sidecars in the mesh and is applied for requests destined for `catalogue.sock-shop.svc.cluster.local`, the VirtualService injects a delay of 10 seconds for all requests prefixed with `/catalogue/3395a43e-2d88-40de-b95f-`

e00e1502085b, and it then forwards them to the catalogue.sock-shop.svc.cluster.local service. Requests that don't have the /catalogue/3395a43e-2d88-40de-b95f-e00e1502085b prefix are forwarded as is to the catalogue.sock-shop.svc.cluster.local service.

Create a namespace for Chapter5 with Istio injection enabled:

```
$ kubectl create ns chapter5
$ kubectl label ns chapter5 istio-injection=enabled
```

Create the Ingress gateway and VirtualService configuration for sockshop.com using the following:

```
$ kubectl apply -f Chapter5/01-sockshop-istio-gateway.yaml
```

After that, apply the VirtualService configuration for the catalog service:

```
$ kubectl apply -f Chapter5/02-faultinjection_delay.yaml
```

After this, open sockshop.com in your browser using the Ingress ELB and custom host headers. Enable the developer tools and search for requests with the /catalogue/3395a43e-2d88-40de-b95f-e00e1502085b prefix. You will notice those particular requests are taking more than 10 seconds to process.

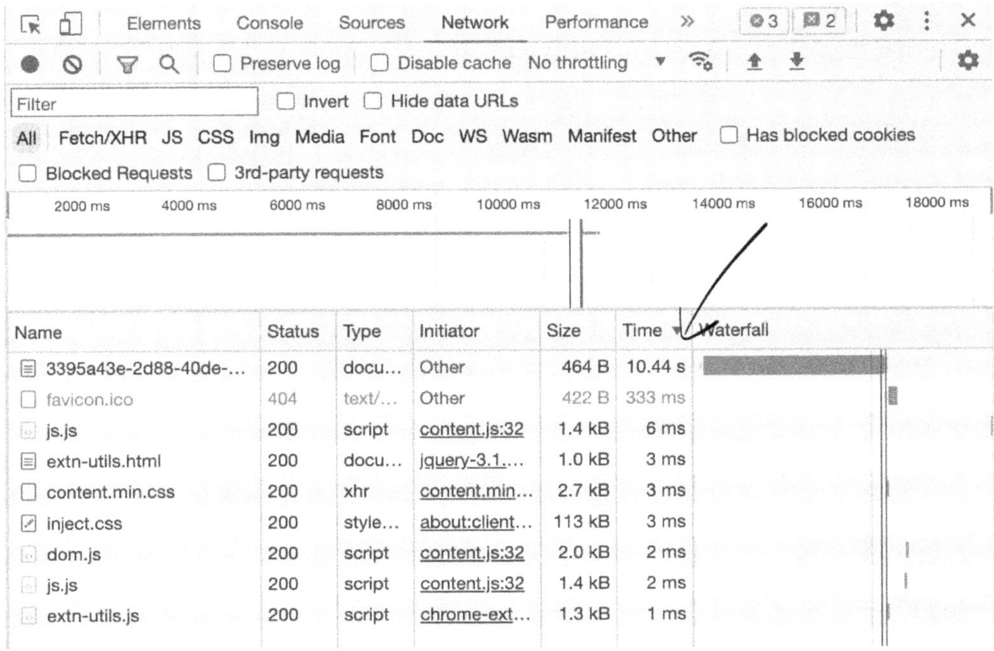

Figure 5.2 – HTTP delay causing requests to take more than 10 seconds

You can also check the sidecar injected in the `front-end` Pod to get the access logs for this request:

```
% kubectl logs -l name=front-end -c istio-proxy -n sock-shop |
grep /catalogue/3395a43e-2d88-40de-b95f-e00e1502085b
[2022-08-27T00:39:09.547Z] "GET /catalogue/3395a43e-2d88-
40de-b95f-e00e1502085b HTTP/1.1" 200 DI via_upstream - "-"
0 286 10005 4 "-" "-" "d83fc92e-4781-99ec-91af-c523e55cdbce"
"catalogue" "10.10.10.170:80" outbound|80||catalogue.sock-
shop.svc.cluster.local 10.10.10.155:59312 172.20.246.13:80
10.10.10.155:40834 - -
```

Highlighted in the preceding code block is the request that, in this instance, took `10005` milliseconds to process.

In this section, we injected the latency of 10 seconds, but you may have also noticed that the front-end web page functioned without any noticeable delays. All the images loaded asynchronously and any latency was limited to the catalogue section of the page. However, by configuring the delay, you are able to test the end-to-end behavior of the application in case of any unforeseen delays in the network or processing in the catalog service.

What is HTTP abort?

HTTP abort is the second type of fault that can be injected using Istio. HTTP abort prematurely aborts the processing of the request; you can also specify the error code that needs to be returned downstream.

The following is the snippet from the `VirtualService` definition with the `abort` configuration for the catalog service. The configuration is available in `Chapter5/03-faultinjection_abort.yaml`:

```
spec:
  hosts:
  - "catalogue.sock-shop.svc.cluster.local"
  gateways:
  - mesh
  http:
  - match:
    - uri:
        prefix: "/catalogue/3395a43e-2d88-40de-b95f-
e00e1502085b"
      ignoreUriCase: true
    fault:
      abort:
        httpStatus: 500
```

```
        percentage:
          value: 100.0
    route:
    - destination:
        host: catalogue.sock-shop.svc.cluster.local
        port:
          number: 80
```

Under `fault`, there is another configuration called `abort` with the following parameters:

- `httpStatus`: Specifies the HTTP status code that needs to be returned downstream
- `percentage`: Specifies the percentage of requests that need to be aborted

The following is a list of the additional configuration that can be applied to the gRPC request:

- `grpcStatus`: The gRPC status code that needs to be returned when aborting gRPC requests

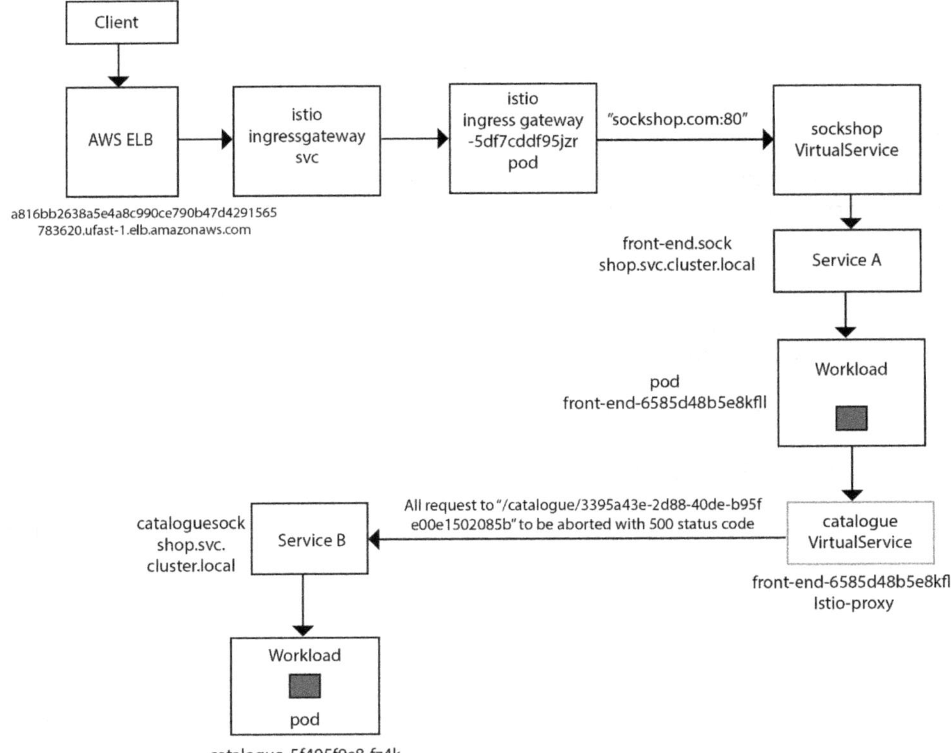

Figure 5.3 – HTTP abort injection in catalog VirtualService

In Chapter5/03-faultinjection_abort.yaml, we have configured a VirtualService rule for all calls originating from within the mesh to http://catalogue.sock-shop.svc.cluster.local/catalogue/3395a43e-2d88-40de-b95f-e00e1502085b to be aborted with an HTTP status code of 500.

Apply the following configuration:

```
$ kubectl apply -f Chapter5/03-faultinjection_abort.yaml
```

When loading sock-shop.com from the browser, you will notice that one image doesn't load. Looking at the istio-proxy access logs for the front-end Pod, you will find the following:

```
% kubectl logs -l name=front-end -c istio-proxy -n sock-shop |
grep /catalogue/3395a43e-2d88-40de-b95f-e00e1502085b
[2022-08-27T00:42:45.260Z] "GET /catalogue/3395a43e-2d88-
40de-b95f-e00e1502085b HTTP/1.1" 500 FI fault_filter_abort -
"-" 0 18 0 - "-" "-" "b364ca88-cb39-9501-b4bd-fd9ea143fa2e"
"catalogue" "-" outbound|80||catalogue.sock-shop.svc.cluster.
local - 172.20.246.13:80 10.10.10.155:57762 - -
```

This concludes fault injection and you have now practiced how to inject delay and abort into a Service Mesh.

> **Reminder**
>
> Please clean up Chapter5/02-faultinjection_delay.yaml and Chapter5/03-faultinjection_abort.yaml to avoid conflict with the upcoming exercises in this chapter.

In this section, we read about how to inject faults into a mesh so that we can then test the microservices for their resiliency and design them to withstand any fault caused by latency and delays due to upstream service communications. In the following section, we will read about implementing timeouts and retries in the Service Mesh.

Application resiliency using timeouts and retries

With communication between multiple microservices, several things can go wrong, network and infrastructure being the most common causes of service degradation and outages. A service too slow to respond can cause cascading failures across other services and have a ripple effect across the whole application. So, microservices design must be prepared for unexpected delays by setting **timeouts** when sending requests to other microservices.

The timeout is the amount of time for which a service can wait for a response from other services; beyond the timeout duration, the response has no significance to the requestor. Once a timeout happens, the microservices will follow contingency methods, which may include servicing the response from the cache or letting the request gracefully fail.

Sometimes, issues are transient, and it makes sense to make another attempt to get a response. This approach is called a **retry**, where a microservice can retry a request based on certain conditions. In this section, we will discuss how Istio enables service timeouts and retries without requiring any code changes for microservices. We will start with timeouts first.

Timeouts

A timeout is the amount of time for which an `istio-proxy` sidecar should wait for replies from a given service. Timeouts help ensure that microservices are not waiting unreasonably long for replies and that calls succeed or fail within a predictable timeframe. Istio lets you easily adjust timeouts dynamically on a per-service basis using `VirtualServices` without having to edit your service code.

As an example, we will configure a timeout of 1 second on the `order` service and we will generate a `delay` of 10 seconds in the `payment` service. The `order` service calls the `payment` service during check out, so we are simulating a slow payment service and implementing resiliency in the frontend service by configuring a timeout during invocation of `order` service:

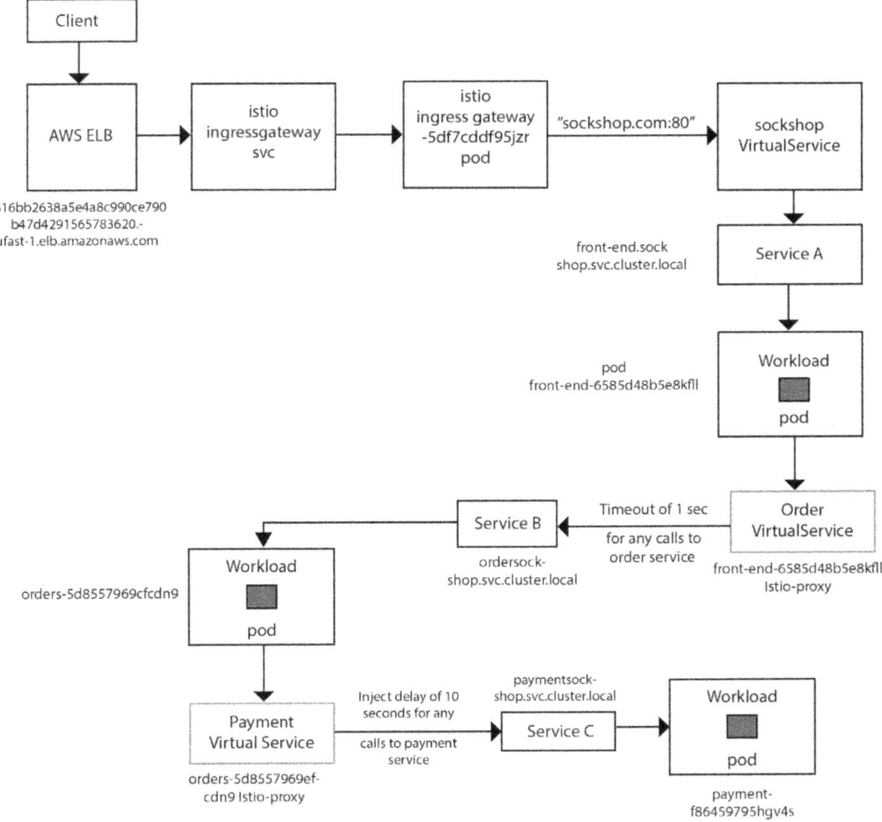

Figure 5.4 – Timeout for order and delay fault in the payment service

We will start by first configuring a timeout in the `order` service, which also happens via the `VirtualService`. You can find the full configuration in `Chapter5/04-request-timeouts.yaml` on GitHub:

```
apiVersion: networking.istio.io/v1alpha3
kind: VirtualService
metadata:
  name: orders
  namespace: chapter5
spec:
  hosts:
  - "orders.sock-shop.svc.cluster.local"
  gateways:
  - mesh
  http:
  - timeout: 1s
    route:
    - destination:
        host: orders.sock-shop.svc.cluster.local
        port:
          number: 80
```

Here, we have created a new VirtualService called `orders` and we have configured a `timeout` of `1` second to any request for `orders.sock-shop.svc.cluster.local` made from within the mesh. The timeout is part of the `http` route configuration.

Following this, we are also injecting a delay of `10` seconds to all requests to the `payment` service. For details, you can refer to the `Chapter5/04-request-timeouts.yaml` file.

Go ahead and apply the changes:

```
% kubectl apply -f Chapter5/04-request-timeouts.yaml
```

From the sockshop website, add any items to the cart and check out. Observe the behavior before and after applying the changes in this section.

Check the logs for the sidecar in the order Pods:

```
% kubectl logs --follow -l name=orders -c istio-proxy -n sock-shop | grep payment
[2022-08-28T01:24:31.968Z] "POST /paymentAuth HTTP/1.1"
200 - via_upstream - "-" 326 51 2 2 "-" "Java/1.8.0_111-internal" "ce406513-fd29-9dfc-b9cd-cb2b3dbd24a6" "payment"
```

```
"10.10.10.171:80" outbound|80||payment.sock-shop.svc.cluster.
local 10.10.10.229:60984 172.20.93.36:80 10.10.10.229:40816 - -
[2022-08-28T01:25:55.244Z] "POST /paymentAuth HTTP/1.1" 200
DI via_upstream - "-" 326 51 10007 2 "-" "Java/1.8.0_111-
internal" "ae00c14e-409c-94b1-8cfb-951a89411246" "payment"
"10.10.10.171:80" outbound|80||payment.sock-shop.svc.cluster.
local 10.10.10.229:36752 172.20.93.36:80 10.10.10.229:52932 - -
```

Notice that the actual request to the `payment` Pod was processed in `2` milliseconds but the overall time taken was `10007` milliseconds due to the injection of the `delay` fault.

Further, check the `istio-proxy` logs in the `front-end` Pods:

```
% kubectl logs --follow -l name=front-end -c istio-proxy -n
sock-shop | grep orders
[2022-08-28T01:25:55.204Z] "POST /orders HTTP/1.1" 504
UT upstream_response_timeout - "-" 232 24 1004 - "-"
"-" "b02ea4a2-b834-95a6-b5be-78db31fabf28" "orders"
"10.10.10.229:80" outbound|80||orders.sock-shop.svc.cluster.
local 10.10.10.155:49808 172.20.11.100:80 10.10.10.155:55974
- -
[2022-08-28T01:25:55.173Z] "POST /orders HTTP/1.1" 504 - via_
upstream - "-" 0 26 1058 1057 "10.10.10.217" "Mozilla/5.0
(Macintosh; Intel Mac OS X 10_15_7) AppleWebKit/537.36 (KHTML,
like Gecko) Chrome/104.0.0.0 Safari/537.36" "a8cb6589-4fd1-
9b2c-abef-f049cd1f6beb" "sockshop.com" "10.10.10.155:8079"
inbound|8079|| 127.0.0.6:36185 10.10.10.155:8079 10.10.10.217:0
outbound_.80_._.front-end.sock-shop.svc.cluster.local default
```

Here, we can see the request was returned after a little over `1` second with an HTTP status code of `504`. Although the underlying request for payment was processed in `10` seconds, the request to the `order` service timed out after `1` second.

In this scenario, we can see that the error was not gracefully handled by the website. Instead of returning a graceful message such as "**Your order has been accepted and we will notify you about its status**" or something similar, the checkout button became unresponsive. We can say that the website has no inbuilt resiliency to any failure in the `order` service.

> **Reminder**
>
> Don't forget to clean up `Chapter5/04-request-timeouts.yaml` to avoid conflict with upcoming exercises.

Retries

Timeouts are good firewalls to stop delays from cascading to other parts of an application, but the root cause of a delay is sometimes transient. In these scenarios, it may make sense to retry the requests at least a couple of times. The number of retries and the interval between them depends on the reason for the delay, which is determined by the error codes returned in the response.

In this section, we will look at how to inject retries into the Service Mesh. To keep things simple so that we can focus on learning about the concepts in this section, we will make use of the `envoydummy` service we created in the previous chapter. Envoy has many filters to simulate various delays, which we will leverage to mimic application failures.

First, configure the `envoydummy` service:

```
$ kubectl create ns utilities
$ kubectl label ns utilities istio-injection=enabled
```

After that deploy the `envoy` service and Pods:

```
$ kubectl apply -f Chapter5/envoy-proxy-01.yaml
```

We then deploy the gateway and VirtualService:

```
$ kubectl apply -f Chapter5/mockshop-ingress_01.yaml
```

Test the service and see whether it is working OK.

After this, we will configure the `envoy` config to abort half of the calls and return an error code of `503`. Notice the similarity between the Istio config and the `envoy` config.

Configure `envoydummy` to abort half of the API calls:

```
  http_filters:
              - name: envoy.filters.http.fault
                typed_config:
                  "@type": type.googleapis.com/envoy.
extensions.filters.http.fault.v3.HTTPFault
                  abort:
                    http_status: 503
                    percentage:
                      numerator: 50
                      denominator: HUNDRED
```

The complete file is available at `Chapter5/envoy-proxy-02-abort-02.yaml`.

Apply the changes:

```
$ kubectl apply -f Chapter5/envoy-proxy-02-abort-02.yaml
```

Conduct a few tests and you will notice that the API calls are working OK, although we have configured `envoydummy` to fail half of the calls. This is because Istio has already enabled default retries. Although the requests are aborted by `envoydummy`, the sidecar retries them for a default of two times and eventually gets a successful response. The interval between retries (over 25 ms) is variable and is determined automatically by Istio, preventing the called service from being overwhelmed by requests. This is made possible by the Envoy proxy in the sidecar, which uses a fully jittered exponential back-off algorithm for retries with a configurable base interval with a default value of 25 ms. If the base interval is C and N is the number of retry attempts, then the back-off for the retry is in the range of [0,(2^N−1)C). For example, for an interval of 25 ms and a retry attempt of 2, then the first retry will be delayed randomly by 0-24 ms and the second by 0-74 ms.

To disable `retries`, make the following changes in the `mockshop` VirtualService:

```
http:
- match:
  - uri:
      prefix: /
  route:
  - destination:
      port:
        number: 80
      host: envoydummy.utilities.svc.cluster.local
  retries:
    attempts: 0
```

Here, we have configured the number of `retries` to `0`. Apply the changes and this time, you will notice that half of the API calls return `503`.

We will be making the changes as depicted in the following illustration:

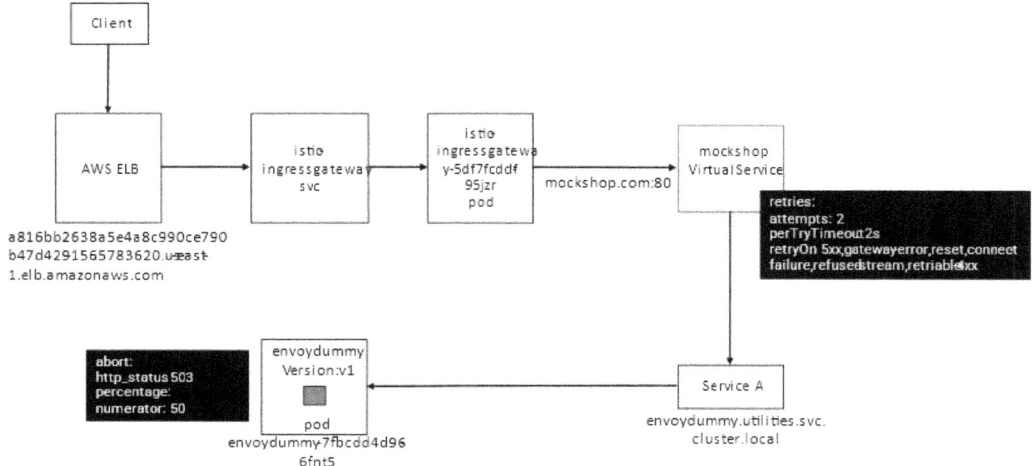

Figure 5.5 – Request retries

Make the following changes to the `retry` block. You can find the full configuration at `Chapter5/05-request-retry.yaml`:

```
    retries:
      attempts: 2
      perTryTimeout: 2s
      retryOn: 5xx,gateway-error,reset,connect-failure,refused-stream,retriable-4xx
```

In the `retry` block, we define the following configurations:

- `attempts`: The number of retries for a given request
- `perTryTimeout`: The timeout for every attempt
- `retryOn`: The condition for which the request should be retried

Apply the configuration and you will notice that the requests are working as usual:

```
$ kubectl apply -f  Chapter5/05-request-retry.yaml
```

Reminder

Please clean up `Chapter5/05-request-retry.yaml` to avoid conflict with upcoming exercises.

This concludes this section, where you have learned how to set timeouts and retries in the Service Mesh to improve application resiliency. In the next section, we will explore various strategies for load balancing.

Building application resiliency using load balancing

Load balancing is another technique to improve application resiliency. Istio load balancing policies help you maximize the availability of your application by distributing network traffic efficiently across the microservices or underlying services. Load balancing uses destination rules. Destination rules define the policy that controls how the traffic should be handled by the service after routing has occurred.

In the previous chapter, we used destination rules for traffic management purposes. In this section, we will go through various load balancing strategies provided by Istio and how to configure them using destination rules.

Deploy another `envoydummy` Pod but with an additional label of `version:v2` this time and an output of `V2----------Bootstrap Service Mesh Implementation with Istio----------V2`. The config is available in `Chapter5/envoy-proxy-02.yaml`:

```
$ kubectl apply -f Chapter5/envoy-proxy-02.yaml
```

Istio supports the following load balancing strategies.

Round-robins

Round robins are one of the simplest ways of distributing loads, where requests are forwarded one by one to underlying backend services. Although simple to use, they don't necessarily result in the most efficient distribution of traffic, because with round-robin load balancing, every upstream is treated the same, as if they are handling the same kind of traffic, are equally performant, and experience similar environmental constraints.

In `Chapter5/06-loadbalancing-roundrobbin.yaml`, we have created a destination rule with `trafficPolicy` as follows:

```
apiVersion: networking.istio.io/v1alpha3
kind: DestinationRule
metadata:
  name: envoydummy
spec:
  host: envoydummy
  trafficPolicy:
```

```
        loadBalancer:
          simple: ROUND_ROBIN
```

In `DestinationRule`, you can define multiple parameters, which we will uncover one by one in this section and subsequent sections.

For round-robin load balancing, we have defined the following destination rules in `trafficPolicy`:

- `simple`: Defines the load balancing algorithms to be used – the possible values are as follows:
  - `UNSPECIFIED`: Istio will select an appropriate default
  - `RANDOM`: Istio will select a healthy host at random.
  - `PASSTHROUGH`: This option allows the client to ask for a specific upstream and the load balancing policy will forward the request to the requested upstream.
  - `ROUND_ROBIN`: Istio will send requests in a round-robin fashion to upstream services.
  - `LEAST_REQUEST`: This distributes the load across endpoints depending on how many requests are outstanding on each endpoint. This policy is the most efficient of all the load balancing policies.

Apply the configuration using this:

```
$ kubectl apply -f Chapter5/06-loadbalancing-roundrobbin.yaml
```

Test the endpoints and you will notice that you are receiving an equal amount of responses from both versions of `envoydummy`.

RANDOM

When using the `RANDOM` load balancing policy, Istio selects a destination host at random. You can find an example of the `RANDOM` load balancing policy in the `Chapter5/07-loadbalancing-random.yaml` file. The following is the destination rule with a `RANDOM` load balancing policy:

```
apiVersion: networking.istio.io/v1alpha3
kind: DestinationRule
metadata:
  name: envoydummy
spec:
  host: envoydummy
  trafficPolicy:
        loadBalancer:
          simple: RANDOM
```

Apply the configuration:

```
$ kubectl apply -f Chapter5/07-loadbalancing-random.yaml
```

Make a few requests to the endpoints and you will notice that the response doesn't have any predictable patterns.

LEAST_REQUEST

As mentioned earlier, in the `LEAST_REQUEST` load balancing policy, Istio routes the traffic that has the least amount of outstanding requests upstream.

To mimic this scenario, we will create another service that specifically sends all requests to an `envoydummy` version 2 Pod. The configuration is available at `Chapter5/08-loadbalancing-leastrequest.yaml`:

```
apiVersion: v1
kind: Service
metadata:
  name: envoydummy2
  labels:
    name: envoydummy2
    service: envoydummy2
  namespace: chapter5
spec:
  ports:
    # the port that this service should serve on
  - port: 80
    targetPort: 10000
  selector:
    name: envoydummy
```

We have also made changes to `DestinationRule` to send the request to the host that has the fewest active connections:

```
  trafficPolicy:
       loadBalancer:
         simple: LEAST_REQUEST
    version: v2
```

Apply the configuration:

```
$ kubectl apply -f Chapter5/08-loadbalancing-leastrequest.yaml
```

Using `kubectl port-forward`, we can send a request to the `envoydummy2` service from `localhost`:

```
$ kubectl port-forward svc/envoydummy2 10000:80 -n utilities
```

After this, we will generate a request targeted for version 2 of the `envoydummy` service using the following command:

```
$ for ((i=1;i<1000000000;i++)); do curl -v "http://
localhost:10000"; done
```

While the load is in progress, access the request using the `mockshop` endpoint and you will notice the majority, if not all the requests, are served by version 1 of the `envoydummy` Pods because of the `LEAST_REQUEST` load balancing policy:

```
$ curl -Hhost:mockshop.com http://
a816bb2638a5e4a8c990ce790b47d429-1565783620.us-east-1.elb.
amazonaws.com/
V1----------Bootstrap Service Mesh Implementation with Istio--
--------V1
V1----------Bootstrap Service Mesh Implementation with Istio--
--------V1
V1----------Bootstrap Service Mesh Implementation with Istio--
--------V1
V1----------Bootstrap Service Mesh Implementation with Istio--
--------V1
```

In the preceding example, you saw how Istio routes all requests for `mockshop` to version 1 of `envoydummy` because `v1` had the fewest active connections.

Defining multiple load balancing rules

Istio has provisions to apply multiple load balancing rules for every subset. In the `Chapter5/09-loadbalancing-multirules.yaml` file, we are defining the default load balancing policy of `ROUND_ROBIN`, the `LEAST_REQUEST` policy for the `v1` subset, and `RANDOM` for the `v2` subset.

The following is the snippet from the configuration defined in `Chapter5/09-loadbalancing-multirules.yaml`:

```
host: envoydummy
trafficPolicy:
     loadBalancer:
       simple: ROUND_ROBIN
subsets:
- name: v1
  labels:
    version: v1
  trafficPolicy:
    loadBalancer:
      simple: LEAST_REQUEST
- name: v2
  labels:
    version: v2
  trafficPolicy:
    loadBalancer:
      simple: RANDOM
```

In the preceding code block, we have applied the `LEAST_REQUEST` load balancing policy to the `v1` subset, the `RANDOM` load balancing policy to the `v2` subset, and `ROUND_ROBBIN` to any other subset not specified in the code block.

Being able to define multiple load balancing rules for workloads enables you to apply fine-grained control on traffic distribution at the level of the destination rule subset. In the next section, we will go over another important aspect of application resiliency called **rate limiting** and how it can be implemented in Istio.

Rate limiting

Another important technique for application resiliency is rate limiting and circuit breaking. Rate limiting helps provide the following controls to handle traffic from consumers without breaking down the provider system:

- Surge protection to prevent a system from being overloaded by a sudden spike in traffic
- Aligning the rate of incoming requests with the available capacity to process requests
- Protecting slow providers from fast consumers

Rate limiting is performed by configuring destination rules with a connection pool for connections to upstream services. Connection pool settings can be applied at the TCP level as well as the HTTP level, as described in the following configuration in `Chapter5/10-connection-pooling.yaml`:

```
apiVersion: networking.istio.io/v1alpha3
kind: DestinationRule
metadata:
  name: envoydummy
  namespace: chapter5
spec:
  host: envoydummy
  trafficPolicy:
      connectionPool:
        http:
          http2MaxRequests: 1
          maxRequestsPerConnection: 1
          http1MaxPendingRequests: 0
```

The following are the key attributes of the connection pool configuration:

- `http2MaxRequests`: The maximum number of active requests to a destination; the default value is `1024`.
- `maxRequestsPerConnection`: The maximum number of requests per connection to upstream. A value of `1` disables keep-alive whereas `0` is unlimited.
- `http1MaxPendingRequests`: The maximum number of requests that will be queued while waiting for a connection from the connection pool; the default value is `1024`.

We have configured a maximum of `1` request per connection to upstream, a maximum of `1` active connection at any point of time, and no queuing for connection requests allowed.

Testing the rate limit, circuit breakers, and outlier detection are not as straightforward as testing other features of application resiliency. Fortunately, there is a very handy load testing utility called `fortio` available at `https://github.com/fortio/fortio` and packaged in the Istio sample directory. We will use `fortio` for generating load and testing rate limit.

Deploy `fortio` from the Istio directory:

```
$ kubectl apply -f samples/httpbin/sample-client/fortio-deploy.
yaml -n utilities
```

Apply one of the load balancing policies to test normal behavior:

```
$ kubectl apply -f Chapter5/06-loadbalancing-roundrobbin.yaml
```

Generate a load using `fortio`:

```
$ kubectl exec -it fortio-deploy-7dcd84c469-xpggh
-n utilities -c fortio -- /usr/bin/fortio load
-qps 0 -c 2 -t 1s -H "Host:mockshop.com" http://
a816bb2638a5e4a8c990ce790b47d429-1565783620.us-east-1.elb.
amazonaws.com/
```

In the previous request, we are configuring `folio` to generate a load test for `1` second with `2` parallel connections with a maximum query per second (`qps`) rate of `0`, meaning no waits/the maximum `qps` rate.

In the output, you will notice that all requests were successfully processed:

```
Code 200 : 486 (100.0 %)
Response Header Sizes: count 486 avg 151.01235 +/- 0.1104 min
151 max 152 sum 73392
Response Body/Total Sizes : count 486 avg 223.01235 +/- 0.1104
min 223 max 224 sum 108384
All done 486 calls (plus 2 warmup) 4.120 ms avg, 484.8 qps
```

In this case, a total of `486 calls` were made with a 100% success rate. Next, we will apply the changes to enforce rate limiting:

```
$ kubectl apply -f Chapter5/10-connection-pooling.yaml
```

Run the test again with `1` connection:

```
kubectl exec -it fortio-deploy-7dcd84c469-xpggh
-n utilities -c fortio -- /usr/bin/fortio load
-qps 0 -c 1 -t 1s -H "Host:mockshop.com" http://
a816bb2638a5e4a8c990ce790b47d429-1565783620.us-east-1.elb.
amazonaws.com/
Code 200 : 175 (100.0 %)
Response Header Sizes : count 175 avg 151.01143 +/- 0.1063 min
151 max 152 sum 26427
Response Body/Total Sizes : count 175 avg 223.01143 +/- 0.1063
min 223 max 224 sum 39027
All done 175 calls (plus 1 warmup) 5.744 ms avg, 174.0 qps
```

Run the test again; this time, with two connections:

```
Code 200 : 193 (66.6 %)
Code 503 : 97 (33.4 %)
Response Header Sizes : count 290 avg 100.55517 +/- 71.29 min 0
max 152 sum 29161
Response Body/Total Sizes : count 290 avg 240.45517 +/- 24.49
min 223 max 275 sum 69732
All done 290 calls (plus 2 warmup) 6.915 ms avg, 288.8 qps
```

You can see that `33.4%` of the calls failed with a 503 error code because the destination rule enforces the rate limiting rules.

In this section, you saw an example of rate limiting, which, in turn, is also circuit breaking based on a rate limiting condition. In the next section, we will read about circuit breaking by detecting outliers.

Circuit breakers and outlier detection

In this section, we will look at outlier detection and circuit breaker patterns. **A circuit breaker** is a design pattern in which you continuously monitor the response processing behavior of upstream systems and when the behavior is unacceptable, you stop sending any further requests upstream until the behavior has become acceptable again.

For example, you can monitor the average response time of the upstream system and when it crosses a certain threshold, you may decide to stop sending any further requests to the system; this is called tripping the circuit breaker. Once the circuit breaker has been tripped, you leave it that way for a certain duration so that the upstream service can heal. After the circuit breaker duration has elapsed, you can reset the circuit breaker to let the traffic pass through again.

While circuit breaking is the part that handles the flow of traffic, **outlier detection** is a set of policies to identify the conditions when the circuit breaker should trip.

We will configure one of `envoy` Pods to return a `503` error at random. We will reuse `Chapter5/envoy-proxy-02-abort-02.yaml`, in which we configured a version of `envoydummy` to return a `503` error for 50% of the requests.

To avoid any confusion, delete all previous deployments of `envoydummy` in the `utilities` namespace and any Istio config we have executed prior to this section.

Perform the following in the same order as shown:

```
Kubectl apply -f Chapter5/envoy-proxy-02.yaml
Kubectl apply -f Chapter5/envoy-proxy-02-abort-02.yaml
```

At this stage, we have two envoydummy Pods. For the Pod labeled `version:v1`, returning `V1----------Bootstrap Service Mesh Implementation with Istio----------V1`, we have modified it to abort 50% of the request and return `503`.

Perform the following command to disable the default automated retries by Istio:

```
$ kubectl apply -f Chapter5/11-request-retry-disabled.yaml
```

Test the request and you will notice that the response is a mixed bag of `v1`, `v2`, and `503`.

Now, the task at hand is to define an outlier detection policy to detect `v1` as the outlier because of its erroneous behavior of returning a `503` error code. We will do that via destination rules as follows:

```
apiVersion: networking.istio.io/v1alpha3
kind: DestinationRule
metadata:
  name: envoydummy
  namespace: utilities
spec:
  host: envoydummy.utilities.svc.cluster.local
  trafficPolicy:
      connectionPool:
        http:
          http2MaxRequests: 2
          maxRequestsPerConnection: 1
          http1MaxPendingRequests: 0
      outlierDetection:
        baseEjectionTime: 5m
        consecutive5xxErrors: 1
        interval: 1s
        maxEjectionPercent: 100
```

In `outlierDetection`, the following parameters are provided:

- `baseEjectionTime`: The minimum ejection duration per ejection, which is then multiplied by the number of times an upstream is found to be unhealthy. For example, if a host is found to be an outlier five times, then it will be ejected from the connection pool for `baseEjectionTime*5`.
- `consecutive5xxErrors`: The number of 5xx errors that need to occur to qualify the upstream to be an outlier.

- `interval`: The time between the checks when Istio scans upstream for health status. The interval is specified in hours, mins, or seconds.
- `maxEjectionPercent`: The maximum number of hosts in the connection pool that can be ejected.

In the destination rule, we have configured Istio to scan upstream at an interval of 1 second. If 1 or more 5xx errors are consecutively returned, then the upstream will be ejected from the connection pool for 5 minutes and if needed, all hosts can be ejected from the connection pool.

The following parameters can also be defined for outlier detection, but we have not used them in our example:

- `splitExternalLocalOriginErrors`: This flag tells Istio whether it should consider the local service behavior to determine whether the upstream service is an outlier. For example, `404` may be returned, which is a valid response, but returning it too frequently can also mean that there may be a problem. Maybe the upstream service has an error but due to bad error handling, the upstream is returning `404`, which, in turn, makes the downstream service return a 5XX error. To summarize, this flag enables outlier detection not just based on the response codes returned upstream but also on how the downstream system perceived the response.
- `consecutiveLocalOriginFailures`: This is the number of consecutive local errors before the upstream service is ejected from the connection pool.
- `consecutiveGatewayErrors`: This is the number of gateway errors before an upstream service is ejected. This might be caused by unhealthy connections or misconfiguration between a gateway and the upstream service. When an upstream host is accessed over HTTP then, HTTP status codes of `502`, `503`, or `504` are usually returned due to communication issues between the gateway and upstream services.
- `minHealthPercent`: This field defines the minimum number of healthy upstream systems available in the load balancing pool for outlier detection to be enabled. Once the number of healthy upstream systems drops below this level, outlier detection is disabled to maintain service availability.

The configuration defined in `Chapter5/12-outlier-detection.yaml` enables us to quickly observe the effects of outlier detection but when deploying this in a non-experimental scenario, the values need to be tuned and configured as per the resiliency requirements.

Apply the updated destination rule:

```
$ kubectl apply -f  Chapter5/12-outlier-detection.yaml
```

After applying the changes, test the request a few times. you will notice that apart from just a few responses with `V1----------Bootstrap Service Mesh Implementation with Istio----------V1`, most of the response contains `V2----------Bootstrap Service Mesh Implementation with Istio----------V2` because Istio detected the `v1` Pod returning `503` and marked it as an outlier in the connection pool.

Summary

In this chapter, we read about how Istio enables application resiliency and testing by providing options to inject delays and faults into request processing. Fault injection assists in validating an application's resiliency when there is unexpected degradation in underlying services, as well as the network and infrastructure. After fault injection, we read about request timeouts and how they improve application resiliency. For transient failures, it might be a wise idea to make a few retries before giving up on the request and hence, we practiced configuring Istio to perform service retries. Fault injection, timeouts, and retries are properties of VirtualServices and are carried out before routing a request to upstream services.

In the second part of the chapter, we read about various load balancing policies and how you can configure load balancing policies based on the dynamic behavior of an upstream service. Load balancing helps to distribute traffic to upstream services, with `LEAST_REQUEST` policy being the most effective policy for distributing traffic based on the number of requests being handled upstream at any point in time. Load balancing is configured in destination rules because it happens as part of the request routing to upstream services. After load balancing, we read about rate limiting and how it is based on the connection pool configuration in the destination rules. In the final part of the chapter, we read about how to configure destination rules to implement outlier detection.

The most noticeable element of everything that we read in this chapter was the ability to implement application resiliency via timeouts, retries, load balancing, circuit breaking, and outlier detection without ever needing to change the application code. Applications benefit from these resiliency strategies just by being part of the Service Mesh. Various types of software and utilities are used by software engineers to perform chaos engineering to understand the resiliency of an application suffering from failures. You can use these chaos engineering tools to test the application resiliency provided by the Service Mesh.

The next chapter is very exciting and intense because we will read about how to make use of Istio to implement iron-clad security for applications running in the mesh.

6 Securing Microservices Communication

Istio secures communication between microservices without microservices requiring any code changes. In *Chapter 4*, we briefly touched upon the topic of security. We configured transport layer security by exposing our sockshop application over HTTPS. We created certificates and configured the Istio Ingress gateway to bind those certificates to hostnames in SIMPLE TLS mode. We also implemented TLS-based security for multiple hosts managed by a single Ingress gateway.

In this chapter, we will dive deeper into some advanced topics of security. We will start by understanding Istio security architecture. We will implement mutual TLS for service communication with other services in the mesh, and we will also implement mutual TLS with downstream clients outside the mesh. We will then perform various hands-on exercises to create custom security policies for authentication and authorization. We will go through these topics in the following order:

- Istio security architecture
- Authenticating using mutual TLS
- How to configure a custom authentication and authorization policy

> **Important note**
> The technical prerequisites for this chapter are the same as *Chapters 4* and *5*.

Understanding Istio security architecture

In *Chapter 3*, we discussed how the Istio control plane is responsible for the injection of sidecars and establishing trust so that sidecars can communicate with the control plane securely and security policies are eventually enforced by the sidecar. When deployed in Kubernetes, Istio relies on Kubernetes service accounts to identify the roles of workloads in a Service Mesh. The Istio CA watches the Kubernetes API server for the addition/deletion/modification of any service accounts in the namespace with Istio injection enabled. It creates a key and certificates for each service account and, during Pod creation, the certificate and key are mounted onto the sidecar. The Istio CA is responsible for managing the life cycle of the certificates distributed to the sidecars, including the rotation and management of private keys. Using the **Secure Production Identity Framework for Everyone** (**SPIFFE**) format identities, Istio provides a strong identity to each service along with service naming, which represents the role that can be taken up by the identity assigned to the service.

SPIFFE is a set of open source standards for software identity. SPIFFE provides platform-agnostic interoperable software identities along with interfaces and documents required to obtain and validate cryptographic identity in a fully automated fashion.

In Istio, each workload is automatically assigned an identity represented in the X.509 certificate format. The creation and signing of **certificate signing request** (**CSRs**) are managed by the Istio control plane, as discussed in *Chapter 3*. The X.509 certificate follows the SPIFFE format.

Let's redeploy the `envoydummy` service and inspect the `envoydummy` Pods:

```
$ kubectl apply -f Chapter6/01-envoy-dummy.yaml
$ istioctl proxy-config all po/envoydummy-2-7488b58cd7-m5vpv -n
utilities -o json | jq -r '.. |."secret"?' | jq -r 'select(.
name == "default")' | jq -r '.tls_certificate.certificate_
chain.inline_bytes' | base64 -d - | step certificate
inspect  --short
X.509v3 TLS Certificate (RSA 2048) [Serial: 3062...1679]
  Subject:      spiffe://cluster.local/ns/utilities/sa/default
  Issuer:
  Valid from:   2022-09-11T22:18:13Z
          to:   2022-09-12T22:20:13Z
```

step CLI

You will need to install step `CLI` to be able to run the preceding command. To install it, please follow the documentation at `https://smallstep.com/docs/step-cli`.

In the output of the preceding command, you will notice that the **Subject Alternative Name** (**SAN**) is `spiffe://cluster.local/ns/utilities/sa/default`. This is the SPIFFE ID, which functions as the unique name:

- `spifee` is the URI scheme
- `cluster.local` is the trust domain
- `/ns/utilities/sa/default` is the URI identifying the service account associated with the workload:

- `ns` stands for namespace
- `sa` stands for service account

The value default for service accounts comes from the service account attached to the workload. In our example of `envoydummy`, we didn't associate any service accounts so, by default, Kubernetes associated the `default` service account. You can find the service account name associated with a Pod using the following command:

```
kubectl get po/envoydummy-2-7488b58cd7-m5vpv -n utilities -o
json | jq .spec.serviceAccountName
"default"
```

You will notice that `default` is the default name for service accounts associated with all Pods in all namespaces, such as `sock-shop`, `utilities`, and so on. Kubernetes creates a service account named `default` in every namespace:

```
% kubectl get sa -n utilities
NAME       SECRETS   AGE
default    1         13d
% kubectl get sa -n sock-shop
NAME       SECRETS   AGE
default    1         27d
```

> **Kubernetes service accounts**
>
> A service account is an identity assigned to workloads in Kubernetes. When processes running inside a workload try to access other Kubernetes resources, they are identified and authenticated as per the details of their service accounts. You can find more details about service accounts at `https://kubernetes.io/docs/tasks/configure-pod-container/configure-service-account/`.

Secure Naming is a technology that decouples the name of services from the identities the services are running as. In the previous example, `spiffe://cluster.local/ns/utilities/sa/default` is the identity of the service presented during mutual TLS presented by the istio-proxy sidecar in `envoydummy-2-7488b58cd7-m5vpv workload`. From the SPIFFE ID, the other party (istio-proxy in another Pod) in the MTLS session can validate that the endpoint has the identity of a service account named `default` in the `utilities` namespace. The Istio control plane propagates the secure naming information to all sidecars in the mesh and during mutual TLS, the sidecar not only verifies that the identity is correct but also that the respective service is assuming the correct identity.

The following diagram summarizes the Istio security architecture:

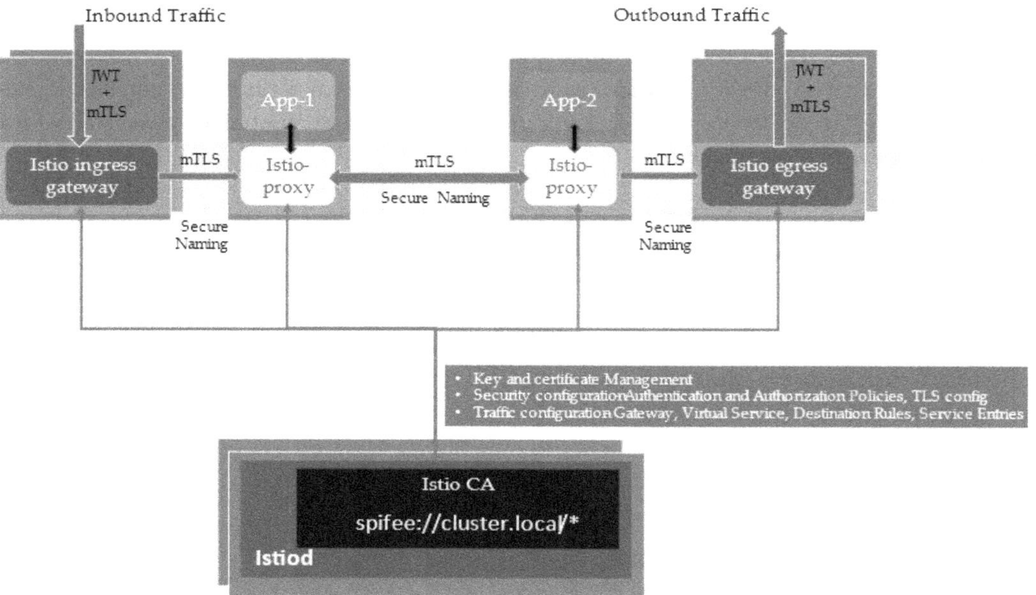

Figure 6.1 – Istio security architecture

These are the key concepts to remember:

- The Istio CA manages keys and certificates and the SANs in certificates are in SPIFFE format
- Istiod distributes authentication and authorization security policies to all sidecars in the mesh
- Sidecars enforce authentication and authorization as per security policies distributed by Istiod

> **Reminder**
> Make sure to clean up `Chapter6/01-envoy-dummy.yaml` to avoid conflict in upcoming exercises.

In the next section, we will read about how to secure data in transit between microservices in a Service Mesh.

Authentication using mutual TLS

Mutual TLS (mTLS) is a technique for authenticating two parties at each end of a network connection. Through mTLS, each party can verify that the other party is what they are claiming to be. Certificate authorities play a critical role in mTLS, and hence we had the previous section on Istio security architecture describing certificate authorities and secure naming in Istio.

mTLS is one of the most frequently used authentication mechanisms for implementing the zero-trust security framework, in which no party trusts another party by default, irrespective of where the other party is placed in the network. Zero trust assumes that there are no traditional network edges and boundaries and hence every party needs to be authenticated and authorized. This helps to eliminate many security vulnerabilities that arise because of the assumption-based trust model.

In the following two subsections, we will look at how Istio helps you implement mTLS for service-to-service authentication inside a mesh, also called east-west traffic, and mTLS between client/downstream systems that are outside the mesh, with services in the mesh called north-south communication.

Service-to-service authentication

Istio provides service-to-service authentication by using mTLS for transport authentication. During traffic processing, Istio performs the following:

- All outbound traffic from Pods is rerouted to istio-proxy.
- istio-proxy starts an mTLS handshake with the server-side istio-proxy. During the handshake, it also does a secure naming check to verify that the service account presented in the server certificate can run the Pod.
- The server-side istio-proxy verifies the client-side istio-proxy in the same fashion and if all is okay, a secure channel is established between the two proxies.

Istio provides the following two options when implementing mTLS:

- **Permissive mode**: In permissive mode, Istio allows traffic in both mTLS and non-mTLS mode. This feature is primarily to improve the onboarding of clients to mTLS. Clients who are not yet ready to communicate over mTLS can continue communicating over TLS with the view that they will eventually migrate to mTLS whenever they are ready.
- **Strict mode**: In strict mode, Istio enforces strict mTLS and any non-mTLS traffic is not allowed.

Mutual TLS traffic can be established between clients outside of a mesh trying to access a workload within the mesh, as well as clients within the mesh trying to access other workloads in the mesh. For the former, we will discuss the details in the next section. For the latter set of clients, we will go through some examples in this section.

Let's set up service-to-service communication using mTLS:

1. Create a namespace called `chapter6` with Istio injection enabled and deploy the `httpbin` service:

   ```
   $ kubectl apply -f Chapter6/01-httpbin-deployment.yaml
   ```

   Most of the config in this deployment is the usual, except that we have also created a default Kubernetes service account called `httpbin` in the `Chapter6` namespace:

   ```
   apiVersion: v1
   kind: ServiceAccount
   metadata:
     name: httpbin
     namespace: chapter6
   ```

   The `httpbin` identity is then assigned to an `httpbin` Pod by following these specs:

   ```
     Spec:
         serviceAccountName: httpbin
         containers:
         - image: docker.io/kennethreitz/httpbin
           imagePullPolicy: IfNotPresent
           name: httpbin
           ports:
           - containerPort: 80
   ```

2. Next, we will create a client in the form of a `curl` Pod to access the `httpbin` service. Create a `utilities` namespace with Istio injection disabled, and create a `curl` Deployment with its own service account:

   ```
   $ kubectl apply -f Chapter6/01-curl-deployment.yaml
   ```

   Make sure the `istio-injection` label is not applied. If it is, you can remove it using the following command:

   ```
   $ kubectl label ns utilities istio-injection-
   ```

3. From the `curl` Pod, try to access the `httpbin` Pod and you should get a response back:

   ```
   $ kubectl exec -it curl -n utilities - curl -v http://
   httpbin.chapter6.svc.cluster.local:8000/get
   {
     "args": {},
     "headers": {
   ```

```
        "Accept": "*/*",
        "Host": "httpbin.chapter6.svc.cluster.local:8000",
        "User-Agent": "curl/7.87.0-DEV",
        "X-B3-Sampled": "1",
        "X-B3-Spanid": "a00a50536c3ec2f5",
        "X-B3-Traceid": "49b6942c85c7c1f2a00a50536c3ec2f5"
    },
    "origin": "127.0.0.6",
    "url": "http://httpbin.chapter6.svc.cluster.local:8000/
get"
```

4. So far, we have the `httpbin` Pod running in the mesh, but by default, in *permissive* TLS mode. We will now create a `PeerAuthentication` policy to enforce `STRICT` mTLS. The `PeerAuthentication` policy defines how traffic will be tunneled via sidecars:

```
apiVersion: security.istio.io/v1beta1
kind: PeerAuthentication
metadata:
  name: "httpbin-strict-tls"
  namespace: chapter6
spec:
  mtls:
    mode: STRICT
  selector:
    matchLabels:
      app: httpbin
```

In the `PeerAuthentication` policy, we defined the following configuration parameters:

- `mtls`: This defines the mTLS setting. If not specified, the value is inherited from the default mesh-wide setting. It has one field called `mode`, which can have the following values:
  - `UNSET`: With this value, the mTLS settings are inherited from the parent and if the parent does not have any settings, then the value is set to `PERMISSIVE`.
  - `MTLS`: With this value, the sidecar accepts both mTLS and non-mTLS connections.
  - `STRICT`: This enforces strict mTLS – any non-mTLS connection will be dropped.
  - `DISABLE`: mTLS is disabled and connections are not tunneled.

- `Selector`: This defines the criteria that need to be satisfied by a workload to be part of this per authentication policy. It has a field named `matchLabels`, which takes label information in the `key:value` format.

To summarize the configuration, we have created `httpbin-strict-tls`, which is a `PeerAuthentication` policy in the `Chapter6` namespace. The policy enforces string mTLS for all workloads that have a label of `app=httpbin`. The configuration is available at `Chapter6/02-httpbin-strictTLS.yaml`.

5. Apply the changes via the following command:

```
$ kubectl apply -f Chapter6/02-httpbin-strictTLS.yaml
peerauthentication.security.istio.io/httpbin-strict-tls
created
```

6. Now try to connect to the `httpbin` service from the `curl` Pod:

```
$ kubectl exec -it curl -n utilities - curl -v http://
httpbin.chapter6.svc.cluster.local:8000/get
* Connected to httpbin.chapter6.svc.cluster.local
(172.20.147.104) port 8000 (#0)
> GET /get HTTP/1.1
> Host: httpbin.chapter6.svc.cluster.local:8000
> User-Agent: curl/7.87.0-DEV
> Accept: */*
>
* Recv failure: Connection reset by peer
* Closing connection 0
curl: (56) Recv failure: Connection reset by peer
command terminated with exit code 56
```

`curl` is not able to connect because the `curl` Pod is running in a namespace with Istio injection disabled, whereas the `httpbin` Pod is running in the mesh with the `PeerAuthentication` policy enforcing `STRICT` mTLS. One option is to manually establish an mTLS connection, which is equivalent to modifying your application code to perform mTLS. In this case, as we are trying to simulate service communication within the mesh, we can simply turn on Istio injection and let Istio take care of client-side mTLS as well.

7. Enable Istio injection for the `curl` Pod using the following steps:

    I. Delete the resource created by `Chapter6/01-curl-deployment.yaml`.

    II. Modify the value of Istio injection to be enabled.

    III. Apply the updated configuration.

Once the `curl` Pod is in the `RUNNING` state, along with the istio-proxy sidecar, you can perform `curl` on the `httpbin` service and you will see the following output:

```
$ kubectl exec -it curl -n utilities -- curl -s http://
httpbin.chapter6.svc.cluster.local:8000/get
{
  "args": {},
  "headers": {
    "Accept": "*/*",
    "Host": "httpbin.chapter6.svc.cluster.local:8000",
    "User-Agent": "curl/7.85.0-DEV",
    "X-B3-Parentspanid": "a35412ed46b7ec46",
    "X-B3-Sampled": "1",
    "X-B3-Spanid": "0728b578e88b72fb",
    "X-B3-Traceid": "830ed3d5d867a460a35412ed46b7ec46",
    "X-Envoy-Attempt-Count": "1",
    "X-Forwarded-Client-Cert": "By=spiffe://cluster.
local/ns/chapter6/sa/
httpbin;Hash=b1b88fe241c557bd1281324b458503274eec3f04b1d-
439758508842d6d5b7018;Subject=\"\";URI=spiffe://cluster.
local/ns/utilities/sa/curl"
  },
  "origin": "127.0.0.6",
  "url": "http://httpbin.chapter6.svc.cluster.local:8000/
get"
}
```

In the response from the `httpbin` service, you will notice all the headers that were received by the `httpbin` Pod. The most interesting header is `X-Forwarded-Client-Cert`, also called `XFCC`. There are two parts of the `XFCC` header value that shed light on mTLS:

- `By`: This is filled with the SAN, which is the SPIFFE ID of the istio-proxy's client certificate of the `httpbin` Pod (`spiffe://cluster.local/ns/chapter6/sa/httpbin`)
- `URI`: This contains the SAN, which is the SPIFFE ID of the `curl` Pod's client certificate presented during mTLS (`spiffe://cluster.local/ns/utilities/sa/curl`)

There is also `Hash`, which is the SHA256 digest of istio-proxy's client certificate of the `httpbin` Pod.

You can selectively apply mTLS configuration at the port level also. In the following configuration, we are implying that mTLS is enforced strictly for all ports except port `8080`, which should allow permissive connections. The configuration is available at `Chapter6/03-httpbin-strictTLSwithException.yaml`:

```
apiVersion: security.istio.io/v1beta1
kind: PeerAuthentication
metadata:
  name: "httpbin-strict-tls"
  namespace: chapter6
spec:
  portLevelMtls:
    8080:
      mode: PERMISSIVE
    8000:
      mode: STRICT
  selector:
    matchLabels:
      app: httpbin
```

So, in this section, we learned how to perform mTLS between services inside the mesh. mTLS can be enabled at the service level as well as at the port level. In the next section, we will read about performing mTLS with clients outside the mesh.

> **Reminder**
>
> Make sure to clean up `Chapter6/01-httpbin-deployment.yaml`, `Chapter6/01-curl-deployment.yaml` and `Chapter6/02-httpbin-strictTLS.yaml` to avoid conflict in upcoming exercises.

Authentication with clients outside the mesh

For clients outside the mesh, Istio supports mTLS with the Istio Ingress gateway. In *Chapter 5*, we configured HTTPS at the Ingress gateway. In this section, we will extend that configuration to also support mTLS.

We will now configure mTLS for the `httpbin` Pod. Notice that the first five steps are very similar to *steps 1-5* of *Exposing Ingress over HTTPS* of *Chapter 5*. The steps are as follows:

1. Create a CA. Here, we are creating a CA with a **Common Name** (**CN**) of `sock.inc`:

    ```
    $ openssl req -x509 -sha256 -nodes -days 365 -newkey
    rsa:2048 -subj '/O=sock Inc./CN=sock.inc' -keyout sock.
    inc.key -out sock.inc.crt
    ```

2. Generate a CSR for `httpbin.org`. Here, we are generating a **Certificate Signing Request** (**CSR**) for `httpbin.org`, which also generates a private key:

    ```
    $ openssl req -out httpbin.org.csr -newkey rsa:2048
    -nodes -keyout httpbin.org.key -subj "/CN=httpbin.org/
    O=sockshop.inc"
    Generating a 2048 bit RSA private key
    .........+++
    ..+++
    writing new private key to 'httpbin.org.key'
    ```

3. Sign the CSR using the CA created in *step 1*:

    ```
    $ openssl x509 -req -sha256 -days 365 -CA sock.inc.crt
    -CAkey sock.inc.key -set_serial 0 -in httpbin.org.csr
    -out httpbin.org.crt
    Signature ok
    subject=/CN=httpbin.org/O=sockshop.inc
    Getting CA Private Key
    ```

4. Load the certificate and private key as a Kubernetes secret along with the CA certificate against which client certificates must be verified:

    ```
    $ kubectl create -n istio-system secret generic httpbin-
    credential --from-file=tls.key=httpbin.org.key --from-
    file=tls.crt=httpbin.org.crt --from-file=ca.crt=sock.inc.
    crt
    secret/httpbin-credential created
    ```

5. Configure the Ingress gateway to enforce mTLS for all incoming connections and use the secret created in *step 4* as the secret containing TLS certificate and the CA certificate:

    ```
      tls:
          mode: MUTUAL
          credentialName: httpbin-credential
    ```

6. Deploy the `httpbin` Pod, the Ingress gateway, and the virtual service:

```
$ kubectl apply -f Chapter6/02-httpbin-deployment-MTLS.
yaml
```

7. To perform mTLS, you also need to generate client certificates that can be used to prove the client's identity. For that, perform the following steps:

```
$ openssl req -out bootstrapistio.packt.com.csr -newkey
rsa:2048 -nodes -keyout bootstrapistio.packt.com.key
-subj "/CN= bootstrapistio.packt.com/O=packt.com"
```

```
$ openssl x509 -req -sha256 -days 365 -CA sock.inc.crt
-CAkey sock.inc.key -set_serial 0 -in bootstrapistio.
packt.com.csr -out bootstrapistio.packt.com.crt
```

8. Test the connection to `httpbin.org` by passing client certificates in the request:

```
% curl -v -HHost:httpbin.org --connect-to "httpbin.
org:443:a816bb2638a5e4a8c990ce790b47d429-1565783620.
us-east-1.elb.amazonaws.com" --cacert sock.inc.crt --cert
bootstrapistio.packt.com.crt --key bootstrapistio.packt.
com.key https://httpbin.org:443/get
```

Reminder

Don't forget to clean up using `kubectl delete -n istio-system secret httpbin-credential` and `kubectl delete -f Chapter6/02-httpbin-deployment-MTLS.yaml`.

Configuring RequestAuthentication

Like service-to-service authentication, Istio can also authenticate an end user or validate that an end user has been authenticated based on assertions presented by the end user. The `RequestAuthentication` policy is used to specify what authentication methods are supported by a workload. This policy identifies the authenticated identity but doesn't enforce whether the request should be allowed or denied. Rather, it provides information about the authenticated identity to the authorization policy, which we will go through in the next section. In this section, we will learn how to make use of the Istio `RequestAuthentication` policy to validate an end user who has been authenticated by Auth0 and is providing a bearer token as security credentials to Istio. If you are not familiar with OAuth then you can read more about it at `https://auth0.com/docs/authenticate/protocols/oauth`.

We will follow the hands-on steps to configure Auth0 and perform an OAuth flow while at the same time demystifying all that is happening under the hood:

1. Sign up for Auth0:

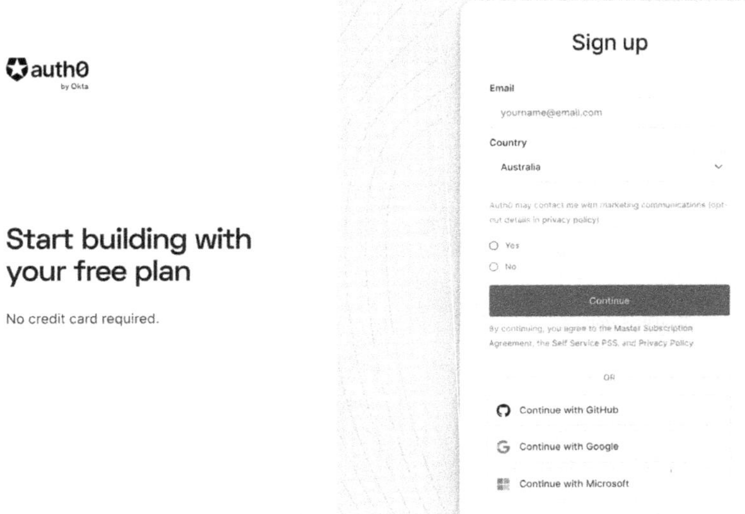

Figure 6.2 – Auth0 signup

2. After signing up, create an application in Auth0:

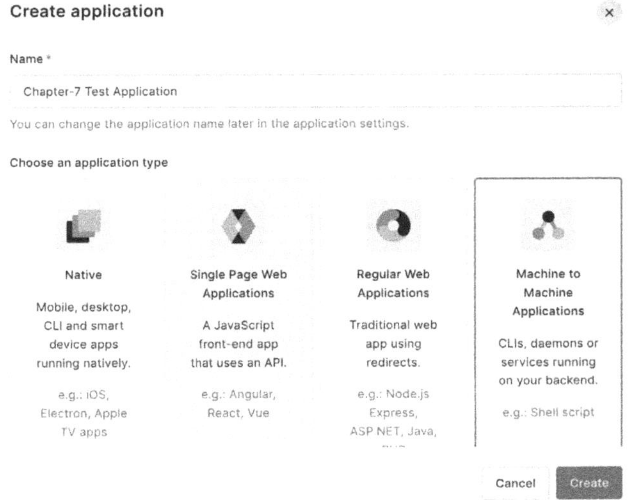

Figure 6.3 – Create application in Auth0

3. After creating the application, you will need to create an API. You can provide the Ingress URL as the identifier:

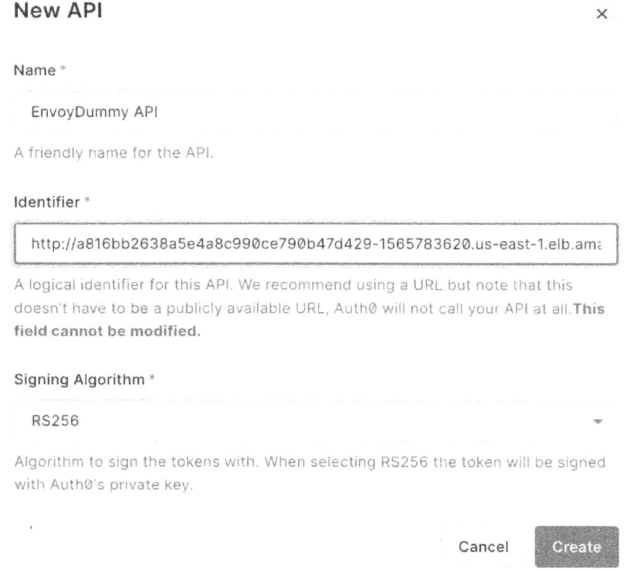

Figure 6.4 – Create an API in Auth0

4. Declare permissions the consumer of this API needs to have to be able to access the API:

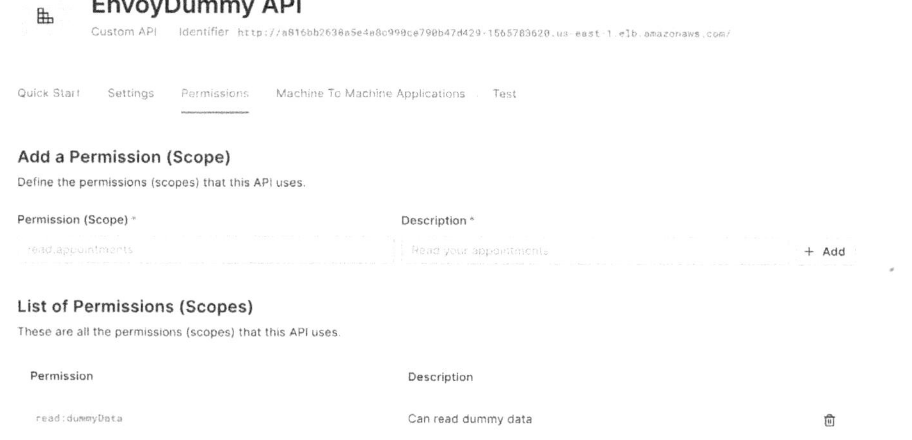

Figure 6.5 – API scopes in Auth0

5. Enable RBAC for the API from **General Settings**:

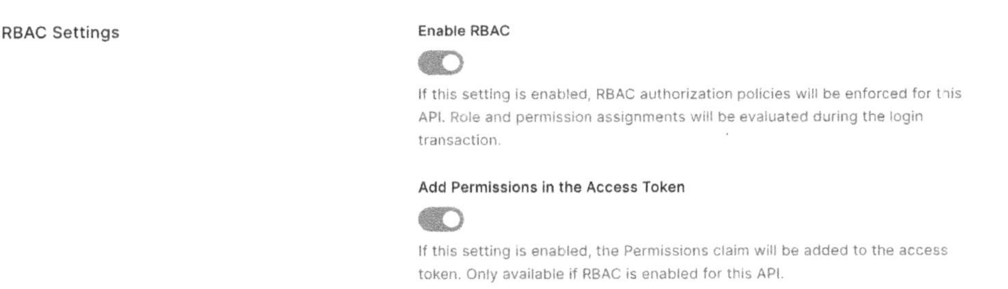

Figure 6.6 – Enable RBAC for the API in Auth0

6. After creating the API, go back to the application and authorize the application for access to the EnvoyDummy API, and while doing so, please also configure scopes:

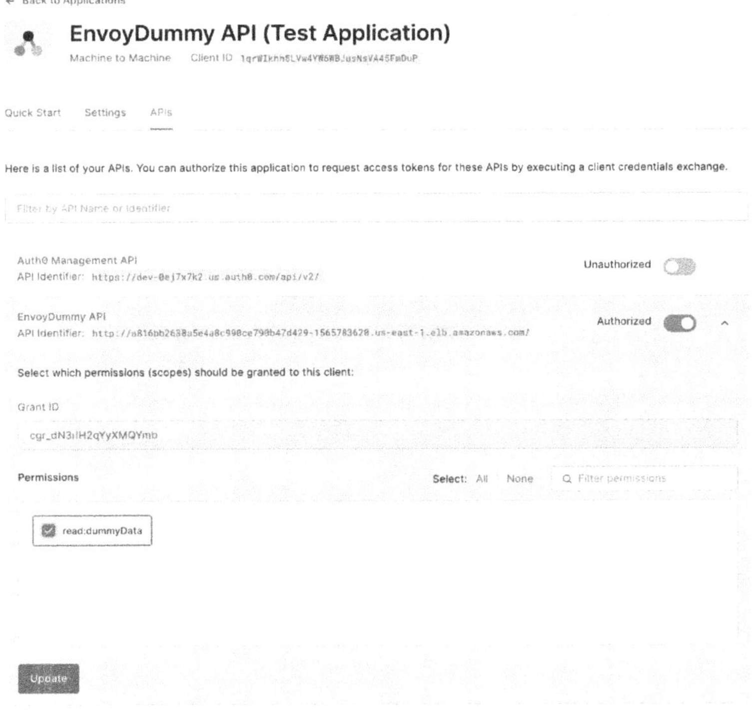

Figure 6.7 – Grant permission to the application in Auth0

7. As the last step, go to the application page to get the request you can use to get the access token:

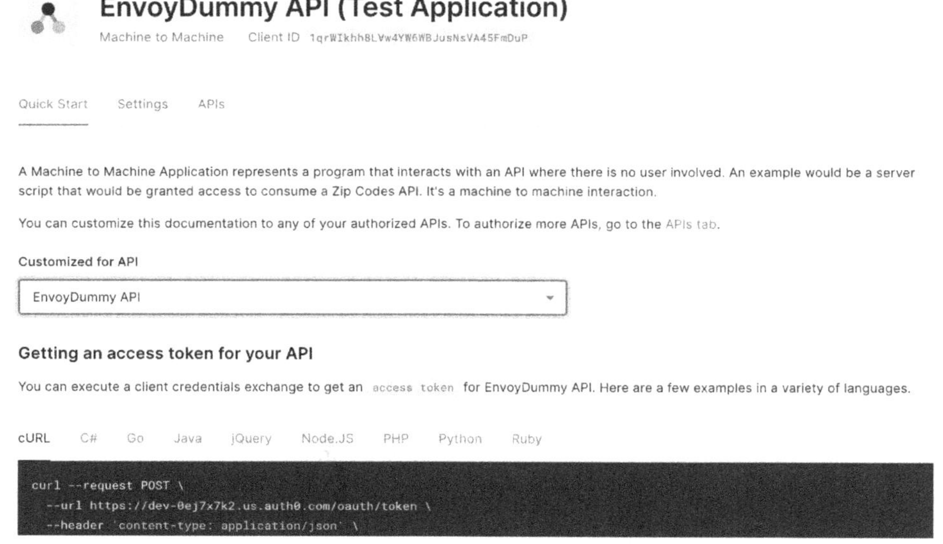

Figure 6.8 – Quickstart example to get the access token in Auth0

Copy the `curl` string, including `client_id`, `client_secret`, and so on, and with this, we have completed all the steps in Auth0.

Now, using the `curl` string you copied in the previous steps, get the access token from the terminal:

```
$ curl --request POST --url https://dev-0ej7x7k2.us.auth0.com/
oauth/token --header 'content-type:application/json' --data
'{"client_id":"XXXXXX-id","client_secret":"XXXXX-secret","
"audience":"http://a816bb2638a5e4a8c990ce790b47d429-1565783620.
us-east-1.elb.amazonaws.com/","grant_type":"client_
credentials"}'

{"access_token":"xxxxxx-accesstoken"
"scope":"read:dummyData","expires_in":86400,"token_
type":"Bearer"}%
```

Once we have received the access token, we will apply the `RequestAuthentication` policy. The `RequestAuthentication` policy specifies the details of how to validate the JWT provided during authentication. Following is the **`RequestAuthentication`** policy:

```
apiVersion: security.istio.io/v1beta1
kind: RequestAuthentication
```

```
metadata:
  name: "auth0"
  namespace: chapter6
spec:
  selector:
    matchLabels:
      name: envoydummy
  jwtRules:
  - issuer: "https://dev-0ej7x7k2.us.auth0.com/"
    jwksUri: "https://dev-0ej7x7k2.us.auth0.com/.well-known/
jwks.json"
```

In the preceding configuration, also available in `Chapter6/01-requestAuthentication.yaml`, we are declaring a **`RequestAuthentication`** policy with the name `auth0` in the `chapter6` namespace with the following specifications:

- `issuer`: This is the value of the domain of the Auth0 application. You can fetch the value from the following screen:

Basic Information

Name *

EnvoyDummy API (Test Application)

Domain

dev-0ej7x7k2.us.auth0.com

Figure 6.9 – Application domain

- `jwksUri`: This is the JWKS endpoint, which can be used by Istio to verify the signature. Auth0 exposes a JWKS endpoint for each tenant, which is found at `https://DOMAIN/.well-known/jwks.json`. This endpoint will contain the JWK used to verify all Auth0-issued JWTs for this tenant. Replace the `DOMAIN` value with the value in the application.

When using the `RequestAuthentication` policy, it is best practice to also configure the `RequestAuthentication` and `AuthorizationPolicy` together and enforce a rule that any request with an empty principal should not be allowed. Following is an example of a sample authorization policy – you will read more about authorization policies in the next section:

```
apiVersion: security.istio.io/v1beta1
kind: AuthorizationPolicy
metadata:
  name: auth0-authz
```

```
  namespace: chapter6
spec:
  action: DENY
  selector:
    matchLabels:
      name: envoydummy
  rules:
  - from:
    - source:
        notPrincipals: ["*"]
```

Configuring RequestAuthorization

In the previous section, we configured a `RequestAuthentication` policy, which verifies a JWT token against the issuer and JWK details as per the JWKS location. We configured Auth0 as the authentication provider and the one that generates the bearer token. In this section, we will learn about how to make use of the information provided by authentication policies such as peer authentication and request authentication to authorize client access to the server (the requested resource, Pod, workload, service, etc.).

We will first focus on implementing an authorization policy in conjunction with the `RequestAuthentication` policy from the previous section.

To let `curl` access the envoy dummy using the access token issued by Auth0, we need to create an `AuthorizationPolicy`:

```
apiVersion: "security.istio.io/v1beta1"
kind: "AuthorizationPolicy"
metadata:
  name: "envoydummy-authz-policy"
  namespace: utilities
spec:
  action: ALLOW
  selector:
    matchLabels:
      name: envoydummy
  rules:
  - when:
    - key: request.auth.claims[permissions]
      values: ["read:profile"]
```

`AuthorizationPolicy` contains the following data:

- `action`: This defines the type of action to be taken when the request matches the defined rule. Possible values for `action` are `ALLOW`, `DENY`, and `AUDIT`.
- `selector`: This defines what workload this policy should be applied to. Here, you provide a set of labels that should match the workload's label to be part of a selection.
- `rules`: Here, we are defining a set of rules that should be matched with the request. Rules contain the following sub-configurations:
    - `source`: This provides the rule about the origin of the request.
    - `to`: This provides rules about the request such as to what host it was addressed, what the method name is, and what resource identified by the URI is requested.
    - `when`: This specifies a list of additional conditions. You can find a detailed list of all parameters at `https://istio.io/latest/docs/reference/config/security/conditions/`.

In this example, we are defining an authorization policy that allows access to Pods with the label `name:envoydummy` if the request contains an authenticated JWT token with a claim of `read:profile`.

Before we apply the changes, make sure that you can access the dummy data and make sure you have the Ingress gateway and `envoydummy` Pods deployed in the `utilities` namespace – if not, you can do that by applying the following commands:

```
$ curl -Hhost:mockshop.com http://
a816bb2638a5e4a8c990ce790b47d429-1565783620.us-east-1.elb.
amazonaws.com/
V1----------Bootstrap Service Mesh Implementation with Istio--
--------V1%
```

Go ahead and apply both of the policies:

```
% kubectl apply -f Chapter6/01-requestAuthentication.yaml
requestauthentication.security.istio.io/auth0 created
% kubectl apply -f Chapter6/02-requestAuthorization.yaml
authorizationpolicy.security.istio.io/envoydummy-authz-policy
created
```

Check that you are able to access `mockshop.com`:

```
% curl -Hhost:mockshop.com http://
a816bb2638a5e4a8c990ce790b47d429-1565783620.us-east-1.elb.
amazonaws.com/
RBAC: access denied
```

The access is denied because we need to provide a valid access token as part of the request. Copy the access token you got from the previous request and try again in the following fashion:

```
$ curl -Hhost:mockshop.com -H "authorization:
Bearer xxxxxx-accesstoken " http://
a816bb2638a5e4a8c990ce790b47d429-1565783620.us-east-1.elb.
amazonaws.com/
RBAC: access denied%
```

Although the JWT verification succeeded, the request failed due to RBAC controls. The error is deliberate because instead of providing `read:dummyData` in `Chapter6/02-requestAuthorization.yaml`, we provide `read:profile`. The changes are updated in `Chapter6/03-requestAuthorization.yaml`. Apply the changes and test the APIs:

```
% kubectl apply -f Chapter6/03-requestAuthorization.yaml
authorizationpolicy.security.istio.io/envoydummy-authz-policy
configured
$ curl -Hhost:mockshop.com -H "authorization:
Bearer xxxxxx-accesstoken " http://
a816bb2638a5e4a8c990ce790b47d429-1565783620.us-east-1.elb.
amazonaws.com/
V2----------Bootstrap Service Mesh Implementation with Istio--
--------V2%
```

To summarize, we did the following, including the previous section:

1. We configured Auth0 as the authentication provider and OAuth server.
2. We created a `RequestAuthentication` policy to validate the bearer token provided in the request.
3. We created an `AuthorizationPolicy` to verify claims presented in the JWT token and whether the claim matched the desired value, then let the request go through the upstream.

Next, we will learn how to configure request authorization in conjunction with `PeerAuthentication`, which we configured in the *Service-to-service authentication* section.

We will modify the `curl` Pod to use a different service account, and let's call it `chapter6sa`:

```
apiVersion: v1
kind: ServiceAccount
metadata:
  name: chapter6sa
  namespace: utilities
```

As you cannot change the service account of an existing Pod, you need to delete the previous deployment and redeploy with a new service account:

```
Kubectl delete -f Chapter6/01-curl-deployment.yaml
kubectl apply -f Chapter6/02-curl-deployment.yaml
```

You can check that the `curl` Pod is running with the identity of the `chapter6sa` service account. After this, let's create an `AuthorizationPolicy` to allow a request to the `httpbin` Pod if the principal of the requestor is `cluster.local/ns/utilities/sa/curl`:

```
apiVersion: "security.istio.io/v1beta1"
kind: "AuthorizationPolicy"
metadata:
  name: "httpbin-authz-policy"
  namespace: chapter6
spec:
  action: ALLOW
  selector:
    matchLabels:
      app: httpbin
  rules:
  - from:
    - source:
        principals: ["cluster.local/ns/utilities/sa/curl"]
    to:
    - operation:
        methods: ['*']
```

Previously, we looked at `AuthorizationPolicy` and you will be familiar with most of the configuration in this example. In this example, we are building `AuthorizationPolicy` on top of peer authentication rather than request authentication. The most interesting part is the `source` field in the `rules` section. In the `source` configuration, we define the source identities of the request. All fields in the source request need to match for the rule to be successful.

The following fields can be defined in the `source` configuration:

- `principals`: This is a list of accepted identities that are derived from the client certificate during mTLS. The values are in the `<TRUST_DOMAIN NAME >/ns/<NAMESPACE NAME>/sa/<SERVICE_ACCOUNT NAME>` format. In this example, the value of principals will be `cluster.local/ns/utilities/sa/curl`.

- `notPrincipals`: This is a list of identities from which the request will not be accepted. The values are derived the same way as `principals`.
- `requestPrincipals`: This is a list of accepted identities where the request principal is derived from JWT and is in the format `<ISS>/<SUB>`.
- `notRequestPrincipals`: This is a list of identities from which the request will not be accepted. The principal is derived from the JWT and is in the format `<ISS>/<SUB>`.
- `namespaces`: This is a list of namespaces from which the request will be accepted. The namespaces are derived from the peer certificate details.
- `notNamespaces`: This is a list of namespaces from which a request will not be allowed. The namespaces are derived from the peer certificate details.
- `ipBlocks`: This is a list of IPs or CIDR blocks from which a request will be accepted. The IP is populated from the source address of the IP packet.
- `notIpBlocks`: This is a list of IP blocks from which a request will be rejected.
- `remoteIpBlocks`: This is a list of IP blocks, populated from the `X-Forwarded-For` header or proxy protocol.
- `notRemoteIpBlocks`: This is a negative list of `remoteIpBlocks`.

Go ahead and apply the configuration and test whether you are able to `curl` to `httpbin`:

```
$ kubectl apply -f Chapter6/04-httpbinAuthorizationForSpecifi
cSA.yaml
authorizationpolicy.security.istio.io/httpbin-authz-policy
configured
$ kubectl exec -it curl -n utilities -- curl -v http://httpbin.
chapter6.svc.cluster.local:8000/headers
*   Trying 172.20.152.62:8000...
* Connected to httpbin.chapter6.svc.cluster.local
(172.20.152.62) port 8000 (#0)
> GET /headers HTTP/1.1
> Host: httpbin.chapter6.svc.cluster.local:8000
> User-Agent: curl/7.85.0-DEV
> Accept: */*
>
* Mark bundle as not supporting multiuse
< HTTP/1.1 403 Forbidden
< content-length: 19
< content-type: text/plain
```

```
< date: Tue, 20 Sep 2022 02:20:39 GMT
< server: envoy
< x-envoy-upstream-service-time: 15
<
* Connection #0 to host httpbin.chapter6.svc.cluster.local left
intact
RBAC: access denied%
```

Istio denies the request from the `curl` Pod to `httpbin` because the peer certificate presented by the `curl` Pod contains `cluster.local/ns/utilities/sa/chapter6sa` instead of `cluster.local/ns/utilities/sa/curl` as the principal. Although the `curl` Pod is part of the mesh and contains a valid certificate, it is not authorized to access the `httpbin` Pod.

Go ahead and fix the problem by assigning the correct service account to the `curl` Pod.

> **Tip**
>
> You can use the following commands to fix the problem:
>
> ```
> $ kubectl delete -f Chapter6/02-curl-deployment.yaml
> $ kubectl apply -f Chapter6/03-curl-deployment.yaml
> ```

We will implement one more authorization policy, but this time the policy will enforce that using the `utilities` or `curl` service account, the requestor can access only `/headers` of `httpbin`.

Following is the authorization policy:

```
apiVersion: "security.istio.io/v1beta1"
kind: "AuthorizationPolicy"
metadata:
  name: "httpbin-authz-policy"
  namespace: chapter6
spec:
  action: ALLOW
  selector:
    matchLabels:
      app: httpbin
  rules:
  - from:
```

```
    - source:
        requestPrincipals: ["cluster.local/ns/utilities/sa/
curl"]
  - to:
    - operation:
        methods: ["GET"]
        paths: ["/get"]
```

In this policy, we have defined the HTTP method and HTTP paths in the `to` field on the rule. The `to` field contains a list of operations on which the rules will be applied. The operation field supports the following parameters:

- `hosts`: This specifies a list of hostnames for which the request will be accepted. If it's not set, then any host is allowed.
- `notHosts`: This is a negative list of hosts.
- `ports`: This is a list of ports for which a request will be accepted.
- `notPorts`: This is a list of negative matches of ports.
- `methods`: This is a list of methods as specified in the HTTP request. If not set, any method is allowed.
- `notMethods`: This is a list of negative matches of methods as specified in the HTTP request.
- `paths`: This is a list of paths as specified in the HTTP request. Paths are normalized as per `https://istio.io/latest/docs/reference/config/security/normalization/`.
- `notPaths`: This is a list of negative matches of paths.

Apply the changes:

```
kubectl apply -f Chapter6/05-httpbinAuthorizationForSpecificP
ath.yaml
```

And then try to access `httpbin`:

```
kubectl exec -it curl -n utilities -- curl -X GET -v http://
httpbin.chapter6.svc.cluster.local:8000/headers
......
RBAC: access denied%
```

Access to the request is denied because the authorization policy only allows the `/get` request made using the HTTP GET method. The following is the correct request:

```
kubectl exec -it curl -n utilities -- curl -X GET -v http://
httpbin.chapter6.svc.cluster.local:8000/get
```

This concludes our lesson on how you can build custom policies for performing request authentication and request authorization. To get more familiar with them, I suggest going through the examples in this chapter a few times and maybe building your own variations to learn how to use these policies effectively.

Summary

In this chapter, we read about how Istio provides authentication and authorization. We also read about how to implement service-to-service authentication using mutual TLS within a Service Mesh using the `PeerAuthentication` policy, as well as mutual TLS with clients external to a Service Mesh by using the *mutual* TLS mode at the Ingress gateway. We then read about end user authentication using the `RequestAuthentication` policy. We configured Auth0 to gain some real-life experience in using authentication and identity providers.

To finish off, we then read about `AuthorizationPolicy` and how it can be used to enforce various authorization checks to ensure that the authenticated identity is authorized to access the requested resources.

In the next chapter, we will read about how Istio helps in making microservices observable and how various observability tools and software can be integrated with Istio.

7
Service Mesh Observability

Distributed systems built using microservice architecture are complex and unpredictable. Irrespective of how diligent you have been in writing code, failures, meltdowns, memory leaks, and so on are highly likely to happen. The best strategy to handle such an incident is to proactively observe systems to identify any failures or situations that might lead to failures or any other adverse behavior.

Observing systems help you understand system behavior and the underlying causes behind faults so that you can confidently troubleshoot issues and analyze the effects of potential fixes. In this chapter, you will read about why observability is important, how to collect telemetry information from Istio, the different types of metrics available and how to fetch them via APIs, and how to enable distributed tracing. We will do so by discussing the following topics:

- Understanding observability
- Metric scraping using Prometheus
- Customizing Istio metrics
- Visualizing telemetry using Grafana
- Implementing distributed tracing

Without further delay, let's start with understanding observability.

> **Important note**
>
> The technical prerequisites for this chapter are the same as the previous chapters.

Understanding observability

The concept of **observability** was originally introduced as part of **control theory**, which deals with the control of self-regulated dynamic systems. Control theory is an abstract concept and has interdisciplinary applications; it basically provides a model governing the application of system inputs to drive a system to a desired state while maximizing its stability and performance.

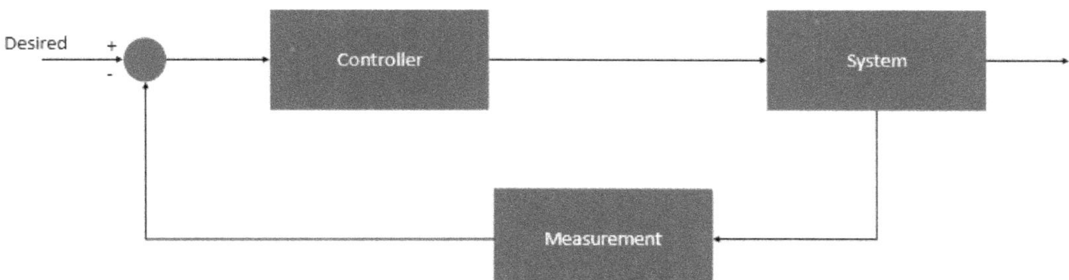

Figure 7.1 – Observability in control theory

The observability of a system is a measure of how well we can understand the internal state of that system, based on the signals and observation of its external outputs. It is then used by controllers to apply compensating control to the system to drive it to the desired state. A system is considered observable if it emits signals, which the controller can use to determine the system's status.

In the world of IT, *systems* are software systems, and *controllers* are operators who are other software systems or sometimes human operators, such as **site reliability engineers** (**SREs**), who rely on measurements provided by observable systems. If you want your software systems to be resilient and self-regulated, then it is important that all parts of your software systems are also observable.

Another concept to remember is **telemetry data**, which is the data transmitted by systems used for the observability of the systems. Usually, they are logs, event traces, and metrics:

- **Logs**: These are information emitted by software systems in verbose format. Logs are usually the data emitted by an application and are premeditated at the time an application is designed. Logs are used heavily by developers to troubleshoot code by correlating logs with the code blocks emitting them. Logs can be structured, meaning that all log entries follow a specific pattern that makes it easier for observability systems to ingest and comprehend them. Logs can also be unstructured, which unfortunately is the case for the majority of the logs. Istio generates a full record of each request, including source and destination metadata.
- **Traces**: In distributed systems or applications, tracing is the means of finding how a request or an activity was processed and executed across multiple components. Traces are made up of spans that describe execution/software processing within a system. Multiple spans are then put together to provide a trace of a request being executed. Traces describe the relationship between various systems and how they partnered together to complete a task. For tracing to work in distributed systems, it is important to share a context between all systems, and those contexts are usually in the form of correlation IDs or something similar, which all participating systems can understand and honor. Istio generates distributed trace spans for each service, providing details of request flows and interdependency between various services.
- **Metrics**: These are numeric measurements of an internal value over a specific period. Metrics are used to aggregate large volumes of data over an interval of time, which are then used as

key performance indicators for the system being observed – for example, CPU consumption percentage over a period of time or the number of requests processed per hour or every second. Istio generates metric data for latency, errors, traffic, and saturation, which are called the **four golden signals** of monitoring. **Latency** is the time it takes to service a request. **Traffic** is the measure of requests being handled by a system – for example, requests per second. The metric for traffic is further broken down into categories corresponding to traffic types. **Error** refers to the rate of request failures – for example, how many requests have the `500` response code. **Saturation** shows how many system resources, such as memory, CPU, network, and storage, are utilized by your system. Istio generates the metric data for both data and control planes.

All this telemetry data is used in conjunction to provide the observability of systems. There are various types of open source and commercial software available for observing software systems; Istio includes various tools out of the box, which we briefly discussed in *Chapter 2*. Prometheus and Grafana are shipped out of the box with Istio; in the next section, we will install Prometheus and Grafana and configure them to collect Istio's metrics data.

Metric scraping using Prometheus

Prometheus is open source system monitoring software, which stores all metric information along with the timestamps of when they were recorded. What differentiates Prometheus from other monitoring software is its powerful multidimensional data model and a powerful query language called **PromQL**. It works by collecting data from various targets and then analyzing and crunching it to produce metrics. Systems can also implement HTTP endpoints that provide metrics data; these endpoints are then called by Prometheus to collect metrics data from the applications. The process of gathering metrics data from various HTTP endpoints is also called **scraping**.

As illustrated in the following figure, the Istio control plane and data plane components expose endpoints that emit metrics, and Prometheus is configured to scrape these endpoints to collect metrics data and store it in a time series database:

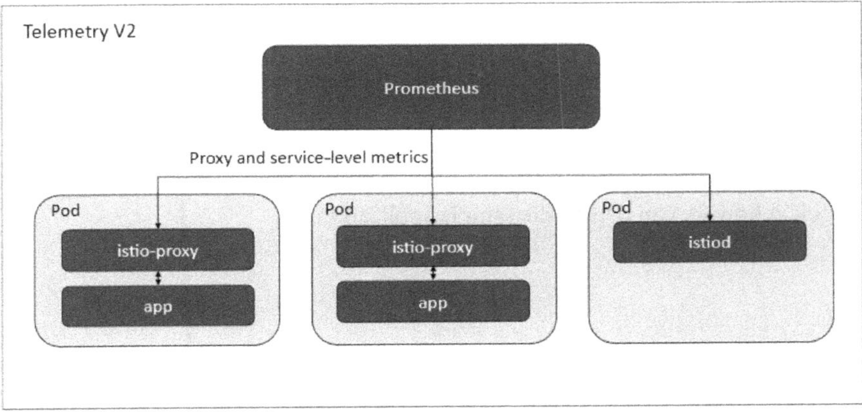

Figure 7.2 – Metric scraping using Prometheus

We will describe the process in detail in the following sections.

Installing Prometheus

Istio already provides a sample installation file available in `/sample/addons/ prometheus.yaml`, which is good enough as a starting point. We have modified the file slightly to cater to applications that support strict mTLS mode only:

```
% kubectl apply -f Chapter7/01-prometheus.yaml
serviceaccount/prometheus created
configmap/prometheus created
clusterrole.rbac.authorization.k8s.io/prometheus created
clusterrolebinding.rbac.authorization.k8s.io/prometheus created
service/prometheus created
deployment.apps/prometheus created
```

The changes in our file, `01-prometheus.yaml`, in comparison to the out-of-the-box file, are that we have provisioned Istio certificates for Prometheus by injecting a sidecar and configuring it to write the certificate to a shared volume, which is then mounted onto the Prometheus container. The sidecar is just for mounting and managing the certificates and doesn't intercept any inbound and outbound requests. You will find the changes in `Chapter7/01-prometheus.yaml`.

You can check what has been installed in the `istio-system` namespace:

```
% kubectl get po -n istio-system
NAME                           READY   STATUS    RESTARTS   AGE
istio-egressgateway-7d75d6f46f-
28r59   1/1     Running   0          48d
istio-ingressgateway-5df7fcddf-
7qdx9   1/1     Running   0          48d
istiod-56fd889679-
ltxg5                   1/1     Running   0          48d
prometheus-7b8b9dd44c-
sp5pc               2/2     Running   0          16s
```

Now, we will look at how we can deploy the sample application.

Deploying a sample application

Let's deploy the `sockshop` application with `istio-injection` enabled.

Modify `sockshop/devops/deploy/kubernetes/manifests/00-sock-shop-ns.yaml` with the following code:

```
apiVersion: v1
kind: Namespace
metadata:
  labels:
    istio-injection: enabled
  name: sock-shop
```

Then, deploy the `sockshop` application:

```
% kubectl apply -f sockshop/devops/deploy/kubernetes/manifests/
```

Finally, we will configure an Ingress gateway:

```
% kubectl apply -f Chapter7/sockshop-IstioServices.yaml
```

Now, make some calls from the browser to send traffic to the frontend service as you've been doing in the previous chapters. We will then check some metrics scraped by Prometheus to access the dashboard, using the following commands:

```
% istioctl dashboard prometheus
http://localhost:9090
```

From the dashboard, we will first check that Prometheus is scraping the metrics. We can do so by clicking on **Status** | **Targets** on the Prometheus dashboard:

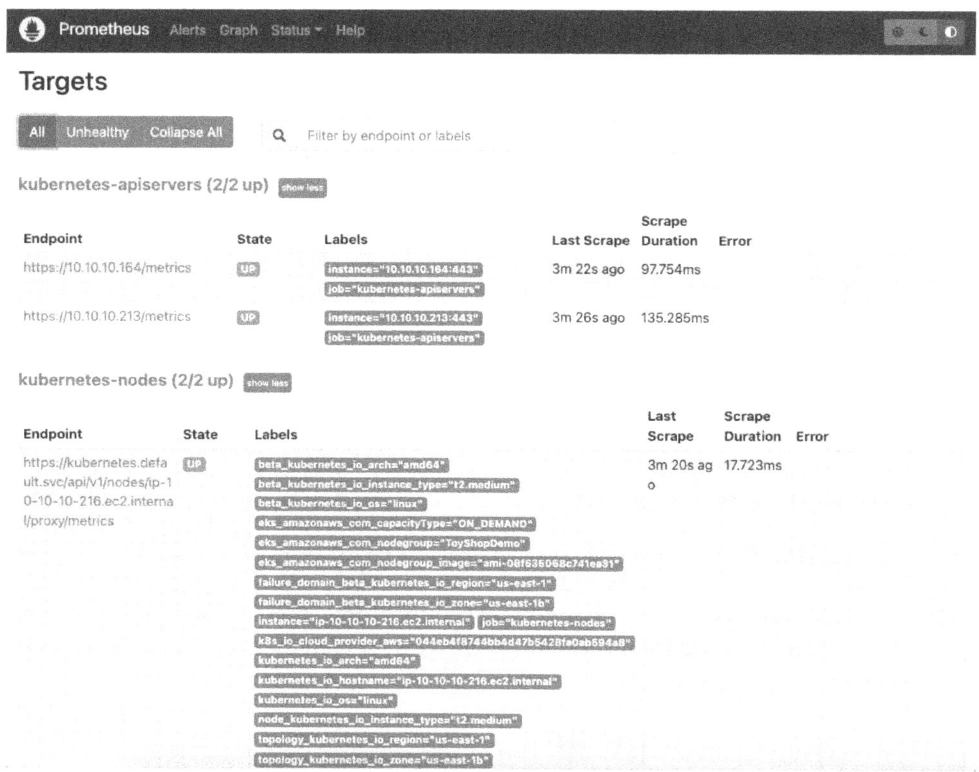

Figure 7.3 – The Prometheus configuration

You will see all targets from which Prometheus is scraping the metrics.

On the dashboard, we will fire up a query to get a total request between the `istio-` Ingress gateway and the frontend service, using the following code:

```
istio_requests_total{destination_service="front-end.sock-shop.
svc.cluster.local",response_code="200",source_app="istio-
ingressgateway",namespace="sock-shop"}
```

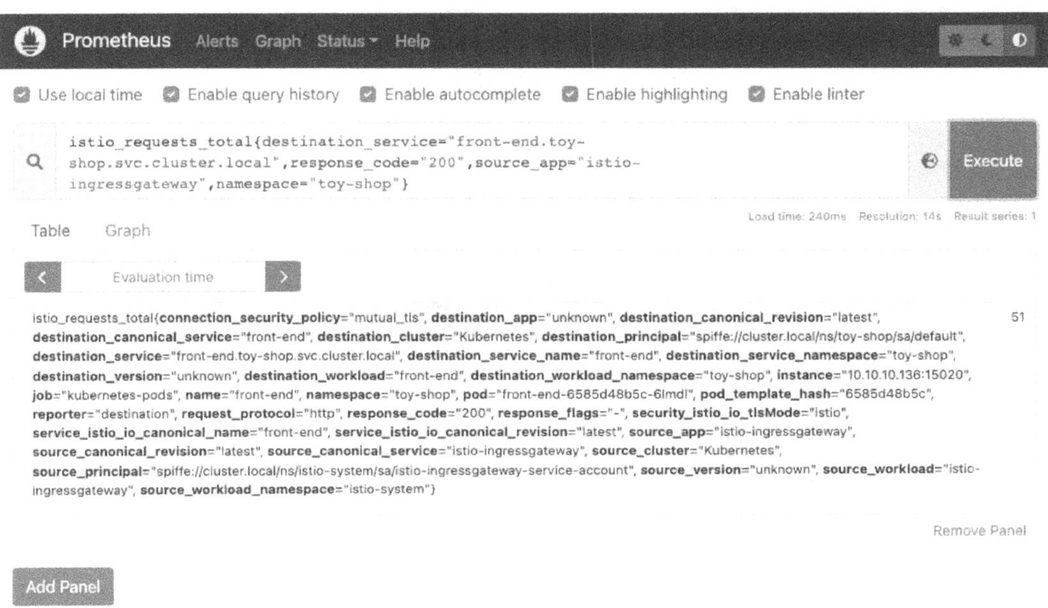

Figure 7.4 – PromQL

In the preceding screenshot, the name of the metric is `istio_requests_total`, and the fields in curly brackets are known as **metric dimensions**. Using the PromQL string, we are specifying that we want the `istio_requests_total` metric whose dimensions are `destination_service`, `response_code`, `source_app`, and `namespace` to match the `front-end.sock-shop.svc.cluster.local`, `200`, `istio-ingressgateway`, and `sock-shop` values respectively.

In response, we receive a metric count of `51` and other dimensions as part of the metric.

Let's make another query to check how many requests to the catalog service have been generated from the frontend service, using the following code:

```
istio_requests_total{destination_service="catalogue.
sock-shop.svc.cluster.local",source_workload="front-
end",reporter="source",response_code="200"}
```

Note in the query how we have provided `reporter = "source"`, which means we want metrics reported by the frontend Pod.

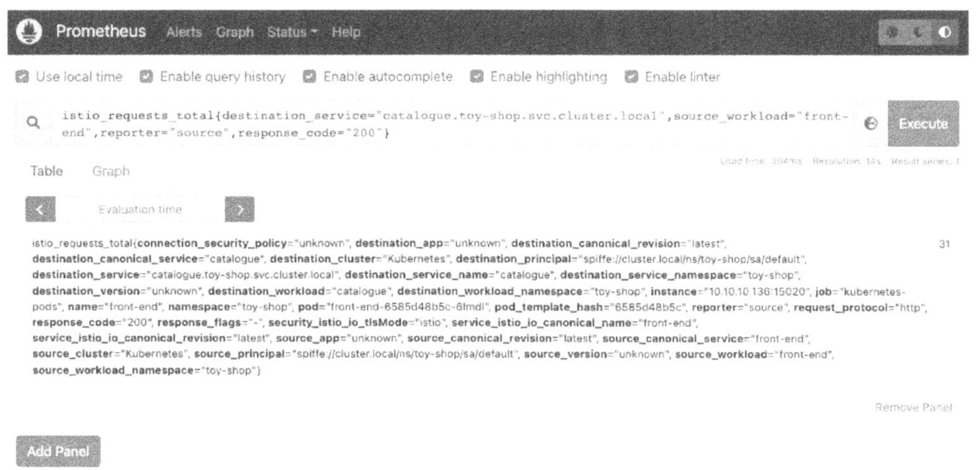

Figure 7.5 – PromQL istio_request_total from the frontend to the catalog

If you change `reporter = "destination"`, you will see similar metrics but reported by the catalog Pod.

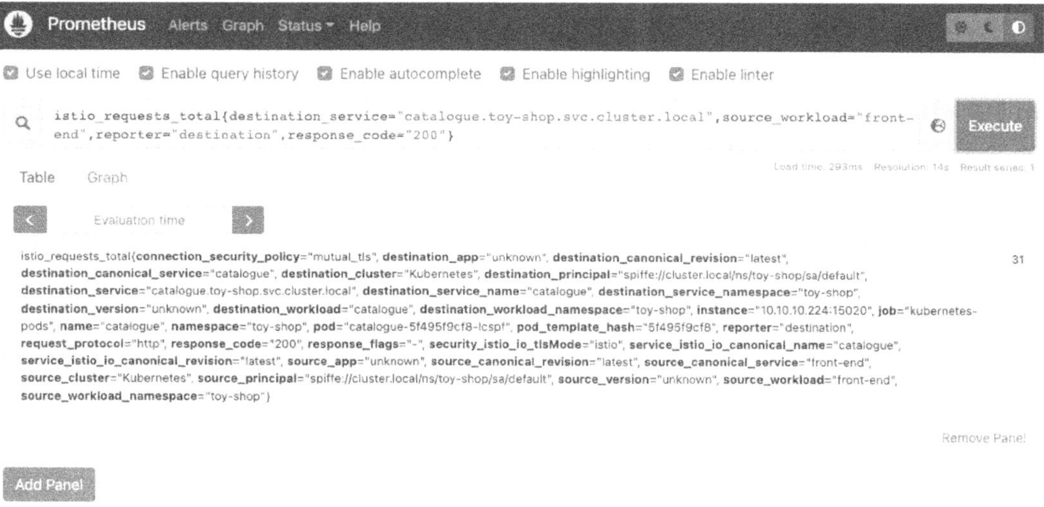

Figure 7.6 – PromQL istio_request_total from the frontend to the catalogue, reported by the catalog sidecar

Let's also check the database connection between the catalog service and the MySQL catalog database, using the following query:

```
istio_tcp_connections_opened_total{destination_canonical_
service="catalogue-db",source_workload="catalogue", source_
workload_namespace="sock-shop}
```

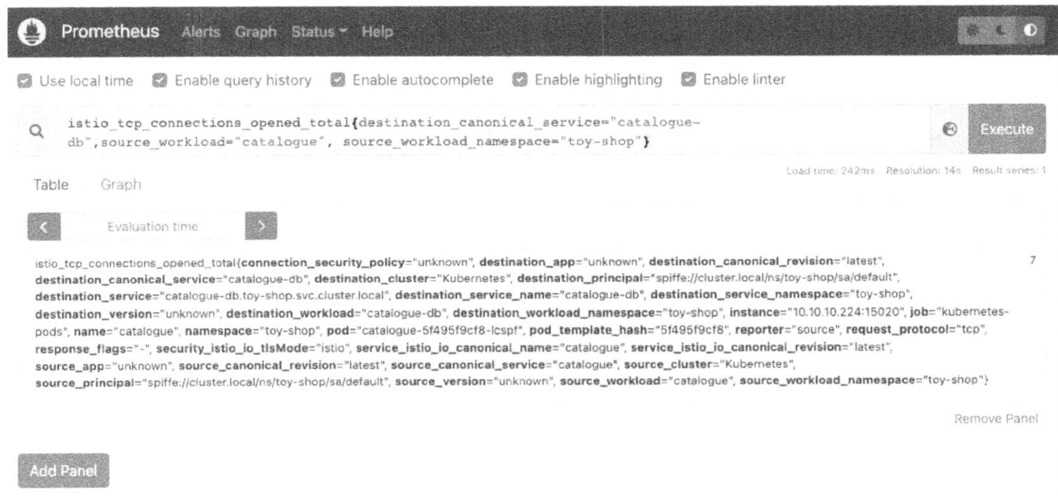

Figure 7.7 – PromQL TCP connections between catalogue and catalogue-db

The metric data shows that the catalog service made seven TCP connections.

So far, we have used default metric configuration. In the next section, we will read about how these metrics are configured and how to customize them by adding new metrics.

Customizing Istio metrics

Istio provides flexibility to observe metrics other than what comes out of the box. This provides flexibility to observe application-specific metrics. With that in mind, let's begin by looking at the `/stats/prometheus` endpoint exposed by the sidecar:

```
% kubectl exec front-end-6c768c478-82sqw -n sock-shop -c istio-
proxy -- curl -sS 'localhost:15000/stats/prometheus' | grep
istio_requests_total
```

The following screenshot shows sample data returned by this endpoint, which is also scraped by Prometheus and is the same data you saw using the dashboard in the previous section:

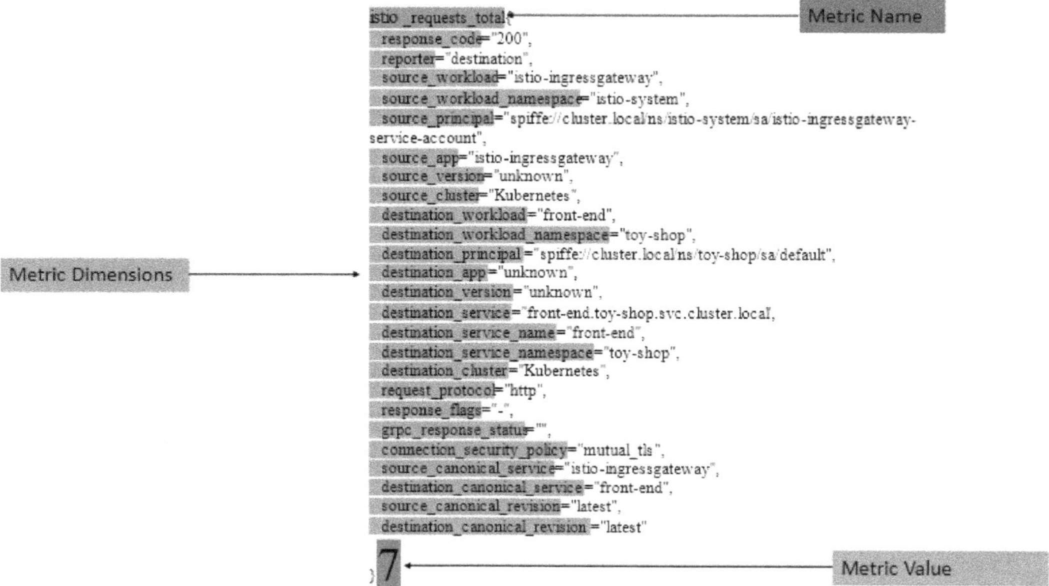

Figure 7.8 – Istio metric, dimensions, and value

The metric is organized in the following structure:

- **Metric name**: This is the name of the metric exported by Istio. Out-of-the-box Istio generates many metric details, which can be found at `https://istio.io/latest/docs/reference/config/metrics/#metrics`.
- **Metric dimensions**: These are the various fields that are part of a metric. These fields are called dimensions in the context of Prometheus and labels in the context of an Istio metric. Details about standard label parts of Istio metrics are available at `https://istio.io/latest/docs/reference/config/metrics/#labels`.
- **Metric value**: This is the value of the metric and can be a counter, gauge, or histogram.
- A **counter** is used to track the occurrence of an event. Counters are continuously increasing values exposed as time series. Some examples of metrics with counter-type values are request counts, bytes received, and TCP connections.
- A **gauge** is a snapshot of a measurement at a single point in time. It is used to measure metrics such as CPU consumption and memory consumption.

- As the name suggests, **histograms** are used for measuring observations spread over a period. They are also the most complex metric to measure.

The telemetry component of Istio is implemented by the `proxy-wasm` plugin. We will read more about this in *Chapter 9*, but for now, just understand it as a means to build extensions for **Envoy**. You can find these filters using the following command:

```
% kubectl get EnvoyFilters -A
NAMESPACE      NAME                              AGE
istio-system   stats-filter-1.16                 28h
istio-system   tcp-stats-filter-1.16             28h
```

The filters run WebAssembly at different points of request execution and collect various metrics. Using the same technique, you can easily customize Istio metrics by adding/removing new dimensions. You can also add new metrics or override any existing metrics. We will discuss how to achieve this in the following sections.

Adding dimensions to the Istio metric

The `istio_request_total` metric doesn't have any dimensions for a request path – that is, we cannot count how many requests we are receiving for individual request paths. We will configure an EnvoyFilter to include `request.url_path` in the `request_total` metric. Please note that `istio_` is a prefix added by Prometheus; the actual metric name in the context of Istio is `request_total`.

We will discuss EnvoyFilter in *Chapter 9*, so if you want to jump to that chapter to understand the various ways of extending Istio, please do so; alternatively, you can also read about this filter at `https://istio.io/latest/docs/reference/config/networking/envoy-filter/#EnvoyFilter-PatchContext`.

In the following configuration, we have created an EnvoyFilter that is applied to frontend Pods, using the condition in `workloadSelector`, in the following code block:

```
apiVersion: networking.istio.io/v1alpha3
kind: EnvoyFilter
metadata:
  name: custom-metrics
  namespace: sock-shop
spec:
  workloadSelector:
    labels:
      name: front-end
```

Next, we apply `configPatch` to `HTTP_FILTER` for inbound traffic flow to the sidecar. Other options are `SIDECAR_OUTBOUND` and `GATEWAY`. The patch is applied to HTTP connection manager filters and, in particular, the `istio.stats` subfilter; this is the filter we discussed in the previous section and is responsible for Istio telemetry:

```
configPatches:
  - applyTo: HTTP_FILTER
    match:
      context: SIDECAR_INBOUND
      listener:
        filterChain:
          filter:
            name: envoy.filters.network.http_connection_manager
            subFilter:
              name: istio.stats
      proxy:
        proxyVersion: ^1\.16.*
```

Note that the proxy version, which is 1.16, must match the Istio version you have installed.

Next, we will replace the configuration of the `istio.stats` filter with the following:

```
patch:
      operation: REPLACE
      value:
        name: istio.stats
        typed_config:
          '@type': type.googleapis.com/udpa.type.v1.TypedStruct
          type_url: type.googleapis.com/envoy.extensions.
filters.http.wasm.v3.Wasm
          value:
            config:
              configuration:
                '@type': type.googleapis.com/google.protobuf.
StringValue
                value: |
                  {
                    "debug": "false",
                    "stat_prefix": "istio",
```

```
                    "metrics": [
                      {
                        "name": "requests_total",
                        "dimensions": {
                          "request.url_path": "request.url_
path"
                        }
                      }
                    ]
                  }
```

In this configuration, we are modifying the `metrics` field by adding a new dimension called `request.url.path` with the same value as the `request.url.path` attribute of Envoy. To remove any existing dimension – for example, `response_flag` – please use the following configuration:

```
"metrics": [
                      {
                        "name": "requests_total",
                        "dimensions": {
                          "request.url_path": "request.url_
path"
                        },
                        "tags_to_remove": [
                          "response_flags"
                        ]
                      }
```

Then, apply the configuration:

```
% kubectl apply -f Chapter7/01-custom-metrics.yaml
envoyfilter.networking.istio.io/custom-metrics created
```

By default, Istio will not include the newly added `request.url.path` dimension for Prometheus; the following annotations need to be applied to include `request.url_path`:

```
spec:
  template:
    metadata:
      annotations:
        sidecar.istio.io/extraStatTags: request.url_path
```

Apply the changes to the frontend deployment:

```
% kubectl patch Deployment/front-end -n sock-shop --type=merge
--patch-file Chapter7/01-sockshopfrontenddeployment_patch.yaml
```

You will now be able to see the new dimension added to the `istio_requests_total` metrics:

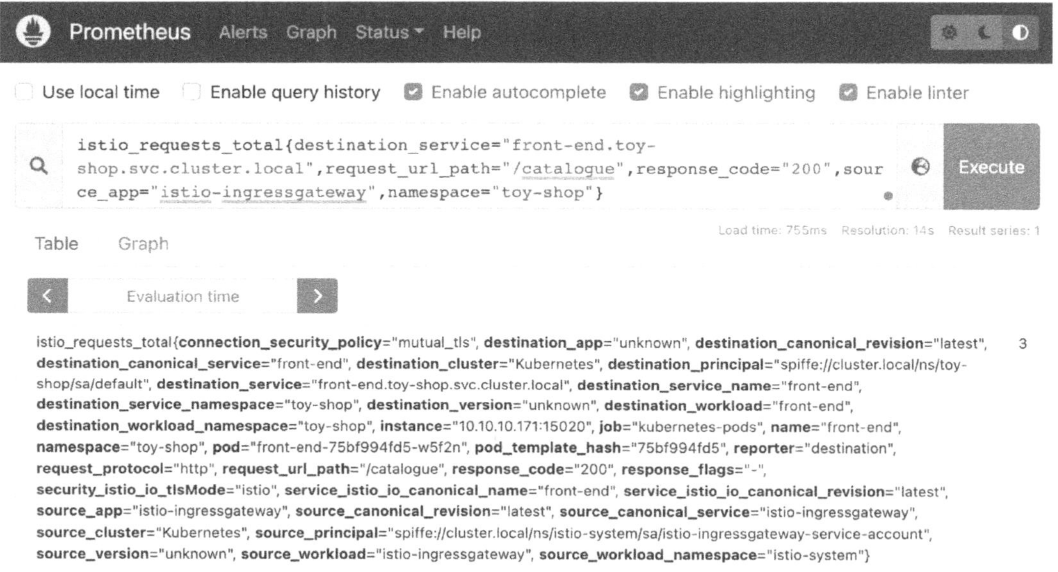

Figure 7.9 – The new metric dimension

You can add any Envoy attributes as a dimension to the metric, and you can find the full list of available attributes at `https://www.envoyproxy.io/docs/envoy/latest/intro/arch_overview/advanced/attributes`.

Creating a new Istio metric

You can also create a new Istio metric using EnvoyFilter, similar to what you used to create custom metrics.

In the following example, we have created new metrics using `definitions` and also added another dimension:

```
                configuration:
                  '@type': type.googleapis.com/google.protobuf.
StringValue
```

```
        value: |
          {
            "debug": "false",
            "stat_prefix": "istio",
            "definitions": [
              {
                "name": "request_total_bymethod",
                "type": "COUNTER",
                "value": "1"
              }
            ],
            "metrics": [
              {
                "name": "request_total_bymethod",
                "dimensions": {
                  "request.method": "request.method"
                }
              }
            ]

          }
```

Next, apply the changes:

```
% kubectl apply -f Chapter7/02-new-metric.yaml
envoyfilter.networking.istio.io/request-total-bymethod
configured
```

We must also annotate the frontend Pod with `sidecar.istio.io/statsInclusionPrefixes` so that the `request_total_bymethod` metric is included for Prometheus:

```
% kubectl patch Deployment/front-end -n sock-shop --type=merge
--patch-file Chapter7/02-sockshopfrontenddeployment_patch.yaml
deployment.apps/front-end patched
```

It would be a good idea to restart the frontend Pod to make sure that the annotation is applied. After applying the changes, you can scrape the Prometheus endpoint using the following code:

```
% kubectl exec front-end-58755f99b4-v59cd -n sock-shop -c
istio-proxy -- curl -sS 'localhost:15000/stats/prometheus' |
grep request_total_bymethod
# TYPE istio_request_total_bymethod counter
istio_request_total_bymethod{request_method="GET"} 137
```

Also, using the Prometheus dashboard, check that the new metric is available:

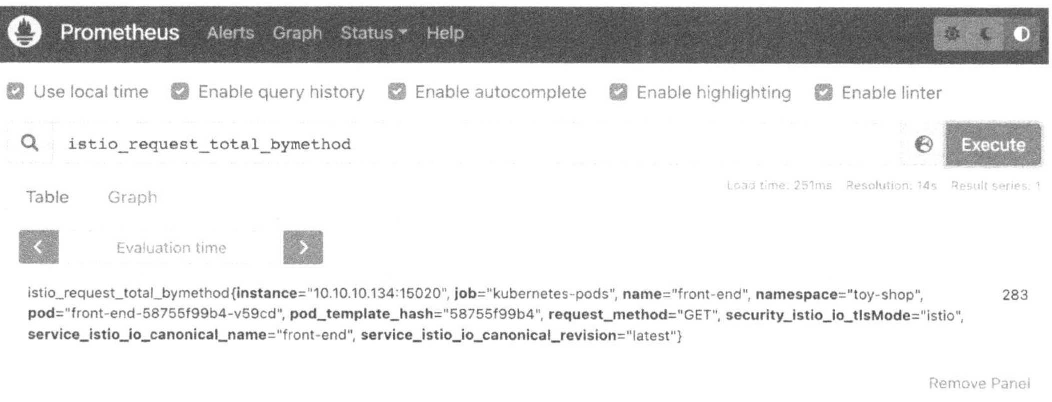

Figure 7.10 – New metrics

With this, you should now be able to create a new Istio metric with dimensions, as well as updating dimensions for any existing metrics. In the next section, we will look at Grafana, which is yet another powerful observability utility.

Visualizing telemetry using Grafana

Grafana is open source software used for the visualization of telemetry data. It provides an easy-to-use and interactive option for visualizing observability metrics. Grafana also helps to unify telemetry data from various systems in a centralized place, providing a unified view of observability across all your systems.

Istio installation provides sample manifests for Grafana, located in `samples/addons`. Install Grafana using the following commands:

```
% kubectl apply -f samples/addons/grafana.yaml
serviceaccount/grafana created
configmap/grafana created
service/grafana created
deployment.apps/grafana created
configmap/istio-grafana-dashboards created
configmap/istio-services-grafana-dashboards created
```

Once you have installed Grafana, you can open the Grafana dashboard using the following commands:

```
% istioctl dashboard grafana
http://localhost:3000
```

This should open the Grafana dashboard, as shown in the following screenshot:

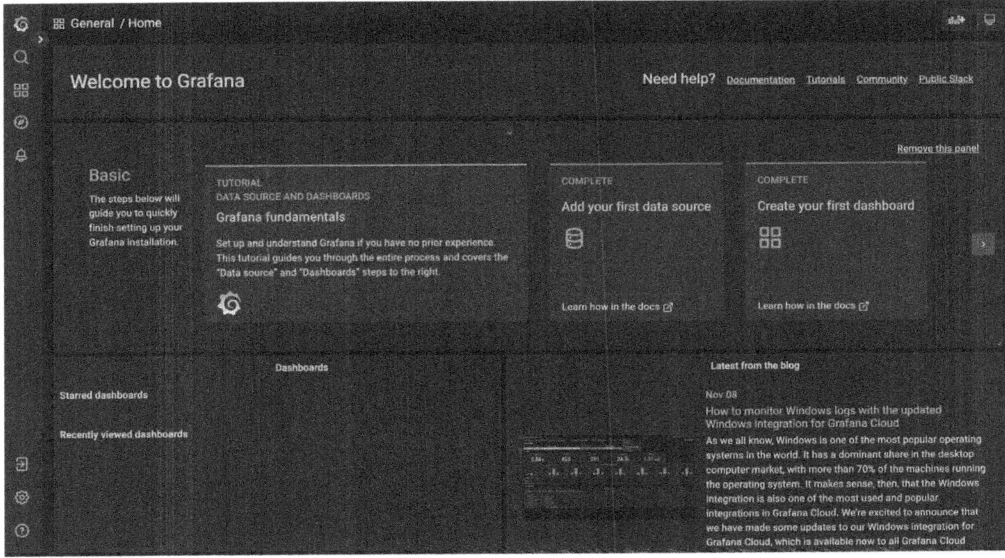

Figure 7.11 – Grafana dashboard

Grafana already includes the following dashboards for Istio:

- **Istio Control Plane Dashboard**: This provides charts showing resource consumption by Istio's control plane components. It also provides metrics on the interaction between the control plane and the data plane, including xDS push, errors during configuration sync, and conflicts in the configuration between the data plane and the control plane.

- **Istio Mesh Dashboard**: This provides a summary view of the mesh. The dashboard provides a summary view of requests, errors, gateways, and policies, as well as details about services and their associated latency during request processing.
- **Istio Performance Dashboard**: This provides charts that show the resource utilization of Istio components.
- **Istio Service and Workload Dashboards**: This provides metrics about the request-response for each service and workload. Using this dashboard, you can find more granular details about how services and workloads are behaving. You can search for a metric based on various dimensions, as discussed in the *Metric scraping using Prometheus* section.

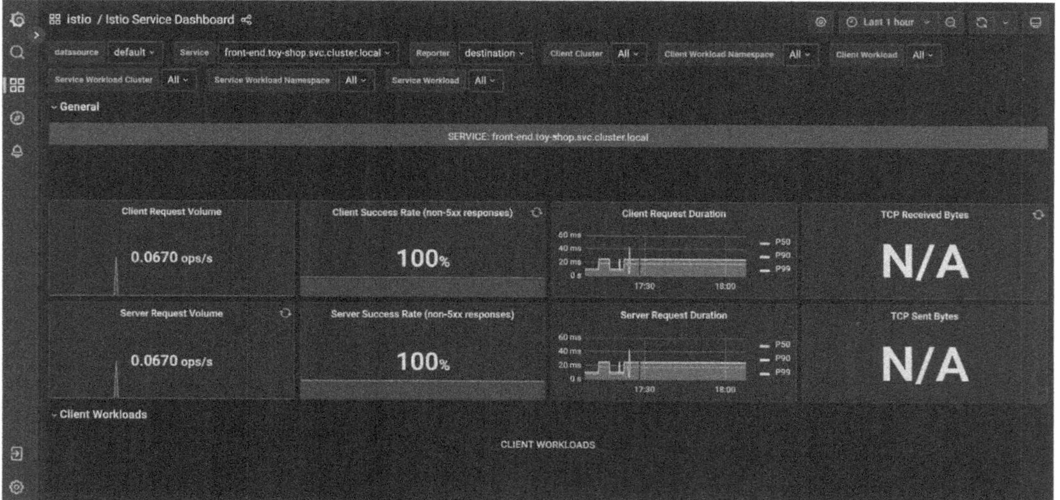

Figure 7.12 – Istio Service Dashboard

Another powerful feature of Grafana is **alerting**, where you can create alerts based on certain kinds of events. In the following example, we will create one such alert:

1. Create an alert when `response_code` is not equal to `200`, based on the `istio_request_total` metric in the last 10 minutes.

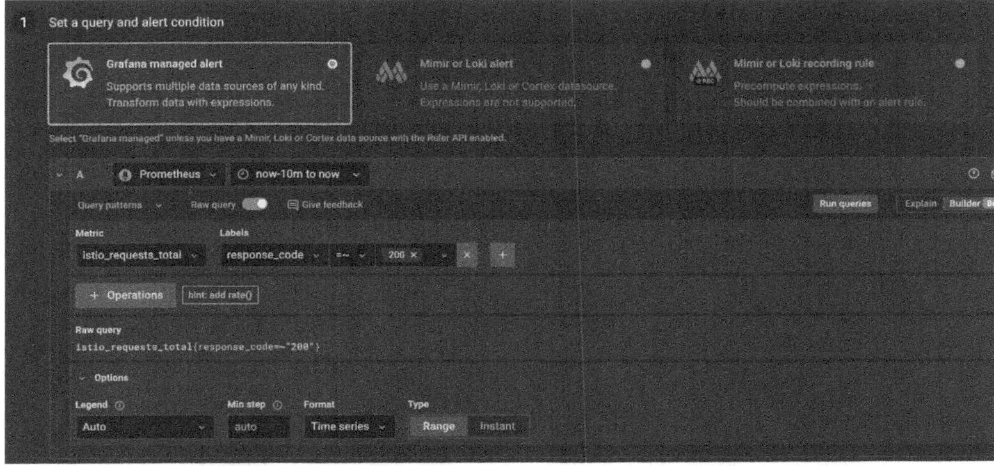

Figure 7.13 – Creating alerts in Grafana

2. Configure an alert to be raised when the count of a request with a `~=200` response code is more than 3 in the past 10 minutes; this is also called the **threshold**. We will also configure the frequency of evaluation for this alert and the threshold for firing the alert. In the following example, we have set the alert to be evaluated every minute but fired after 5 minutes. By adjusting these parameters, we can prevent the alert from being fired too soon or too late.

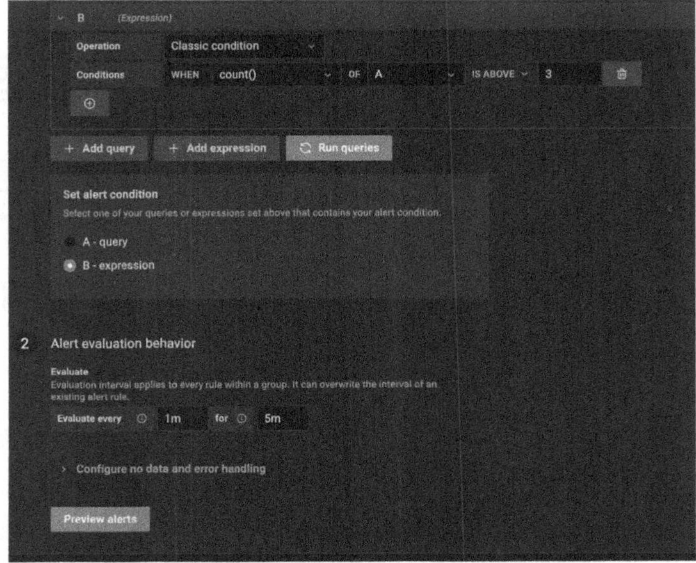

Figure 7.14 – Configuring the threshold to raise an alert

3. Next, you configure the name of the alert rule and where the alert should be stored in Grafana:

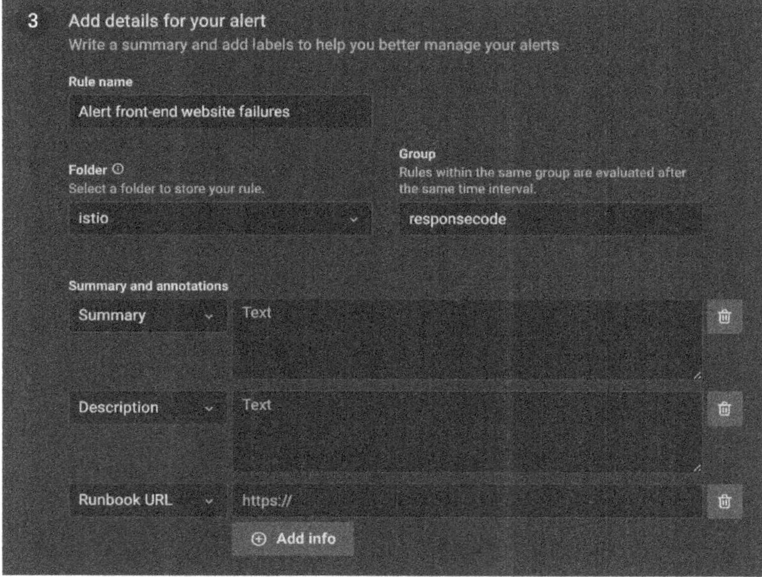

Figure 7.15 – Adding details about the alert

4. After configuring the name of the rule, you configure labels, which are a way to associate alerts with notification policies:

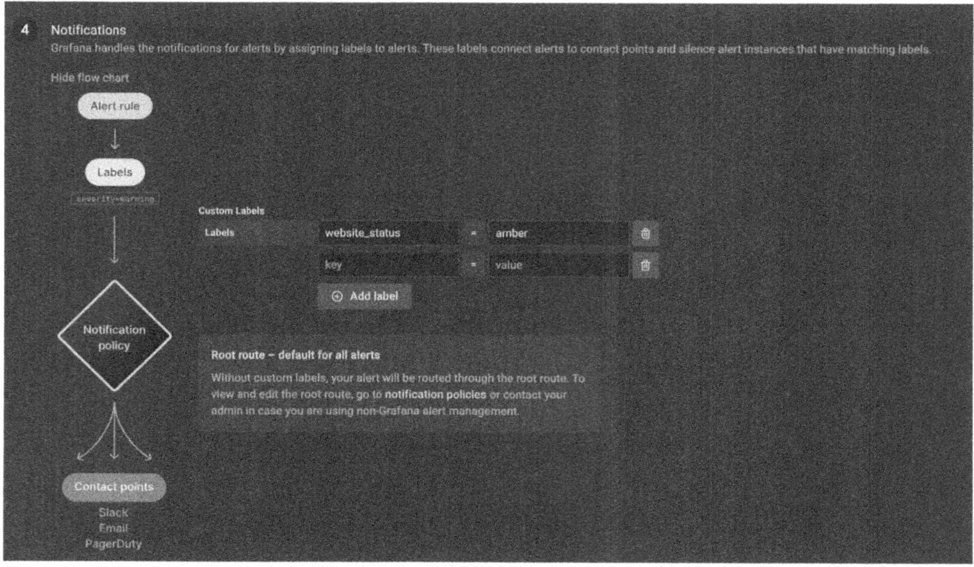

Figure 7.16 – Alert notifications

5. Next, you configure contact points that need to be notified when an alert is raised:

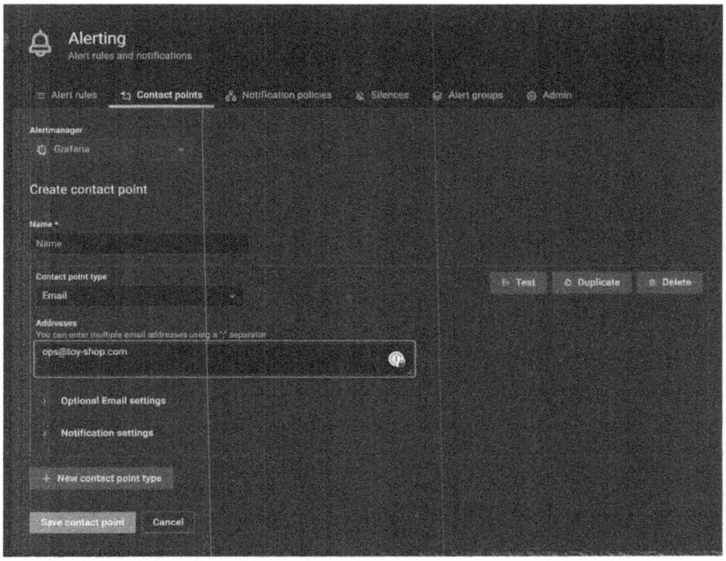

Figure 7.17 – Configure contact points

6. And finally, you create a notification policy, which specifies the contact points that will be notified about the alert.

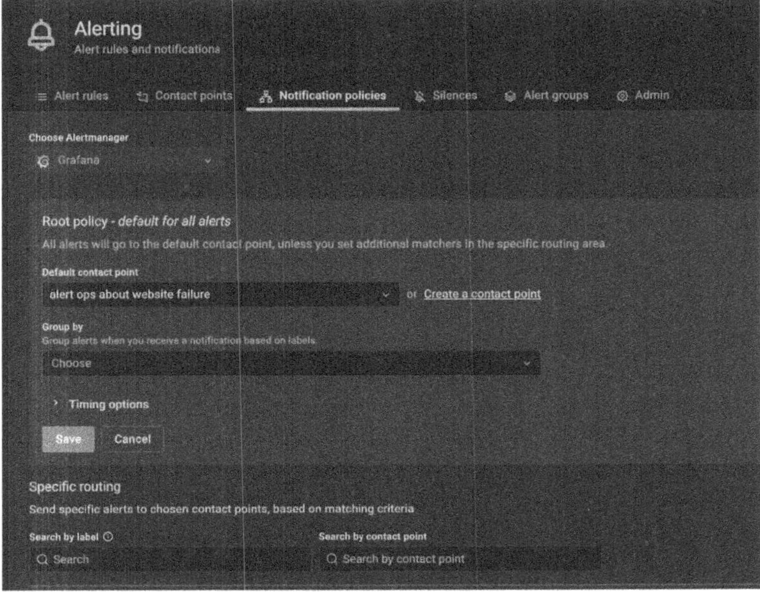

Figure 7.18 – Configuring a notification policy

You finally have your alert configured. Now, go ahead and disable the catalog service in `sockshop.com`, make a few requests from the website, and you will see the following alert fired in Grafana:

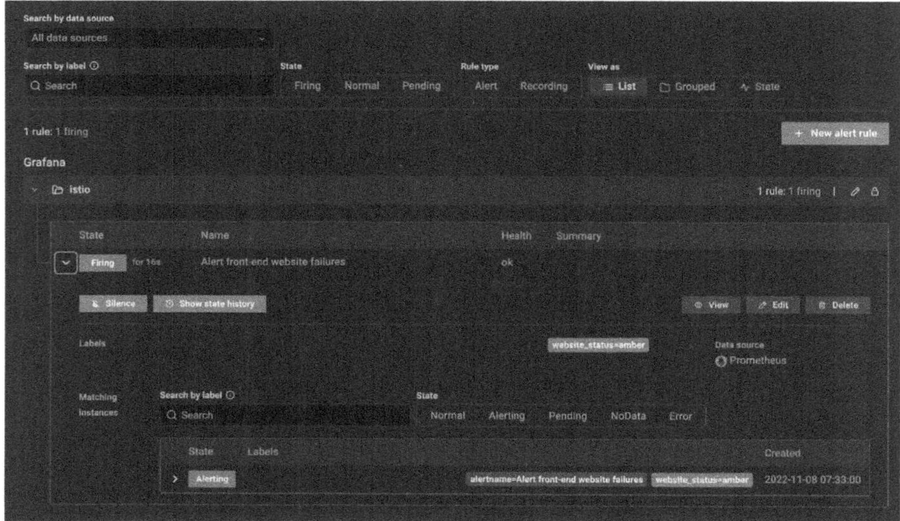

Figure 7.19 – Alerts raised due to a failure caused by catalog service outage

In this section, we saw an example of how we can use Grafana to visualize various metrics produced by Istio. Grafana provides comprehensive tooling to visualize data, which helps in uncovering new opportunities as well as unearthing any issues occurring within your system.

Implementing distributed tracing

Distributed tracing helps you understand the journey of a request through various IT systems. In the context of microservices, distributed tracing helps you understand the flow of requests through various microservices, helps you to diagnose any issues a request might be encountering, and helps you quickly diagnose any failure or performance issues.

In Istio, you can enable distributed tracing without needing to make any changes in application code, provided your application forwards all tracing headers to upstream services. Istio supports integrations with various distributed tracing systems; Jaeger is one such supported system, which is also provided as an add-on with Istio. Istio distributed tracing is built upon Envoy, where tracing information is sent directly to the tracing backend from Envoy. The tracing information comprises `x-request-id`, `x-b3-trace-id`, `x-b3-span-id`, `x-b3-parent-spanid`, `x-b3-sampled`, `x-b3-flags`, and `b3`. These custom headers are created by Envoy for every request that flows through Envoy. Envoy forwards these headers to the associated application container in the Pod. The application container then needs to ensure that these headers are not truncated and, rather, forwarded to any upstream services in the mesh. The proxied application then needs to propagate these headers in all outbound requests from the application.

You can read more about headers at `https://www.envoyproxy.io/docs/envoy/latest/intro/arch_overview/observability/tracing`.

In the following section, we will learn how to install Jaeger and enable distributed tracing for the `sockshop` example.

Enabling distributed tracing with Jaeger

Jaeger is open source distributed tracing software, originally developed by Uber Technologies and later donated to CNCF. Jaeger is used to monitor and troubleshoot microservices-based systems. It is used primarily for the following:

- Distributed context propagation and transaction monitoring
- Microservice dependency analysis and troubleshooting
- Understanding bottlenecks in distributed architectures

Yuri Shkuro, the creator of Jaeger, published a book called *Mastering Distributor Tracing* (`https://www.shkuro.com/books/2019-mastering-distributed-tracing`) that explains many aspects of Jaeger design and operations. You can read more about Jaeger at `https://www.jaegertracing.io/`.

Next, we will install and configure Jaeger in Istio.

The Kubernetes manifest file for deploying Jaeger is already available in `samples/addons/jaeger.yaml`:

```
% kubectl apply -f samples/addons/jaeger.yaml
deployment.apps/jaeger created
service/tracing created
service/zipkin created
service/jaeger-collector created
```

This code block installs Jaeger in the `istio-system` namespace. You can open the dashboard using the following command:

```
$ istioctl dashboard jaeger
```

Unfortunately, the `sockshop` application wasn't designed to propagate the headers, so for this scenario, we will make use of the `bookinfo` application as an example with Istio. But before that, we will deploy the `httpbin` application to understand the Zipkin tracing headers injected by Istio:

```
% kubectl apply -f  Chapter7/01-httpbin-deployment.yaml
```

Let's make a request to `httpbin` and check the response headers:

```
% curl -H "Host:httpbin.com"  http://
a858beb9fccb444f48185da8fce35019-1967243973.us-east-1.elb.
amazonaws.com/headers
{
  "headers": {
    "Accept": "*/*",
    "Host": "httpbin.com",
    "User-Agent": "curl/7.79.1",
    "X-B3-Parentspanid": "5c0572d9e4ed5415",
    "X-B3-Sampled": "1",
    "X-B3-Spanid": "743b39197aaca61f",
    "X-B3-Traceid": "73665fec31eb46795c0572d9e4ed5415",
    "X-Envoy-Attempt-Count": "1",
    "X-Envoy-Internal": "true",
    "X-Forwarded-Client-Cert": "By=spiffe://cluster.local/
ns/Chapter7/sa/default;Hash=5c4dfe997d5ae7c853efb8b-
81624f1ae5e4472f1cabeb36a7cec38c9a4807832;Subject=\"\";URI=s
piffe://cluster.local/ns/istio-system/sa/istio-ingressgate-
way-service-account"
  }
}
```

In the response, note the headers injected by Istio – `x-b3-parentspanid`, `x-b3-sampled`, `x-b3-spanid`, and `x-b3-traceid`. These headers are also called B3 headers, which are used for trace context propagation across a service boundary:

- **x-b3-sampled**: This denotes whether a request should be traced or not. A value of `1` means yes and `0` means prohibited.
- **x-b3-traceid**: This is 8 or 16 bytes in length, indicating the overall ID of the trace.
- **x-b3-parentspanid**: This is 8 bytes in length and represents the position of the parent operation in the trace tree. Every span will have a parent span unless it is the root itself.
- **x-b3-spanid**: This is 8 bytes in length and represents the position of the current operation in the trace tree.

In the response from `httpbin`, the request is traversed to the Ingress gateway and then to `httpbin`. The B3 headers were injected by Istio as soon as the request arrived at the Ingress gateway. The span ID generated by the Ingress gateway is `5c0572d9e4ed5415`, which is a parent of the `httpbin` span that has a span ID of `743b39197aaca61f`. Both the Ingress gateway and `httpbin` spans will have the same trace ID because they are part of the same trace. As the Ingress gateway is the root span, it will have no `parentspanid`. In this example, there are only two hops and, thus, two spans. If there were more, they all would have generated the B3 headers because the value of `x-b3-sampled` is `1`.

You can find out more about these headers at `https://www.envoyproxy.io/docs/envoy/latest/configuration/http/http_conn_man/headers.html`.

Now that you are familiar with `x-b3` headers injected by Istio, let's deploy the sample `bookinfo` application and configure the Ingress. If you have not created a `Chapter7` namespace, then please do so with Istio injection enabled:

```
% kubectl apply -f samples/bookinfo/platform/kube/bookinfo.ya
ml -n Chapter7
% kubectl apply -f Chapter7/bookinfo-gateway.yaml
```

Note that `Chapter7/bookinfo-gateway.yaml` configures `bookshop.com` as the host; we did it so that it can run along with `sock-shop.com`. Once the Ingress configuration is deployed, you can access `bookinfo` using the external IP of the `istio-ingress` gateway service. Please use `/productpage` as the URI. Go ahead and make some requests to the `bookinfo` app, after which you can check the Jaeger dashboard and select **productpage.Chapter7** as the service. Once you have selected the service, you can click on **Find Traces**, which will then show a detailed view of the latest traces for the service:

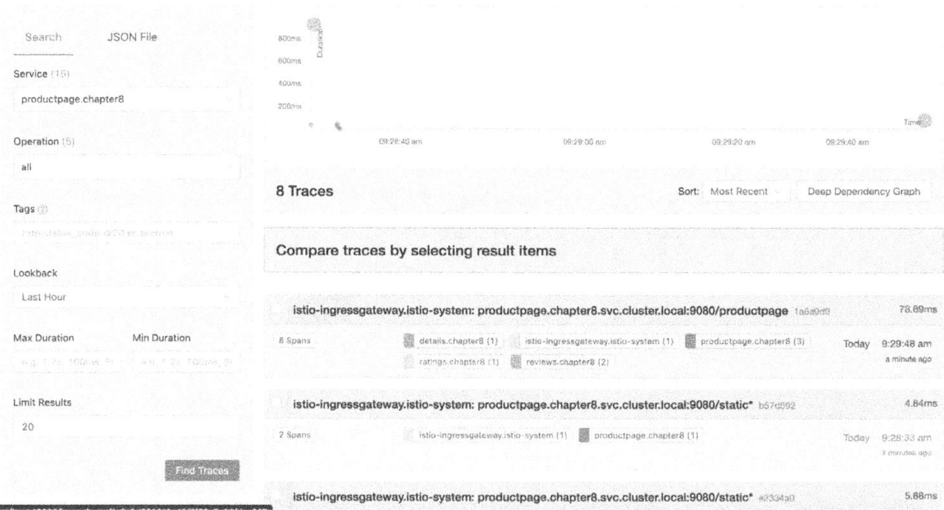

Figure 7.20 – The Jaeger dashboard

A **trace** in Jaeger is a representation of a request execution and is composed of multiple **spans**; the trace records the path taken and traversed by a request. A trace is made up of multiple spans; a span represents a unit of work and is used to track specific operations made by a request. The first span represents the **root span**, which is a request from start to finish; each subsequent span provides a more in-depth context of what has happened in that part of the request execution.

You can click on any of the traces on the dashboard. The following is an example of a trace with eight spans:

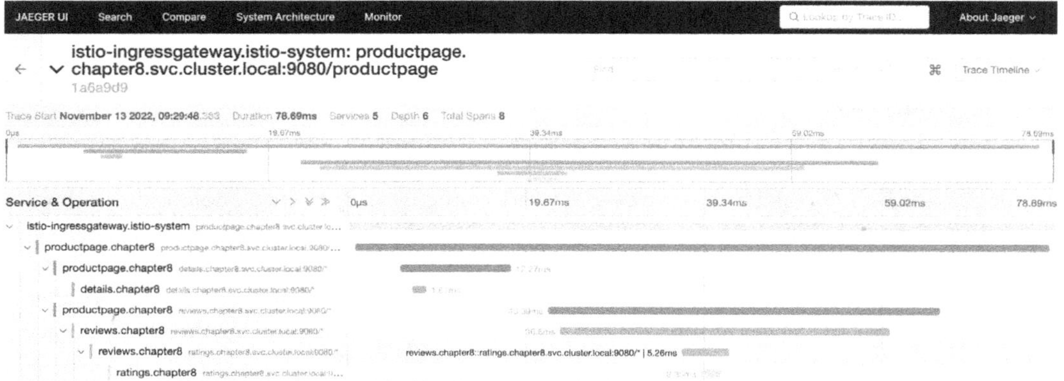

Figure 7.21 – Trace and spans in Jaeger

In the following screenshots, you can observe the following:

- The request took 78.69 milliseconds in `istio-ingressgateway`, which is also the root span. The request was then forwarded to the `productpage` upstream service at port `9080`. If you look at the next child span, you will see that the time taken in `istio-ingressgateway` is 78.69 – 76.73 milliseconds = 1.96 milliseconds.

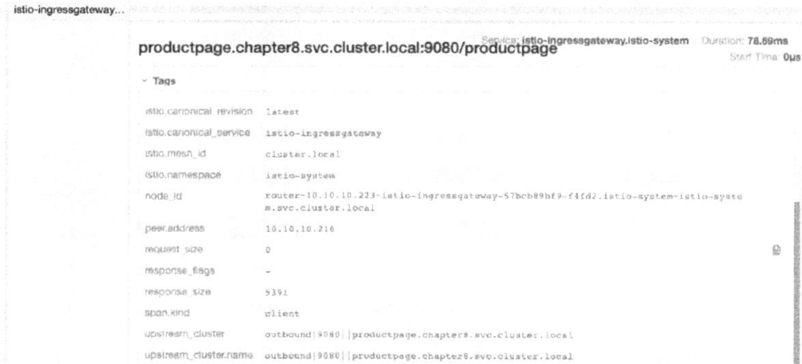

Figure 7.22 – The root span of BookInfo

- The request was then received by the `productpage` service at 867 microseconds in the overall processing timeline. It took 76.73 milliseconds to process the request.

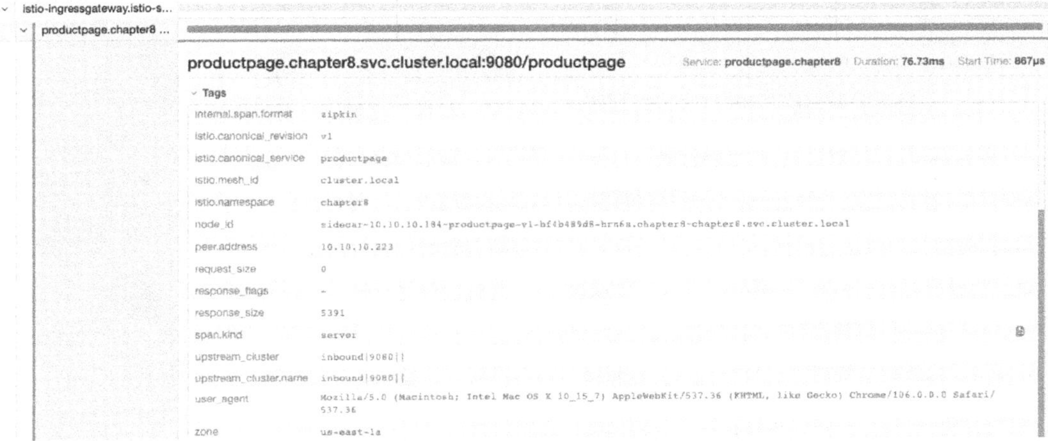

Figure 7.23 – The request arriving on the product page

- The `productpage` service did some processing between 867 microseconds and 5.84 milliseconds, and after that, it invoked the `details` service at port `9080`. It took 12.27 milliseconds to make the round trip to the `details` service.

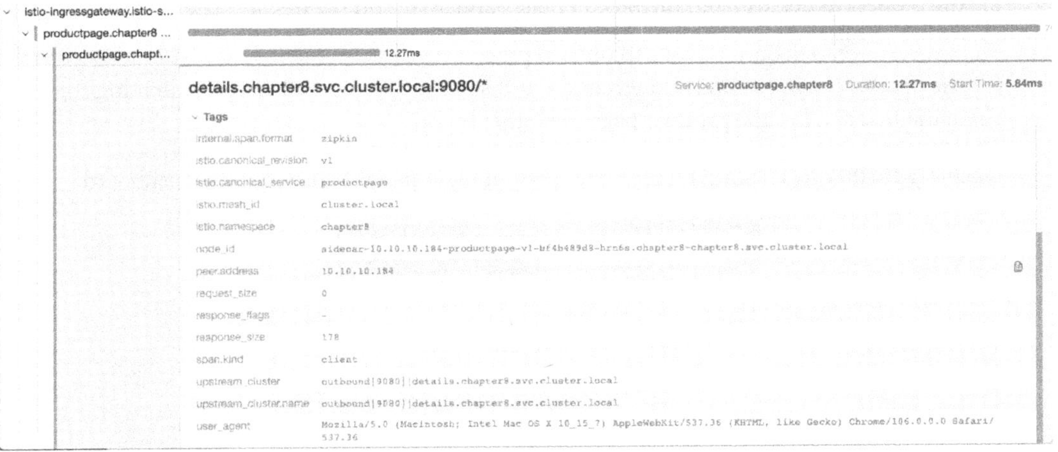

Figure 7.24 – The request from the product page to the details service

- The request was then received by the `details` service after 7.14 milliseconds, and it took 1.61 milliseconds to process the request.

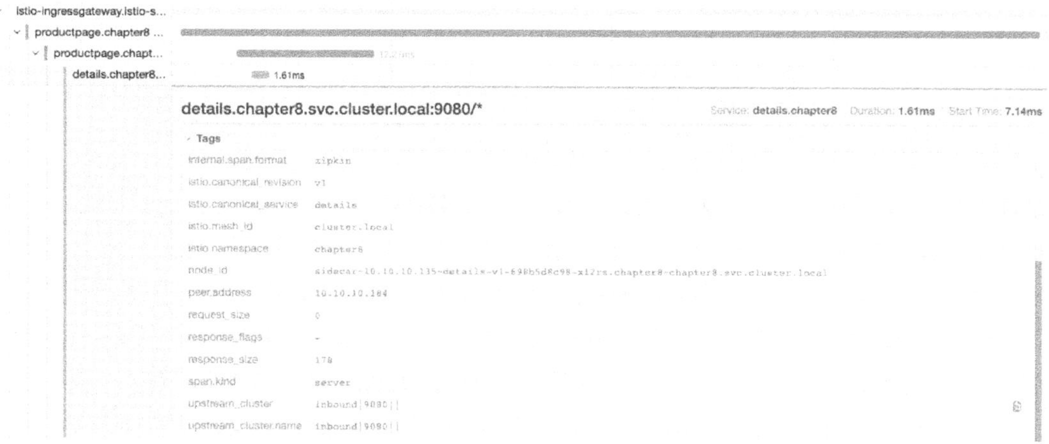

Figure 7.25 – The request arriving at the details service

I have not illustrated the rest of the spans, but I hope you get an idea of the benefits of doing this exercise. The example we just went through raises some intriguing observations:

- By comparing the start time of the third and fourth spans, it is clear that it took 1.3 milliseconds for the request to depart from the product page and arrive on the details page
- The details page took only 1.6 milliseconds to process the request, but it took 12.27 milliseconds for the product page to receive the request and send it to the details page, which highlights some inefficiencies in the product page implementation

You can explore further by clicking on the dropdown in the top-right corner of the dashboard.

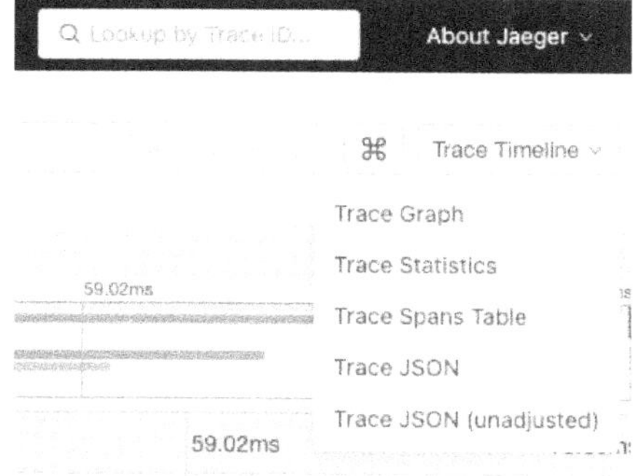

Figure 7.26 – Other options to see the tracing details

The **Trace Spans Table** option is very useful in presenting a summarized view of the time taken by multiple spans to process requests:

Service Name	Operation	ID	Duration
istio-ingressgateway.istio-system	productpage.chapter8.svc.cluster.local:9080/productpage	7f56e5a06fdec307	78.689ms
productpage.chapter8	productpage.chapter8.svc.cluster.local:9080/productpage	2454be88e319126c	76.725ms
productpage.chapter8	details.chapter8.svc.cluster.local:9080/*	cb7c6d2190474871	12.273ms
details.chapter8	details.chapter8.svc.cluster.local:9080/*	62e4ad2e3e086f07	1.608ms
productpage.chapter8	reviews.chapter8.svc.cluster.local:9080/*	62c3e3f1bcbd7c5b	43.394ms
reviews.chapter8	reviews.chapter8.svc.cluster.local:9080/*	a148d22bea2ee476	36.501ms
reviews.chapter8	ratings.chapter8.svc.cluster.local:9080/*	9a5452fa8e216070	5.262ms
ratings.chapter8	ratings.chapter8.svc.cluster.local:9080/*	f83b5a365d198da0	2.353ms

Figure 7.27 – The trace span table in Jaeger

Tracing comes at the cost of performance, and it is not ideal to trace all requests because they will cause performance degradation. We installed Istio in a demo profile, which by default samples all requests. This can be controlled by the following configuration map:

```
% kubectl get cm/istio -n istio-system -o yaml
```

You can control the sampling rate by providing the correct value for `sampling` in `tracing`:

```
apiVersion: v1
data:
  mesh: |-
..
      tracing:
       sampling: 10
        zipkin:
          address: zipkin.istio-system:9411
    enablePrometheusMerge: true
..
kind: ConfigMap
```

This can also be controlled at the deployment level – for example, we can configure the product page to sample only 1% of the requests by adding the following to `bookinfo.yaml`:

```
template:
  metadata:
    annotations:
      proxy.istio.io/config: |
        tracing:
          sampling: 1
          zipkin:
            address: zipkin.istio-system:9411
```

The entire configuration is available in `Chapter7/bookinfo-samplingdemo.yaml`.

In this section, we learned about how distributed tracing can be performed using Jaeger without making any changes in the application code, provided your application can forward `x-b3` headers.

Summary

In this chapter, we read about how Istio makes systems observable by generating various telemetry data. Istio provides various metrics that can then be used by Istio operators to fine-tune and optimize a system. This is all achieved by Envoy, which generates various metrics that are then scraped by Prometheus.

Istio allows you to configure new metrics as well as add new dimensions to existing metrics. You learned how to use Prometheus to query various metrics using PromQL and build queries that can provide insight into your system, as well as business operations. We later installed Grafana to visualize the metrics collected by Prometheus, even though there are a number of out-of-the-box dashboards provided for Istio, and you can easily add new dashboards, configure alerts, and create policies on how these alerts should be distributed.

Finally, we installed Jaeger to perform distributed tracing to understand how a request is processed in a distributed system, and we did all this without needing to modify the application code. This chapter provides the foundational understanding of how Istio makes systems observable, resulting in systems that are not only healthy but also optimal.

In the next chapter, we will learn about the various issues you may face when operating Istio and how to troubleshoot them.

Part 3: Scaling, Extending, and Optimizing

This part takes you into advanced topics of Istio. You will read about various architectures for deploying Istio to production environments. You will also explore various options to extend the Istio data plane and learn why it is a very useful and powerful feature of Istio. Istio provides great flexibility for virtual machine-based workloads so, in this part, you will read about how to extend Istio to virtual machines. Toward the end of this part, you will read various tips for troubleshooting Istio and best practices to operate and configure Istio in production. We will finish the book by summarizing what we have learned and applying this to another sample application, along with discussing eBPF and how you can learn more about Istio. The appendix provides you with valuable details about other Service Mesh technologies and will help you get comparable knowledge of Istio in regard to other options.

This part contains the following chapters:

- *Chapter 8, Scaling Istio to Multiple Clusters Deployments A Kubernetes*
- *Chapter 9, Extending Istio Data Plane*
- *Chapter 10, Deploying Istio Service Mesh for Non-Kubernetes Workloads*
- *Chapter 11, Troubleshooting and Operating Istio*
- *Chapter 12, Summarizing What We Have Learned and Next Steps*

8
Scaling Istio to Multi-Cluster Deployments Across Kubernetes

Containers have changed not only how applications are developed but also how applications connect. Application networking, that is, networking between applications, is critical for production deployments and must be automated, elastically scalable, and secure. Real-world applications are deployed across on-premises, multiple clouds, Kubernetes clusters, and namespaces within clusters. Hence, there is a need to provide a Service Mesh across legacy and modern environments, providing seamless connectivity between applications.

In *Chapter 3*, we briefly discussed deployment models for Istio control planes. We discussed a single cluster with a local control plane, a primary and remote cluster with a single control plane, and a single cluster with an external control plane. A single-mesh/single-cluster deployment is the simplest but is also a non-practical deployment model because your production workload will include multiple Kubernetes clusters, possibly spread across multiple data centers.

In this chapter, we will go through the following topics to learn how to deploy Istio across multiple Kubernetes clusters and then how to federate them:

- Establishing mutual trust in multi-cluster deployments
- Primary-remote on multi-network
- Primary-remote on the same network
- Multi-primary on different networks
- Multi-primary on the same network

This chapter is extremely hands-on, so please pay special attention to the *Technical requirements* section. Also, in each section, please pay special attention to the instructions for setting up clusters and configuring Istio.

Technical requirements

In this chapter, we will be using **Google Cloud** for hands-on activities. If you are a first-time user, you may be eligible for free credits, as described at `https://cloud.google.com/free`. You will need a Google account to sign up; once you are signed up, please follow the Google documentation to install the **Google CLI**, as described at `https://cloud.google.com/sdk/docs/install`. After installing the Google CLI, you will need to initialize it using the steps described at `https://cloud.google.com/sdk/docs/initializing`. The init steps will make the necessary configurations so that you can interact with your Google Cloud account using the CLI.

Setting up Kubernetes clusters

Once you have the account set up, we will create two Kubernetes clusters using the **Google Kubernetes Engine service**. To do this, follow these steps:

1. Create cluster 1 using the following commands:

```
% gcloud beta container --project "istio-book-370122"
clusters create "cluster1" --zone "australia-
southeast1-a" --no-enable-basic-auth --cluster-version
"1.23.12-gke.100" --release-channel "regular" --machine-
type "e2-medium" --image-type "COS_CONTAINERD" --disk-
type "pd-standard" --disk-size "30" --num-nodes "3"
```

In this example, we are creating a cluster named `cluster1` in the `australia-southeast1-a` zone in the `australia-southeast1` region. The machine type to be used is `e2-medium` with a default pool size of `3`. You can change the regions to whatever is closest to your location. You can change the instance type and other parameters, but be conscious of any costs it may incur.

2. Next, create cluster 2. The process is the same as that in the previous step, but we are using a different region and different subnets:

```
% gcloud beta container --project "istio-book-370122"
clusters create "cluster2" --zone "australia-
southeast2-a" --no-enable-basic-auth --cluster-version
"1.23.12-gke.100" --release-channel "regular" --machine-
type "e2-medium" --image-type "COS_CONTAINERD" --disk-
type "pd-standard" --disk-size "30" --max-pods-per-node
"110" --num-nodes "3"
```

3. Now, set up the environment variables to reference the created clusters. Find the cluster reference name from the `kubectl` config:

```
% kubectl config view -o json | jq '.clusters[].name'
"gke_istio-book-370122_australia-southeast1-a_primary-
cluster"
```

```
"gke_istio-book-370122_australia-southeast2-a_primary2-
cluster"
"minikube"
```

Set up the following context variables in every terminal window you will be using in this chapter:

```
export CTX_CLUSTER1="gke_istio-book-370122_australia-
southeast1-a_primary-cluster"
export CTX_CLUSTER2="gke_istio-book-370122_australia-
southeast2-a_primary2-cluster"
```

This completes the setup of the Kubernetes cluster in Google Cloud. In the next section, you will set up OpenSSL on your workstation.

Setting up OpenSSL

We will be using OpenSSL to generate a root and intermediate **certificate authority** (**CA**). You will need OpenSSL 3.0 or higher. Mac users can follow the instructions at `https://formulae.brew.sh/formula/openssl@3`.

You may see the following response:

```
openssl@3 is keg-only, which means it was not symlinked into /
opt/homebrew,
because macOS provides LibreSSL.
```

In this case, manually add OpenSSL to `PATH`:

```
% export PATH="/opt/homebrew/opt/openssl@3/bin:$PATH"
% openssl version
OpenSSL 3.0.7 1
```

Please make sure that the path reflects the terminals from where you will be performing certificate-related commands.

Additional Google Cloud steps

The following steps are useful for establishing connectivity between the two Kubernetes clusters. Please do not perform the steps in this section yet. We will refer to these steps while carrying out the practical exercises in the subsequent sections:

1. Calculate the **Classless Inter-Domain Routing** (**CIDR**) block of clusters 1 and 2:

```
% function join_by { local IFS="$1"; shift; echo "$*"; }
ALL_CLUSTER_CIDRS=$(gcloud container clusters list -
```

```
format='value(clusterIpv4Cidr)' | sort | uniq)
ALL_CLUSTER_CIDRS=$(join_by , $(echo "${ALL_CLUSTER_
CIDRS}"))
```

The value of `ALL_CLUSTER_CIDR` will be something similar to `10.124.0.0/14,10.84.0.0/14`.

2. Get the `NETTAGS` of clusters 1 and 2:

```
% ALL_CLUSTER_NETTAGS=$(gcloud compute instances list -
format='value(tags.items.[0])' | sort | uniq)
ALL_CLUSTER_NETTAGS=$(join_by , $(echo "${ALL_CLUSTER_
NETTAGS}"))
```

The value of `ALL_CLUSTER_NETTAGS` will be something similar to `gke-primary-cluster-9d4f7718-node`, `gke-remote-cluster-c50b7cac-node`.

3. Create a firewall rule to allow all traffic between clusters 1 and 2:

```
% gcloud compute firewall-rules create primary-remote-
shared-network \
  --allow=tcp,udp,icmp,esp,ah,sctp \
  --direction=INGRESS \
  --priority=900 \
  --source-ranges="${ALL_CLUSTER_CIDRS}" \
  --target-tags="${ALL_CLUSTER_NETTAGS}" -quiet
```

4. Delete Google Cloud Kubernetes clusters and firewall rules by performing the following steps:

   - Delete the firewall:

```
% gcloud compute firewall-rules delete primary-remote-
shared-network
```

   - Use the following command to delete `cluster1`:

```
% gcloud container clusters delete cluster1 -zone
"australia-southeast1-a"
```

   - Use the following command to delete `cluster2`:

```
%gcloud container clusters delete cluster2 -zone
"australia-southeast2-a"
```

This concludes all the steps required to prepare for the upcoming sections. In the next section, we will start with the fundamentals required for multi-cluster deployments.

Establishing mutual trust in multi-cluster deployments

When setting up multi-cluster deployments, we must also establish trust between the clusters. The Istio architecture is based on the zero-trust model, where the network is assumed to be hostile and there is no implicit trust for services. Thus, Istio authenticates each service communication to establish the authenticity of the workload. Every workload in the cluster is assigned an identity and service-to-service communication is performed over mTLS by sidecars. Also, all communication between the sidecar and control plane happens over mTLS. In the previous chapters, we used an Istio CA with a self-signed root certificate. When setting up multi-clusters, we must ensure that the workload is assigned identities that can be understood and trusted by all other services in the mesh. Istio does this by distributing a CA bundle to all workloads, which contains a chain of certificates that can then be used by sidecars to identify the sidecar at the other end of the communication. In a multi-cluster environment, we need to ensure that the CA bundle contains the correct certificate chain to validate all services in the data plane.

There are two options to achieve this:

- **Plugin CA certificate**: Using this option, we create root and intermediate certificates outside Istio and configure Istio to use the created intermediate certificate. This option allows you to make use of a known CA or even your own internal CA as a root CA to generate an intermediate CA for Istio. You provide the intermediate CA certificate and keys to Istio along with root CA certificates. Istio then makes use of the intermediate CA and key to sign workloads and embeds root CA certificates as a root of trust.

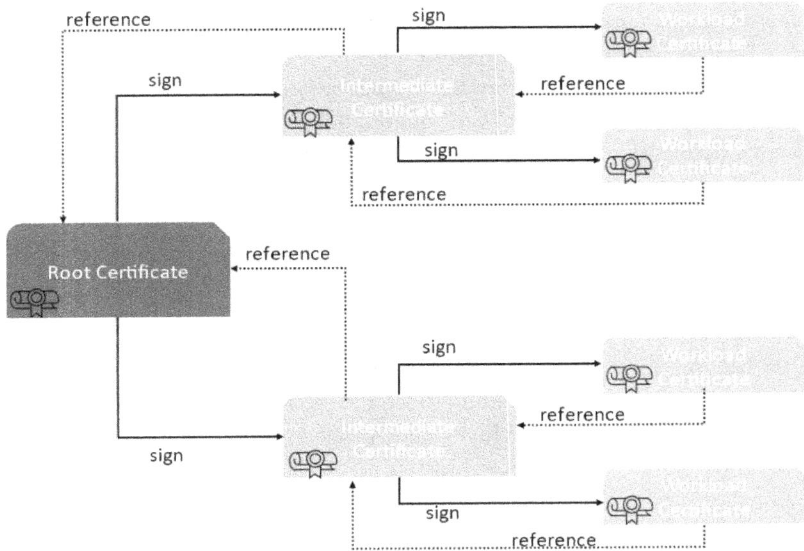

Figure 8.1 – Intermediate CA as a plugin CA to Istio

- **CA external to Istio**: We make use of an external CA that can sign certificates without needing to store private keys inside the Kubernetes cluster. When Istio uses a self-signed certificate, it stores its self-signed private keys as a Secret in the Kubernetes cluster. If using a plugin CA, it still has to save its intermediate keys in the cluster. Storing private keys in the Kubernetes cluster is not a secure option if access to Kubernetes is not restricted. In such cases, we can make use of an external CA to act as a CA for signing certificates.

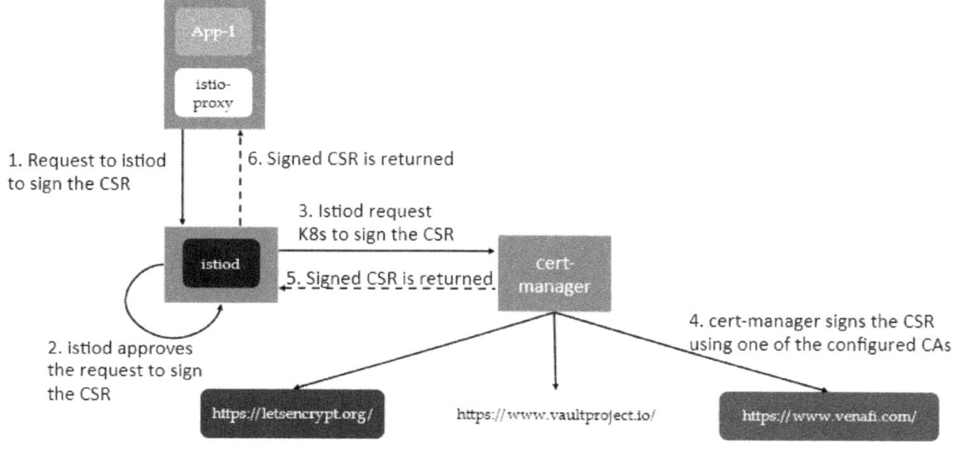

Figure 8.2 – cert-manager

One such certificate management software is **cert-manager**. It adds external certificates and certificate issuers as resource types in Kubernetes clusters and simplifies the process of obtaining, renewing, and using those certificates. It can integrate with a variety of supported sources, including Let's Encrypt and HashiCorp Vault. It ensures certificates are valid and up to date, and it attempts to renew certificates at a configured time before expiry. The cert-manager software integrates with Kubernetes via the **Kubernetes CSR API**; you can read about it at `https://kubernetes.io/docs/reference/access-authn-authz/certificate-signing-requests/`. When using cert-manager, Istio approves the CSR from the service workload and forwards the request to cert-manager for signing. cert-manager signs the request and returns the certificates to istiod, which are then passed on to istio-agent.

In this chapter, we will make use of the plugin CA certificate option, which is the simpler and easier option to use, so that we can focus on the multi-cluster setup of Istio. In the following sections, we will go through setting up Istio in various cluster configurations.

Primary-remote on multi-network

In the primary-remote configuration, we will install the Istio control plane on cluster 1. Clusters 1 and 2 are on different networks with no direct connectivity between the Pods. Cluster 1 will host the Istio control plane as well as a data plane. Cluster 2 will only host the data plane and uses the control plane from cluster 1. Clusters 1 and 2 both use an intermediate CA signed by a root CA. In cluster 1, istiod observes the API server in clusters 1 and 2 for any changes to Kubernetes resources. We will create an Ingress gateway in both clusters, which will be used for cross-network communications between the workloads. We will call this Ingress gateway the east-west gateway because it is used for east-west communication. The east-west gateway takes care of authentication workloads between clusters 1 and 2 and acts as a hub for all traffic traveling between the two clusters. In the following diagram, the dashed arrows in data plane traffic represent service requests from cluster 1 to cluster 2 traversing via the east-west gateway. In cluster 2, the dotted arrows represent data plane traffic traveling from cluster 2 to cluster 1.

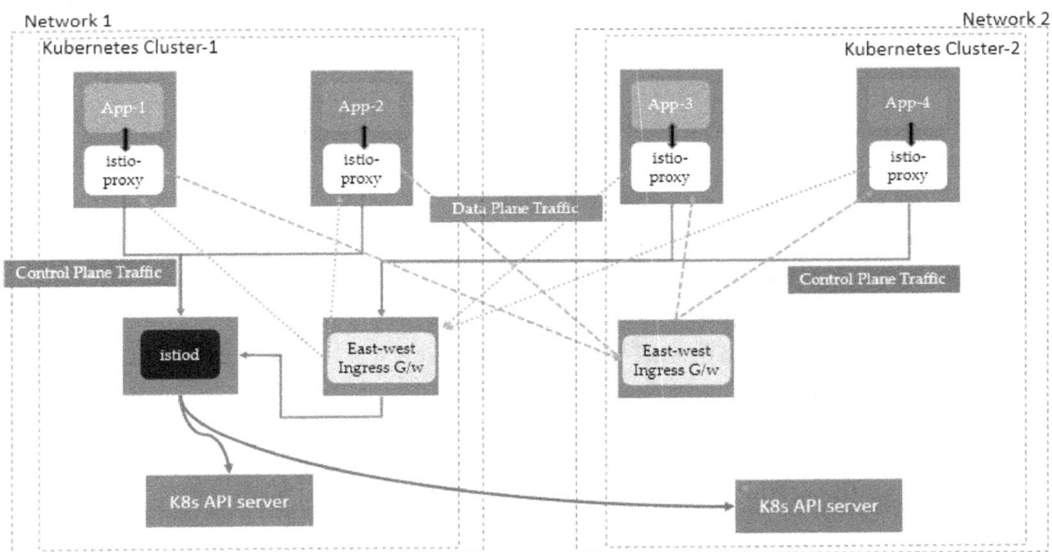

Figure 8.3 – Primary-remote cluster on different networks

In the next section, we will start with configuring mutual trust between the two Kubernetes clusters. We will make use of the plugin CA option, as described in the previous section, *Establishing mutual trust in multi-cluster deployments*.

Establishing trust between the two clusters

We need to configure Istio CAs on both clusters to trust each other. As described in the previous section, we will do that by creating a root CA and using it to generate intermediate CAs for both clusters.

Go to the Istio installation directory and create a folder called `certs` to hold the generated certificates. Then, perform the following instructions from the `cert` directory:

1. Generate the root certificate:

```
% mkdir -p certs
% cd certs
% make -f ../tools/certs/Makefile.selfsigned.mk root-ca
generating root-key.pem
generating root-cert.csr
generating root-cert.pem
Certificate request self-signature ok
subject=O = Istio, CN = Root CA
```

   This will generate `root-key.pem`, which is the private key, and `root-cert.pem`, which is the root certificate.

2. Generate the intermediate CA for `cluster1`:

```
% make -f ../tools/certs/Makefile.selfsigned.mk cluster1-
cacerts
generating cluster1/ca-key.pem
generating cluster1/cluster-ca.csr
generating cluster1/ca-cert.pem
Certificate request self-signature ok
subject=O = Istio, CN = Intermediate CA, L = cluster1
generating cluster1/cert-chain.pem
Intermediate inputs stored in cluster1/
done
rm cluster1/cluster-ca.csr cluster1/intermediate.conf%
% ls cluster1
ca-cert.pem      ca-key.pem cert-chain.pem  root-cert.pem
```

   This will generate an intermediate CA for `cluster1`, with the CA key in `ca-key.pem`, the certificate in `ca-cert.pem`, and the chain in `cert-chain.pem`.

3. Generate an intermediate CA for `cluster2`:

```
% % make -f ../tools/certs/Makefile.selfsigned.mk
cluster2-cacerts
generating cluster2/ca-key.pem
generating cluster2/cluster-ca.csr
```

```
generating cluster2/ca-cert.pem
Certificate request self-signature ok
subject=O = Istio, CN = Intermediate CA, L = cluster2
generating cluster2/cert-chain.pem
Intermediate inputs stored in cluster2/
done
rm cluster2/cluster-ca.csr cluster2/intermediate.conf
% ls cluster2
ca-cert.pem     ca-key.pem  cert-chain.pem  root-cert.pem
```

This will generate an intermediate CA for `cluster2`, with the CA key in `ca-key.pem`, the certificate in `ca-cert.pem`, and the chain in `cert-chain.pem`.

4. Set the environment variables as described in the third and fourth steps of the *Setting up Kubernetes clusters* subsection in the *Technical requirements* section. This helps to run commands targeting multiple Kubernetes clusters:

```
 % export CTX_CLUSTER1=" gke_istio-book-370122_australia-
southeast1-a_primary-cluster"
% export CTX_CLUSTER2=" gke_istio-book-370122_australia-
southeas1-b_remote-cluster"
```

5. Create namespaces in the primary and remote clusters. We will install Istio in this namespace:

```
% kubectl create ns istio-system --context="${CTX_
CLUSTER1}"
namespace/istio-system created
% kubectl create ns istio-system --context="${CTX_
CLUSTER2}"
namespace/istio-system created
```

6. Create `secret` in `cluster1`, which will be used by Istio as an intermediate CA:

```
% kubectl create secret generic cacerts -n istio-system \
      --from-file=cluster1/ca-cert.pem \
      --from-file=cluster1/ca-key.pem \
      --from-file=cluster1/root-cert.pem \
      --from-file=cluster1/cert-chain.pem
--context="${CTX_CLUSTER1}"
secret/cacerts created
```

7. Create `secret` in `cluster2`, which will be used by Istio as an intermediate CA:

```
% kubectl create secret generic cacerts -n istio-system \
      --from-file=cluster2/ca-cert.pem \
      --from-file=cluster2/ca-key.pem \
      --from-file=cluster2/root-cert.pem \
      --from-file=cluster2/cert-chain.pem
--context="${CTX_CLUSTER2}"
secret/cacerts created
```

8. Label the `namespace` in `cluster1` and `cluster2` with the network name:

```
% kubectl --context="${CTX_CLUSTER1}" label namespace
istio-system topology.istio.io/network=network1
namespace/istio-system labeled
% kubectl --context="${CTX_CLUSTER2}" label namespace
istio-system topology.istio.io/network=network2
namespace/istio-system labeled
```

9. Configure `cluster1` as follows:

   I. Create the Istio operator config:

```
apiVersion: install.istio.io/v1alpha1
kind: IstioOperator
spec:
  values:
    global:
      meshID: mesh1
      multiCluster:
        clusterName: cluster1
      network: network1
```

   The file is available in `Chapter08/01-Cluster1.yaml` on GitHub.

   II. Install Istio:

```
% istioctl install --set values.pilot.env.EXTERNAL_
ISTIOD=true --context="${CTX_CLUSTER1}" -f Chapter08/01-
Cluster1.yaml"
This will install the Istio 1.16.0 default profile with
["Istio core" "Istiod" "Ingress gateways"] components
into the cluster. Proceed? (y/N) y
```

```
✓ Istio core installed
✓ Istiod installed
✓ Ingress gateways installed
✓ Installation complete
Making this installation the default for injection and
validation.
```

III. In this step, we will install the east-west gateway in `cluster1`, which will expose all services in the mesh in `cluster1` to services in `cluster2`. This gateway is accessible to all services in `cluster2` but it can only be accessed by services with a trusted mTLS certificate and workload ID, that is, services that are part of a mesh:

```
% samples/multicluster/gen-eastwest-gateway.sh \
    --mesh mesh1 --cluster cluster1 --network network1 |
\
    istioctl --context="${CTX_CLUSTER1}" install -y -f -
 Ingress gateways installed
 Installation complete
```

IV. The east-west gateway is also used to expose istiod endpoints to `cluster2`. These endpoints are used by the mutating webhooks and `istio-proxy` in `cluster2`. The following configuration creates a gateway named `istiod-gateway` and exposes ports `15012` and `15017` over TLS:

```
apiVersion: networking.istio.io/v1alpha3
kind: Gateway
metadata:
  name: istiod-gateway
spec:
  selector:
    istio: eastwestgateway
  servers:
    - port:
        name: tls-istiod
        number: 15012
        protocol: tls
      tls:
        mode: PASSTHROUGH
```

```
    hosts:
      - "*"
  - port:
      name: tls-istiodwebhook
      number: 15017
      protocol: tls
    tls:
      mode: PASSTHROUGH
    hosts:
      - "*"
```

V. The following virtual service routes inbound traffic on ports `15012` and `15017` to `15012` and `443` to the `istiod.istio-system.svc.cluster.local` service on `cluster1`:

```
tls:
- match:
  - port: 15012
    sniHosts:
    - "*"
  route:
  - destination:
      host: istiod.istio-system.svc.cluster.local
      port:
        number: 15012
- match:
  - port: 15017
    sniHosts:
    - "*"
  route:
  - destination:
      host: istiod.istio-system.svc.cluster.local
      port:
        number: 443
```

The configuration is available in `samples/multicluster/expose-istiod.yaml` in the Istio installation folder. Apply the configuration using the following commands:

```
% kubectl apply --context="${CTX_CLUSTER1}" -n istio-
system -f "samples/multicluster/expose-istiod.yaml"
gateway.networking.istio.io/istiod-gateway created
virtualservice.networking.istio.io/istiod-vs created
```

10. Create another gateway for exposing workload services to `cluster2`. The configuration is very similar to `istiod-gateway` except that we are exposing port `15443`, which is specifically dedicated to traffic designated for services in the mesh:

```
apiVersion: networking.istio.io/v1alpha3
kind: Gateway
metadata:
  name: cross-network-gateway
spec:
  selector:
    istio: eastwestgateway
  servers:
    - port:
        number: 15443
        name: tls
        protocol: TLS
      tls:
        mode: AUTO_PASSTHROUGH
      hosts:
        - "*.local"
```

11. A sample file is available in `samples/multicluster/expose-services.yaml` in the Istio installation directory:

```
% kubectl --context="${CTX_CLUSTER1}" apply -n istio-
system -f samples/multicluster/expose-services.yaml
gateway.networking.istio.io/cross-network-gateway created
```

12. In this step, we will configure `cluster2`. To do this, you will need to note down the external IP for the east-west gateway created in the previous step. In the following steps, we will first prepare the configuration file for Istio and then use that to install Istio in `cluster2`:

    I. Configure the Istio operator configuration. The following is the sample configuration, and it has two noteworthy configurations:

    - `injectionPath`: Constructed as `/inject/cluster/CLUSTER_NAME OF_REMOTE_CLUSTER/net/ NETWORK_NAME OF_REMOTE_CLUSTER`
    - `remotePilotAddress`: IP of the east-west gateway exposing ports `15012` and `15017` and the network reachable to `cluster2`

    The sample file is available in `Chapter08/01-Cluster2.yaml` on GitHub:

```
apiVersion: install.istio.io/v1alpha1
kind: IstioOperator
spec:
  profile: remote
  values:
    istiodRemote:
      injectionPath: /inject/cluster/cluster2/net/
network2
    global:
      remotePilotAddress: 35.189.54.43
```

    II. Label and annotate the namespace. Setting the `topology.istio.io/controlPlaneClusters` namespace annotation to `cluster1` instructs istiod running on `cluster1` to manage `cluster2`, when it is attached as a remote cluster:

```
% kubectl --context="${CTX_CLUSTER2}" annotate
namespace istio-system topology.istio.io/
controlPlaneClusters=cluster1
namespace/istio-system annotated
```

    III. Set the network for `cluster2` by adding a label to the `istio-system` namespace. The network name should be the same as you configured in the `01-Cluster2.yaml` file in the previous step:

```
% kubectl --context="${CTX_CLUSTER2}" label namespace
istio-system topology.istio.io/network=network2
namespace/istio-system labeled
```

IV. Install Istio in `cluster2`:

```
istioctl install --context="${CTX_CLUSTER2}" -f "
Chapter08/01-Cluster2.yaml"
This will install the Istio 1.16.0 remote profile with
["Istiod remote"] components into the cluster. Proceed?
(y/N) y
 Istiod remote installed
 Installation complete
Making this installation the default for injection and
validation.
```

13. Provide primary cluster access to the API server of the remote cluster:

```
% istioctl x create-remote-secret \
  --context="${CTX_CLUSTER2}" \
  --name=cluster2 \
  --type=remote \
  --namespace=istio-system \
  --create-service-account=false | \
  kubectl apply -f - --context="${CTX__CLUSTER1}"
secret/istio-remote-secret-cluster2 created
```

After performing this step, istiod in `cluster1` will be able to communicate with the Kubernetes API server in `cluster2`, giving it visibility of services, endpoints, and namespaces in `cluster2`. As soon the API server is accessible to istiod, it will patch the certificates in webhooks in `cluster2`. Now perform the following before and after *step 13*:

```
% kubectl get mutatingwebhookconfiguration/istio-sidecar-
injector --context="${CTX_CLUSTER2}" -o json
```

You will notice that the following has been updated in the sidecar injector:

```
                "caBundle":
"..MWRNPQotLS0tLUVORCBDRVJUSUZJQ0FURS0tLS0tCg==",
                "url": https://
a5bcd3e72e1f04379a75247f8f718bb1-689248335.us-east-1.elb.
amazonaws.com:15017/inject/cluster/cluster2/net/network2
```

14. Create the east-west gateway to handle traffic Ingress from the primary cluster to the remote cluster:

```
% samples/multicluster/gen-eastwest-gateway.sh --mesh
mesh1 --cluster "${CTX_CLUSTER2}" --network network2 >
eastwest-gateway-1.yaml
% istioctl manifest generate -f eastwest-gateway-remote.
yaml --set values.global.istioNamespace=istio-system |
kubectl apply --context="${CTX_CLUSTER2}" -f -
```

15. Install CRDs so that you can configure traffic rules:

```
% kubectl apply -f manifests/charts/base/crds/crd-all.
gen.yaml --context="${CTX_CLUSTER2}"
```

16. Expose all services in the remote cluster:

```
% kubectl --context="${CTX_CLUSTER2}" apply -n istio-
system -f samples/multicluster/expose-services.yaml
```

This completes the installation and configuration of Istio in both clusters.

Deploying the Envoy dummy application

In this section, we will first deploy two versions of the Envoy dummy application and then test the traffic distribution of the dummy application. Let's get started with deploying two versions of the Envoy dummy application:

1. Create namespaces and enable `istio-injection`:

```
% kubectl create ns chapter08 --context="${CTX_CLUSTER1}"
% kubectl create ns chapter08 --context="${CTX_CLUSTER2}"
% kubectl label namespace chapter08 istio-
injection=enabled --context="${CTX_CLUSTER1}"
% kubectl label namespace chapter08 istio-
injection=enabled --context="${CTX_CLUSTER2}"
```

2. Create config maps:

```
% kubectl create configmap envoy-dummy --from-
file=Chapter3/envoy-config-1.yaml -n chapter08
--context="${CTX_CLUSTER1}"
% kubectl create configmap envoy-dummy --from-
file=Chapter4/envoy-config-2.yaml -n chapter08
--context="${CTX_CLUSTER2}"
```

3. Deploy the Envoy application:

```
% kubectl create -f "Chapter08/01-envoy-proxy.yaml"
--namespace=chapter08 --context="${CTX_CLUSTER1}"
% kubectl create -f "Chapter08/02-envoy-proxy.yaml"
--namespace=chapter08 --context="${CTX_CLUSTER2}"
```

4. Expose Envoy using a gateway and virtual services. You can use any cluster as `context`; istiod will propagate the configuration to another cluster:

```
% kubectl apply -f "Chapter08/01-istio-gateway.yaml" -n
chapter08 --context="${CTX_CLUSTER2}"
```

We have now successfully deployed the `envoydummy` application across both clusters. Now, let's move on to testing the traffic distribution of the dummy application:

1. The IP is the external IP of the Ingress gateway. Please note that it is different from the east-west gateway. The east-west gateway is used for inter-cluster communication between services workloads, whereas the Ingress gateway is used for north-south communication. As we are using `curl` from outside the cluster, we will make use of the north-south gateway:

```
% kubectl get svc -n istio-system --context="${CTX_
CLUSTER1}"
NAME                    TYPE                CLUSTER-IP
   EXTERNAL-IP    PORT(S)       AGE
istio-eastwestgateway    LoadBal-
ancer    10.0.7.123    35.189.54.43    15021:30141/
TCP,15443:32354/TCP,15012:30902/TCP,15017:32082/TCP    22h
istio-ingressgateway    LoadBal-
ancer    10.0.3.75    34.87.233.38    15021:30770/
TCP,80:30984/TCP,443:31961/TCP                             22h
istiod   ClusterIP       10.0.6.149    <none>        15010/
TCP,15012/TCP,443/TCP,15014/
TCP                                22h
```

2. Go ahead and call the Envoy dummy using the following commands:

```
% for i in {1..10}; do curl -Hhost:mockshop.com -s
"http://34.87.233.38";echo "\\n"; done
V2----------Bootstrap Service Mesh Implementation with
Istio----------V2
Bootstrap Service Mesh Implementation with Istio
Bootstrap Service Mesh Implementation with Istio
V2----------Bootstrap Service Mesh Implementation with
```

```
Istio----------V2
Bootstrap Service Mesh Implementation with Istio
V2----------Bootstrap Service Mesh Implementation with
Istio----------V2
V2----------Bootstrap Service Mesh Implementation with
Istio----------V2
Bootstrap Service Mesh Implementation with Istio
V2----------Bootstrap Service Mesh Implementation with
Istio----------V2
Bootstrap Service Mesh Implementation with Istio
```

As you must have observed in the output, the traffic is distributed across both clusters. The Ingress gateway in `cluster1` has awareness of `v2` of the Envoy dummy in `cluster2` and is able to route traffic between `v1` and `v2` of the Envoy dummy services.

This concludes the setup of primary-remote on separate networks. In the next section, we will set up primate-remote on the same network.

Primary-remote on the same network

In primary-remote on the same network cluster, the services can access other inter-cluster services because they are on the same network. That means we don't need an east-west gateway for inter-cluster communication between services. We will make `cluster1` the primary cluster and `cluster2` the remote cluster. We still need an east-west gateway to proxy istiod services. All control plane-related traffic from cluster 2 to cluster 1 will traverse via the east-west gateway.

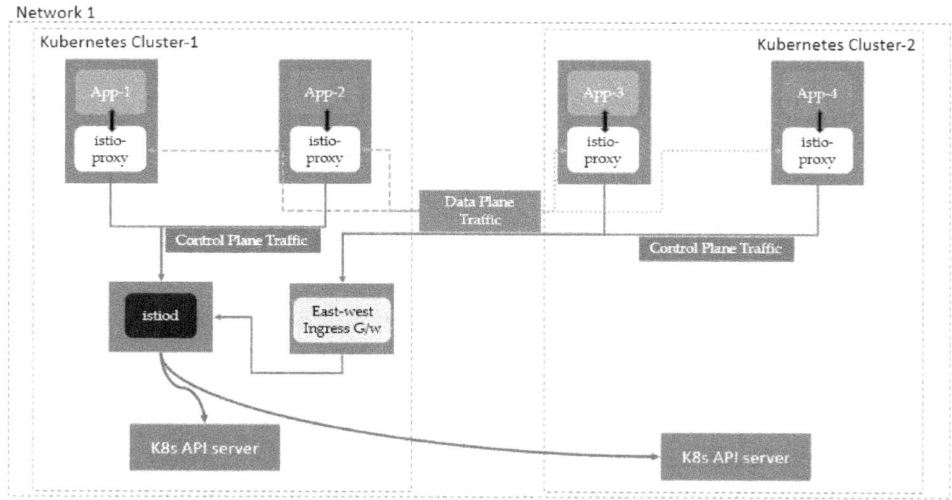

Figure 8.4 – Primary remote cluster sharing the same network

Here, we will make use of the infrastructure we set up in the previous section, *Primary-remote on multi-network*, but if you want, you can also create a separate infrastructure.

Let's get started!

1. If you are using the Kubernetes cluster from the previous section, you will need to first uninstall Istio on the remote cluster using the following code block:

```
$ istioctl uninstall --purge --context="${CTX_CLUSTER2}"
All Istio resources will be pruned from the cluster
Proceed? (y/N) y
 ..::istio-reader-clusterrole-istio-system.
 Uninstall complete
% kubectl delete ns istio-system --context="${CTX_
CLUSTER2}"
namespace "istio-system" deleted
```

2. We will then open the firewall between the two clusters by following the steps provided in the *Additional Google Cloud steps* section:

```
% function join_by { local IFS="$1"; shift; echo "$*"; }
ALL_CLUSTER_CIDRS=$(gcloud container clusters list
--format='value(clusterIpv4Cidr)' | sort | uniq)
ALL_CLUSTER_CIDRS=$(join_by , $(echo "${ALL_CLUSTER_
CIDRS}"))
```

Step 1 will assign a value of something similar to `10.124.0.0/14,10.84.0.0/14`, as seen in the following snippet:

```
% ALL_CLUSTER_NETTAGS=$(gcloud compute instances list
--format='value(tags.items.[0])' | sort | uniq)
ALL_CLUSTER_NETTAGS=$(join_by , $(echo "${ALL_CLUSTER_
NETTAGS}"))
```

Step 2 will assign a value of something similar to `gke-primary-cluster-9d4f7718-node,gke-remote-cluster-c50b7cac-node`.

Step 3 creates a firewall rule to allow all traffic between clusters 1 and 2:

```
% gcloud compute firewall-rules create primary-remote-
shared-network \
  --allow=tcp,udp,icmp,esp,ah,sctp \
  --direction=INGRESS \
  --priority=900 \
  --source-ranges="${ALL_CLUSTER_CIDRS}" \
```

```
  --target-tags="${ALL_CLUSTER_NETTAGS}" -quiet
Creating firewall...⁚Created
Creating firewall...done.
NAME  NETWORK  DIRECTION  PRIORITY
  ALLOW                          DENY  DISABLED
primary-remote-shared--network   default   INGRESS
     900       tcp,udp,icmp,esp,ah,sctp        False
```

After performing these steps, both `cluster1` and `cluster2` will have bidirectional network access.

3. Perform *step 7* from the *Primary-remote on multi-network* section. The step creates `secret` in `cluster2`, which will be used by Istio as an intermediate CA. We will also need to annotate the `istio-system` namespace using the following steps:

```
% kubectl --context="${CTX_CLUSTER2}" annotate
namespace istio-system topology.istio.io/
controlPlaneClusters=cluster1
```

4. Install Istio in `cluster2`. We will be using the following configuration in the installation:

```
apiVersion: install.istio.io/v1alpha1
kind: IstioOperator
spec:
  profile: remote
  values:
    istiodRemote:
      injectionPath: /inject/cluster/cluster2/net/
network1
    global:
      remotePilotAddress: 35.189.54.43
```

Note that `injectionPath` has the value `network1` instead of `network2`. `remotePilotAddress` is the external IP of the east-west gateway of `cluster1`. You will find this configuration in `Chapter08/02-Cluster2.yaml`. The following command will install Istio in cluster 2 using the configuration file:

```
% istioctl install --context="${CTX_CLUSTER2}"
-f  Chapter08/02-Cluster2.yaml -y
✓ Istiod remote installed
✓ Installation complete
             Making this installation the default for
injection and validation.
Thank you for installing Istio 1.16
```

This will complete the installation of Istio in `cluster2`.

5. Next, we will create a remote Secret that provides istiod in `cluster1` with access to the Kubernetes API server in `cluster2`:

   ```
   % istioctl x create-remote-secret --context="${CTX_
   CLUSTER2}" --name=cluster2 |    kubectl apply -f -
   --context="${CTX_CLUSTER1}"
   secret/istio-remote-secret-cluster2 configured
   ```

This concludes the setup of the primary-remote cluster in the same network.

Next, we will test the setup by deploying the Envoy dummy application as we did earlier in the *Primary-remote on multi-network* section. Follow *steps 1–4* of the *Deploying the Envoy dummy application* sub-section of Primary-remote on multi-network section to install the `envoydummy` app. Once it's deployed, we can test whether the Envoy dummy service traffic is distributed across the two clusters:

```
% for i in {1..10}; do curl -Hhost:mockshop.com -s
"http://34.129.4.32";echo '\n'; done
Bootstrap Service Mesh Implementation with Istio
Bootstrap Service Mesh Implementation with Istio
V2----------Bootstrap Service Mesh Implementation with Istio--
--------V2
Bootstrap Service Mesh Implementation with Istio
V2----------Bootstrap Service Mesh Implementation with Istio--
--------V2
V2----------Bootstrap Service Mesh Implementation with Istio--
--------V2
Bootstrap Service Mesh Implementation with Istio
Bootstrap Service Mesh Implementation with Istio
V2----------Bootstrap Service Mesh Implementation with Istio--
--------V2
V2----------Bootstrap Service Mesh Implementation with Istio--
--------V2
```

From the response, you can observe that traffic is distributed across both `cluster1` and `cluster2`. Both clusters are aware of each other's services and the Service Mesh is able to distribute traffic across the two clusters.

This concludes the setup of the primary remote cluster over the shared network. As we have made several changes to our Kubernetes cluster, it is recommended that you delete them, as well as the firewall rule, to get a clean slate before performing the tasks described in the subsequent sections. The following is an example of how you can delete clusters. Please change the parameter values as per your configuration:

```
% gcloud container clusters delete remote-cluster --zone
"australia-southeast2"
% gcloud container clusters delete primary-cluster --zone
"australia-southeast1-a"
% gcloud firewall delete primary-remote-shared-network
```

In the next section, we will perform a primary-primary setup of clusters on separate networks.

Multi-primary on different networks

The control plane has high availability in a multi-primary setup. In the architecture options discussed in the previous sections, we had one primary cluster and the rest of the clusters didn't use istiod, risking a loss of control if the primary control plane suffers an outage due to unforeseen circumstances. In a multi-primary cluster, we have multiple primary control planes providing uninterrupted access to the mesh even if one of the control planes suffers a temporary outage.

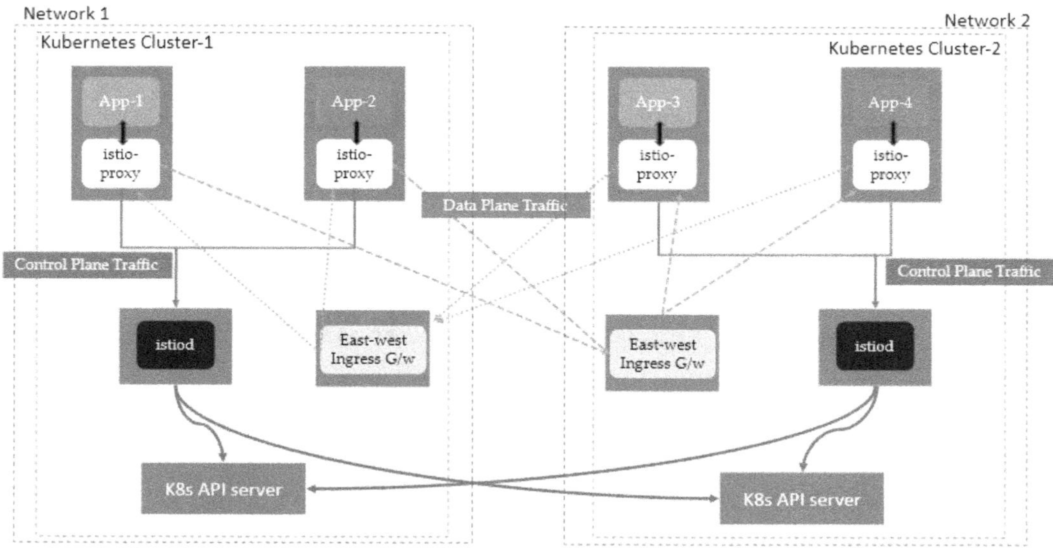

Figure 8.5 – Primary-primary on separate networks

We will start by first setting up the clusters, followed by establishing trust between the two clusters. Perform the following steps to establish a multi-primary cluster:

1. Set up the two clusters as per the initial set of steps in the *Setting up Kubernetes clusters* section. This will complete the creation of the cluster as well as setting up the context variables. Because both clusters are primary, let's call them `primary1` and `primary2` when creating the cluster in Google Cloud.
2. Perform *steps 1–7* of the *Primary-remote on multi-network* section to establish trust between the cluster. These steps will create the certificates, create namespaces, and then create a Secret in the namespaces.
3. Label the `istio-system` namespaces in both clusters with their network names.
4. First, apply the `topology.istio.io/network` label with the value `network1` to the `istio-system` namespace in `cluster1`:

```
% kubectl --context="${CTX_CLUSTER1}" get namespace
istio-system && kubectl --context="${CTX_CLUSTER1}" label
namespace istio-system topology.istio.io/network=network1
NAME           STATUS   AGE
istio-system   Active   3m38s
namespace/istio-system labelled
```

5. Next, apply the `topology.istio.io/network` label with the value `network2` to the `istio-system` namespace in `cluster2`:

```
% kubectl --context="${CTX_CLUSTER2}" get namespace
istio-system && kubectl --context="${CTX_CLUSTER2}" label
namespace istio-system topology.istio.io/network=network2
NAME           STATUS   AGE
istio-system   Active   3m45s
namespace/istio-system labeled
```

6. The Istio operator configuration for `cluster1` is similar to the primary remote configuration, so we will be using the `01-cluster1.yaml` file to install Istio in `cluster1`:

```
% istioctl install --context="${CTX_CLUSTER1}" -f "
Chapter08/01-Cluster1.yaml" -n istio-system -y
 Istio core installed
 Istiod installed
 Ingress gateways installed
 Installation complete
```

7. Install the east-west gateway in `cluster1`:

```
% samples/multicluster/gen-eastwest-gateway.sh --mesh
mesh1 --cluster cluster1 --network network1 | istioctl
--context="${CTX_CLUSTER1}" install -y -f -
 Ingress gateways installed
 Installation complete
```

8. Create a gateway configuration to expose all services in `cluster1` via the east-west gateway:

```
% kubectl --context="${CTX_CLUSTER1}" apply -n istio-
system -f samples/multicluster/expose-services.yaml
gateway.networking.istio.io/cross-network-gateway created
```

9. We will be using the following Istio operator configuration to configure `cluster2` and install Istio:

```
apiVersion: install.istio.io/v1alpha1
kind: IstioOperator
spec:
  values:
    global:
      meshID: mesh1
      multiCluster:
        clusterName: cluster2
      network: network2
```

Note that we are not providing a profile in the configuration, which means a default configuration profile will be selected. In the default configuration, `istioctl` installs the Ingress gateway and istiod. To learn more about the configuration profile and what is included in each profile, please use the following command:

```
% istioctl profile dump default
```

The sample file is available in `Chapter08/03-Cluster3.yaml`. Install Istio in `cluster2` using the following command:

```
% istioctl install --context="${CTX_CLUSTER2}" -f "
Chapter08/03-Cluster3.yaml" -y
 Istio core installed
 Istiod installed
 Ingress gateways installed
 Installation complete
```

10. Install the east-west gateway and expose all services:

```
% samples/multicluster/gen-eastwest-gateway.sh --mesh
mesh1 --cluster cluster2 --network network2 | istioctl
--context="${CTX_CLUSTER2}" install -y -f -
 Ingress gateways installed
 Installation complete
% kubectl --context="${CTX_CLUSTER2}" apply -n istio-
system -f samples/multicluster/expose-services.yaml
gateway.networking.istio.io/cross-network-gateway created
```

11. Create a remote Secret for `cluster1` to be able to access the API server in `cluster2`:

```
% istioctl x create-remote-secret --context="${CTX_
CLUSTER2}" --name=cluster2 | kubectl apply -f -
--context="${CTX_CLUSTER1}"
secret/istio-remote-secret-cluster2 created
```

12. Finally, create a remote Secret for `cluster2` to be able to access the API server in `cluster1`:

```
% istioctl x create-remote-secret --context="${CTX_
CLUSTER1}" --name=cluster1 | kubectl apply -f -
--context="${CTX_CLUSTER2}"
secret/istio-remote-secret-cluster1 configured
```

Now it's time to deploy and test our setup.

Deploying and testing via Envoy dummy services

Next, we will test the setup by deploying an Envoy dummy application as we did in previous sections. Follow *steps 1–4* of the *Deploying the Envoy dummy application* section under the *Primary-remote on multi-network* sub-section.

Test the Envoy dummy application:

```
% for i in {1..5}; do curl -Hhost:mockshop.com -s
"http://34.129.4.32";echo '\n'; done
Bootstrap Service Mesh Implementation with Istio
V2----------Bootstrap Service Mesh Implementation with Istio--
--------V2
Bootstrap Service Mesh Implementation with Istio
```

```
V2----------Bootstrap Service Mesh Implementation with Istio--
--------V2
V2----------Bootstrap Service Mesh Implementation with Istio--
--------V2
```

Another test to perform is for the high availability of the control plane. You can shut down istiod in any of the clusters but that will not impact control plane operations. You will still be able to publish new services into the mesh.

Perform the following tests to validate the high availability of the control plane. I have left out the command instructions because we have performed those steps several times in this book:

1. Shut down istiod in `cluster1`.
2. Delete the Envoy dummy application from `cluster1` using `01-envoy-proxy.yaml`.
3. Test the Envoy dummy application.
4. Deploy the Envoy dummy application in `cluster1` using `01-envoy-proxy.yaml`.
5. Test the Envoy dummy application.

Because we have set up a multi-primary cluster, there should be no interruptions to control plane operations even if the `cluster1` control plane is not available.

In the next section, we will set up a multi-primary control plane where `cluster1` and `cluster2` share the same network.

Multi-primary on the same network

In this section, we will set up a multi-primary Istio cluster with a shared network. In this architecture, workloads in `cluster1` can directly access services in `cluster2` and vice versa. In multi-primary clusters, we don't need an east-west gateway because of the following:

- Services can directly communicate with each other across cluster boundaries
- Each control plane observes the API servers in both clusters

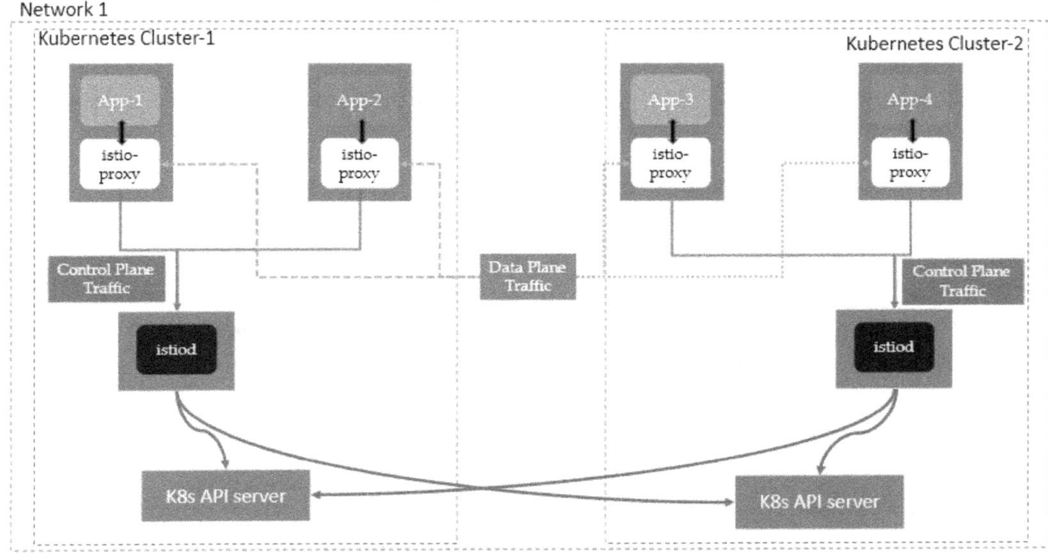

Figure 8.6 – Multi-primary on the same network

As we set up multi-primary in a separate network in the previous section, we will first need to do some cleanup to set up the environment. To do this, we will need to perform the following steps:

1. Uninstall Istio in both the primary and remote clusters:

```
% istioctl uninstall --purge --context="${CTX_CLUSTER2}"
-y
  Uninstall complete
% istioctl uninstall --purge --context="${CTX_CLUSTER1}"
-y
  Uninstall complete
```

2. Remove all the labels from the `istio-system` namespace in `cluster2`:

```
% kubectl label namespace istio-system topology.istio.io/
network- --context="${CTX_CLUSTER2}"
namespace/istio-system unlabeled
```

3. We will then open the firewall between the two clusters following the initial set of deployment steps in the *Additional Google Cloud steps* section.

After the preceding steps, the two clusters are ready for the next steps to install Istio. Perform the following steps to install Istio on both clusters:

1. Install Istio using `01-Cluster1.yaml`. The setup for the primary `cluster1` is the same as other architectures:

```
% istioctl install --context="${CTX_CLUSTER1}"
-f  "Chapter08/01-Cluster1.yaml"

This will install the Istio 1.16.0 default profile with
["Istio core" "Istiod" "Ingress gateways"] components
into the cluster. Proceed? (y/N) y
✓ Istio core installed
✓ Istiod installed
✓ Ingress gateways installed
✓ Installation complete
             Making this installation the default for
injection and validation.
```

2. For `cluster2`, we will be using the default profile, which will install istiod and the Ingress gateway. As `cluster2` is sharing the network with `cluster1`, we will use `cluster2` and `network1` values for the `clusterName` and `network` parameters, respectively:

```
apiVersion: install.istio.io/v1alpha1
kind: IstioOperator
spec:
  values:
    global:
      meshID: mesh1
      multiCluster:
        clusterName: cluster2
      network: network1
```

The sample file is available in `Chapter08/04-Cluster2.yaml`. Install Istio using the sample file using the following commands:

```
% istioctl install --context="${CTX_CLUSTER2}" -f
"Chapter08/04-Cluster2.yaml"
This will install the Istio 1.16.0 default profile with
["Istio core" "Istiod" "Ingress gateways"] components
into the cluster. Proceed? (y/N) y
✓ Istio core installed
```

```
✓ Istiod installed
✓ Ingress gateways installed
✓ Installation complete
```

3. Create a remote Secret so that the `cluster1` control plane can access the Kubernetes API server in `cluster2`:

```
% istioctl x create-remote-secret --context="${CTX_
CLUSTER2}" --name=cluster2 | kubectl apply -f -
--context="${CTX_CLUSTER1}"
secret/istio-remote-secret-cluster2 created
```

4. Create a remote Secret so that the `cluster2` control plane can access the Kubernetes API server in `cluster1`:

```
% istioctl x create-remote-secret --context="${CTX_
CLUSTER1}" --name=cluster1 | kubectl apply -f -
--context="${CTX_CLUSTER2}"
secret/istio-remote-secret-cluster1 created
```

Next, we will test the setup by deploying the Envoy dummy application as we did in previous sections. To install the `envoydummy` app, follow *steps 1–4* in the *Deploying the Envoy dummy application* section under the *Primary-remote on multi-network* section. Similarly, follow *steps 5–6* to perform the testing. The following code block demonstrates the distribution of traffic across both clusters:

```
% for i in {1..5}; do curl -Hhost:mockshop.com -s
"http://34.129.4.32";echo '\n'; done
V2----------Bootstrap Service Mesh Implementation with
Istio----------V2
Bootstrap Service Mesh Implementation with Istio
V2----------Bootstrap Service Mesh Implementation with
Istio----------V2
Bootstrap Service Mesh Implementation with Istio
V2----------Bootstrap Service Mesh Implementation with
Istio----------V2
```

5. Also, test the dummy application by shutting down istiod in the primary cluster and redeploying the application to verify whether mesh operations are uninterrupted even if one of the primary cluster's control planes is not available.

This concludes the setup of multi-primary on separate networks. Multi-primary on a shared network is arguably the simplest Istio setup, which doesn't need an east-west gateway to coordinate traffic between various Kubernetes clusters.

> **Reminder**
>
> Delete the clusters and firewall rule. The following is an example of how you can delete clusters. Please change the parameter values as per your configuration:

```
% gcloud container clusters delete primary1-cluster --zone
"australia-southeast2"
% gcloud container clusters delete primary2-cluster --zone
"australia-southeast1-a"
% gcloud firewall delete primary1-primary2-shared-network
```

Summary

This chapter was very hands-on, but we hope you learned how to set up Istio in various cluster configurations. Every section used two clusters as an example to demonstrate the setup, but we would recommend that you extend each of the examples by adding more clusters. Practice various scenarios by deploying an Envoy dummy application and `curl` Pod from the utilities namespace and then applying virtual and destination rules and testing the behavior of the services in a multi-cluster environment. Practice east-west traffic scenarios by configuring the east-west gateway to only be accessible cross-cluster and see how that plays out using the instructions in this chapter.

Although we discussed four deployment options, choosing the right deployment model depends on specific use cases, and you should consider your underlying infrastructure provider, application isolation, network boundaries, and service-level agreement requirements to consider which architecture is the best fit for you. By having multi-cluster deployments, you get better availability of the Service Mesh and restricted fault boundaries so that an outage doesn't bring down the whole cluster. In an enterprise environment, multiple teams working together may need to isolate data planes but it might be OK to have a shared control plane to save operation costs. In that case, multi-cluster environments such as primary-remote provide isolation and centralized control.

In the next chapter, we will read about web assembly and how it can be used to extend Istio data planes.

9
Extending Istio Data Plane

Istio provides various APIs to manage data plane traffic. There is one API called `EnvoyFilter` that we have not yet used. The `EnvoyFilter` API provides a means to customize the istio-proxy configuration generated by the Istio control plane. Using the `EnvoyFilter` API, you can directly use Envoy filters even if they are not directly supported by Istio APIs.

There is another API called `WasmPlugins`, which is another mechanism to extend the istio-proxy functionality **WebAssembly** (**Wasm**) support is becoming common for proxies such as Envoy to enable developers to build extensions.

In this chapter, we will discuss these two topics; however, the content on `EnvoyFilter` will be brief, as you have already learned about filters and plugins for Envoy in *Chapter 3*. Rather, we will focus on how to invoke Envoy plugins from Istio configurations. However, we will delve deeper into Wasm with hands-on activities as usual.

In this chapter we will be covering the following topics:

- Why extensibility?
- Customizing the data plane using EnvoyFilter
- Understanding the fundamentals of Wasm
- Extending the Istio data plane using Wasm

Technical requirements

To keep it simple, we will be using minikube to perform the hands-on exercises in this chapter. By now, you must be familiar with installing and configuring minikube, and if not, please refer to the *Technical requirements* section of *Chapter 4*.

In addition to minikube, it is good to have Go and TinyGo installed on your workstation. If you are new to Go, then follow the instructions at `https://go.dev/doc/install` to install it. Install TinyGo for your host OS by following the instructions at `https://tinygo.org/getting-started/install/macos/`. Then validate the installation by using the following command:

```
% tinygo version
tinygo version 0.26.0 darwin/amd64 (using go version go1.18.5
and LLVM version 14.0.0)
```

Why extensibility

As with any good architecture, **extensibility** is very important because there is no *one size fits all* approach to technology that can adapt to every application. Extensibility is important in Istio as it provides options to users to build corner cases and extend Istio as per their individual needs. In the early days of Istio and Envoy, the projects took different approaches to build extensibility. Istio took the approach of building a generic out-of-process extension model called **Mixer** (`https://istio.io/v1.6/docs/reference/config/policy-and-telemetry/mixer-overview/`), whereas Envoy focused on in-proxy extensions (`https://www.envoyproxy.io/docs/envoy/latest/extending/extending`). Mixer is now deprecated; it was a plugin-based implementation used for building extensions (also called adaptors) for various infrastructure backends. Some examples of adapters are Bluemix, AWS, Prometheus, Datadog, and SolarWinds. These adapters allowed Istio to interface with various kinds of backend systems for logging, monitoring, and telemetry, but the adapter-based extension model suffered from significant resource inefficiencies that impacted tail latencies and resource utilization. This model was also intrinsically limited and had limited application. The Envoy extension approach required users to write filters in C++, which is also Envoy's native language. Extensions written in C++ are then packaged along with Envoy's code base, compiled, and tested to make sure that they are working as expected. The in-proxy extension approach for Envoy imposed a constraint of writing extensions in C++ followed by a monolithic build process and the fact that you must now maintain the Envoy code base yourself. Some bigger organizations were able to manage their own copy of the Envoy code base, but most of the Envoy community found this approach impractical. So, Envoy adopted other approaches for building extensions, one being **Lua-based filters** and the other being Wasm extensions. In Lua-based extensions, users can write inline Lua code in an existing Envoy HTTP Lua filter. The following is an example of a Lua filter; the Lua script has been highlighted:

```
http_filters:
name: envoy.filters.http.lua
typed_config:
  "@type": type.googleapis.com/envoy.extensions.filters.http.
lua.v3.Lua
  default_source_code:
    inline_string: |
```

```
    function envoy_on_request(request_handle)
...... -- Do something on the request path.
      request_handle:headers():add("NewHeader", "XYZ")
    end
    function envoy_on_response(response_handle)
      -- Do something on the response path.
     response_handle:logInfo("Log something")
    response_handle:headers:add("response_size",response_
handle:body():length())
     response_handle:headers:remove("proxy")
   end
```

In this example, we are using the **HTTP Lua filter**. The HTTP Lua filter allows Lua scripts to be run during both the request and response cycle. Envoy runs the Lua script as a coroutine; LuaJIT is used as the Lua runtime environment and is allocated per Envoy worker thread. The Lua scripts should contain the `envoy_on_request` and/or `envoy_on_response` functions, which are then executed as coroutines on the request and response cycles, respectively. You can write Lua code in these functions to perform the following during request/response processing:

- Inspection and modification of headers, body, and trailers of request and response flows
- Asynchronous HTTP invocation of upstream systems
- Performing direct response and skipping further filter iteration

You can read more about Envoy HTTP Lua filters at `https://www.envoyproxy.io/docs/envoy/latest/configuration/http/http_filters/lua_filter.html?highlight=lua%20filter`. This approach is great for simple logic, but when writing complex processing instructions then writing inline Lua code is not practical. Inline code cannot be easily shared with other developers or easily aligned with best practices of software programming. The other drawback is the lack of flexibility, as developers are obliged to only use Lua, which inhibits non-Lua developers from writing these extensions.

To provide extensibility to Istio, an approach that imposed fewer tradeoffs was needed. As Istio's data plane comprises Envoy, it made sense to converge on a common approach for extensibility for Envoy and Istio. This can decouple Envoy releases from their extension ecosystem, enables Istio consumers to build data plane extensions using their languages of choice, using best-of-breed programming languages and practices, and then deploy these extensions without causing any downtime risk to their Istio deployments in production. Based on this common effort, Wasm support for Istio was introduced. In the upcoming sections, we will discuss Wasm. But before that, let's quickly touch on Istio support for running Envoy filters in the next section.

Customizing the data plane using Envoy Filter

Istio provides an `EnvoyFilter` API, which provides options to modify configurations created via other Istio **custom resource definitions** (**CRDs**). Essentially one of the functions performed by Istio is translating high-level Istio CRDs into low-level Envoy configurations. Using the `EnvoyFilter` CRD, you can change those low-level configurations directly. This is a very powerful feature but also should be used cautiously as it has the potential to make things worse if not used correctly. Using `EnvoyFilter`, you can apply configurations that are not directly available in Istio CRDs and perform more advanced Envoy functions. The filter can be applied at the namespace level as well as selective workload levels identified by labels.

Let's try to understand this further via an example.

We will pick one of the hands-on exercises we performed in *Chapter 7* to route a request to `hhtppbin.org`. Do not forget to create the `Chapter09` folder and turn on `istio-injection`. The following commands will deploy the `httpbin` Pod as described in `Chapter09/01-httpbin-deployment.yaml`:

```
kubectl apply -f Chapter09/01-httpbin-deployment.yaml
curl -H "Host:httpbin.org" http://
a816bb2638a5e4a8c990ce790b47d429-1565783620.us-east-1.elb.
amazonaws.com/get
```

Carefully check all the response fields containing all the headers passed in the request.

Using `EnvoyFilter`, we will add a custom header to the request before sending it to the `httpbin` Pod. For this example, let's pick the `ChapterName` header name and set its value to `ExtendingIstioDataPlane`. The configuration in `Chapter09/02-httpbinenvoyfilter-httpbin.yaml` adds the custom header to the request.

Apply the following configuration using `EnvoyFilter`:

```
$ kubectl apply -f Chapter09/02-httpbinenvoyfilter-httpbin.yaml
envoyfilter.networking.istio.io/updateheaderhorhttpbin
configured
```

Let's go through `Chapter09/02-httpbinenvoyfilter-httpbin.yaml` in two parts:

```
apiVersion: networking.istio.io/v1alpha3
kind: EnvoyFilter
metadata:
  name: updateheaderforhttpbin
  namespace: chapter09
```

```
spec:
  workloadSelector:
    labels:
      app: httpbin
  configPatches:
  - applyTo: HTTP_FILTER
    match:
      context: SIDECAR_INBOUND
      listener:
        portNumber: 80
        filterChain:
          filter:
            name: "envoy.filters.network.http_connection_
manager"
            subFilter:
              name: "envoy.filters.http.router"
```

In this part, we will create an EnvoyFilter named `updateheaderforhttpbin` in the `chapter09` namespace, which will be applied to the workload which has the `app` label with a `httpbin` value. For that configuration, we are applying a configuration patch to all inbound traffic to the Istio sidecar aka istio-proxy aka Envoy for port `80` of the `httpbin` Pod. The configuration patch is applied to `HTTP_FILTER` and, in particular, to the HTTP router filter of the `http_connection_manager` network filter.

In the next part of the EnvoyFilter configuration, we apply configuration before the existing route configuration and, in particular, we are appending a Lua filter with inline code as specified in the `inlineCode` section. The Lua code runs during the `envoy_on_request` phase and adds a request header with the `X-ChapterName` name and the `ExtendingIstioDataPlane` value:

```
  patch:
      operation: INSERT_BEFORE
      value:
        name: envoy.lua
        typed_config:
          "@type": "type.googleapis.com/envoy.extensions.
filters.http.lua.v3.Lua"
          inlineCode: |
            function envoy_on_request(request_handle)
              request_handle:logInfo(" ========= XXXXX
==========");
```

```
                request_handle:headers():add("X-ChapterName",
"ExtendingIstioDataPlane");
              end
```

Now, go ahead and test the endpoint using the following command:

```
% curl -H "Host:httpbin.org" http://
a816bb2638a5e4a8c990ce790b47d429-1565783620.us-east-1.elb.
amazonaws.com/get
```

You will receive the added headers in the response.

You can see the final Envoy config applied using the following commands. To find the exact name of the `httpbin` Pod, you can make use of `proxy-status`:

```
% istioctl proxy-status | grep httpbin
httpbin-7bffdcffd-152sh.chapter09
Kubernetes     SYNCED      SYNCED      SYNCED      SYNCED
     NOT SENT     istiod-56fd889679-ltxg5     1.14.3
```

This is followed by the `proxy-config` details for listeners:

```
% istioctl proxy-config listener httpbin-7bffdcffd-152sh.
chapter09  -o json
```

In the output, look for `envoy.lua`, which is the name of the patch and the filter we applied via the config. In the output, look for `filterChainMatch` and for `destinationPort` set to `80`:

```
"filterChainMatch": {
                    "destinationPort": 80,
                    "transportProtocol": "raw_buffer"
                },
```

We applied the config via EnvoyFilter:

```
  {
                                    "name": "envoy.lua",
                                    "typedConfig": {
                                        "@type": "type.
googleapis.com/envoy.extensions.filters.http.lua.v3.Lua",
                                        "inlineCode":
"function envoy_on_request(request_handle)\n  request_
handle:logInfo(\" ========= XXXXX ==========\");\n  request_
```

```
handle:headers():add(\"X-ChapterName\",
\"ExtendingIstioDataPlane\");\nend \n"
                    }
                }
```

Hopefully, that gave you an idea of EnvoyFilter and how the overall mechanism works. In the hands-on exercise for this chapter, another example applies the same changes but at the Ingress gateway level. You can find the example at `Chapter09/03-httpbinenvoyfilter-httpbiningress.yaml`. Make sure that you delete the `Chapter09/02-httpbinenvoyfilter-httpbin.yaml` file before applying the Ingress gateway changes.

For more details about the various configurations of EnvoyFilter, please refer to the Istio documentation at `https://istio.io/latest/docs/reference/config/networking/envoy-filter/#EnvoyFilter-EnvoyConfigObjectPatch`.

> **Important note**
>
> For cleanup, use this command: `kubectl delete ns chapter09`.

In the next section, we will read about Wasm fundamentals, followed by how to use Wasm to extend the Istio data plane.

Understanding the fundamentals of Wasm

Wasm is a portable binary format designed to run on **virtual machines (VMs)**, allowing it to run on various computer hardware and digital devices, and is very actively used to improve the performance of web applications. It is a virtual **instruction set architecture (ISA)** for a stack machine designed to be portable, compact, and secure with a smaller binary file size to reduce download times when executed on web browsers. A modern browser's JavaScript engines can parse and download the Wasm binary format in order of magnitude faster than JavaScript. All major browser vendors have adopted Wasm, and as per the Mozilla Foundation, Wasm code runs between 10% and 800% faster than the equivalent JavaScript code. It provides faster startup time and higher peak performance without memory bloat.

Wasm is also a preferred and practical choice for building extensions for Envoy for the following reasons:

- Wasm extensions can be delivered at runtime without needing to restart istio-proxy. Furthermore, the extension can be loaded to istio-proxy through various means without needing any changes to istio-proxy. This allows the delivery of changes to the extension and changes to proxy behavior in the form of extensions without any outages.

- Isolated from the host and executed in a sandbox/VM environment, Wasm communicates with the host machine via an **application binary interface** (**ABI**). Through ABIs, we can control what can and cannot be modified and what is visible to the extension.
- Another benefit of running Wasm in a sandbox environment is the isolation and defined fault boundaries. If anything goes wrong with Wasm execution, then the scope of disruption is limited to the sandbox and won't spread to the host process.

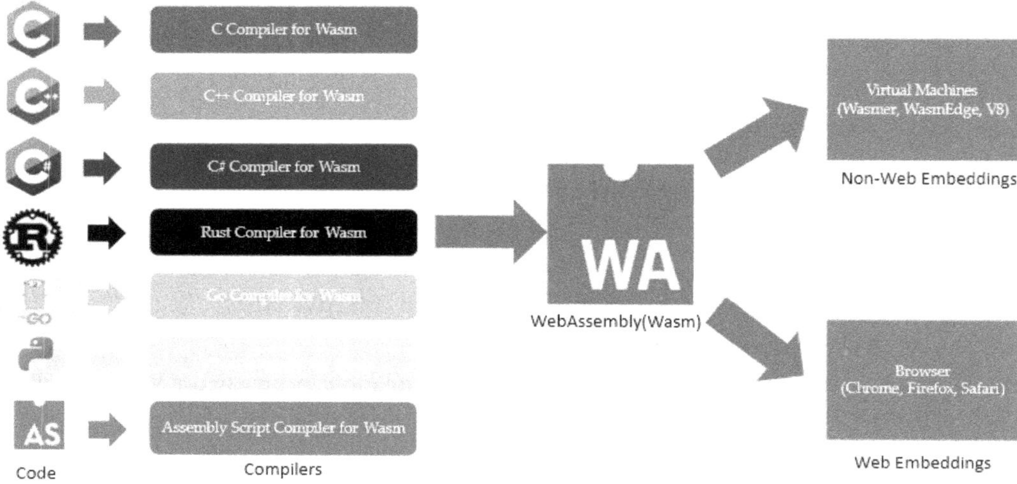

Figure 9.1 – An overview of Wasm

There are over thirty programming languages that support compilation to Wasm modules. Some examples are C, Java, Go, Rust, C++, and TypeScript. This allows most developers to build Istio extensions using the programming language of their choice.

To get familiar with Wasm, we will build a sample application using Go. The source code is available in the `Chapter09/go-Wasm-example` folder.

The problem statement is to build an HTML page that takes a string in lowercase and provides the output in uppercase. We assume that you have some experience working with Go and that it is installed in your hands-on environment. If you don't want to use Go, then try implementing the example using the language of your choice:

1. Copy the code from `Chapter09/go-Wasm-example` and reinitialize the Go module:

```
% go mod init Bootstrap-Service-Mesh-Implementations-
with-Istio/Chapter09/go-Wasm-example
% go mod tidy
```

First, let's check `Chapter09/go-Wasm-example/cmd/Wasm/main.go`:

```
package main
import (
    "strings"
    "syscall/js"
)
func main() {
    done := make(chan struct{}, 0)
    js.Global().Set("WasmHash", js.FuncOf(convertToUpper))
    <-done
}
func convertToUpper(this js.Value, args []js.Value)
interface{} {
    strings.ToUpper(args[0].String())
    return strings.ToUpper(args[0].String())
}
```

`done := make(chan struct{}, 0)` and `<-done` is a Go channel. A Go channel is used for communication between concurrent functions.

`js.Global().Set("WasmHash", hash)` exposes the Go hash function to JavaScript.

The `convertToUpper` function takes a string as an argument, which is then typecasted using the `.String()` function from the `syscall/js` package. The `strings.ToUpper(args[0].String())` line converts all arguments provided by JavaScript into an uppercase string and returns it as output of the function.

2. The next step is to compile `Chapter09/go-Wasm-example/cmd/Wasm/main.go` using the following command:

   ```
   % GOOS=js GOARCH=Wasm go build -o static/main.Wasm cmd/
   Wasm/main.go
   ```

   The secret recipe here is `GOOS=js GOARCH=Wasm`, which tells the Go compiler to compile for JavaScript as the target host and Wasm as the target architecture. Without this, the Go compiler will compile for the target OS and architecture as per your workstation specifications. You can find more about the possible values of `GOOS` and `GOARCH` at `https://gist.github.com/asukakenji/f15ba7e588ac42795f421b48b8aede63`.

   The command will then produce the Wasm file with the `main.Wasm` name in the static folder.

3. We also need to fetch and execute Wasm in the browser. Luckily, Go makes that possible with `Wasm_exec.js`.

   The JavaScript file can be found in the `GOROOT` folder. To copy it to the static directory, use the following command:

   ```
   % cp "$(go env GOROOT)/misc/Wasm/Wasm_exec.js" ./static
   ```

4. We have Wasm and JavaScript to load and execute Wasm in the browser. We need to create an HTML page and then load JavaScript from there. You will find the sample HTML page at `Chapter09/go-Wasm-example/static/index.html`. You will find the following snippet in the HTML to load JavaScript and instantiate Wasm:

   ```
   <script src="Wasm_exec.js"></script>
   <script>
       const go = new Go();
       WebAssembly.instantiateStreaming(fetch("main.Wasm"),
   go.importObject).then((result) => {
           go.run(result.instance);
       });
   </script>
   ```

5. As the last step, we need a web server. You can use `nginx` or a sample HTTP server package with the sample code at `Chapter09/go-Wasm-example/cmd/webserver/main.go`. Run the server using the following command:

   ```
   % go run ./cmd/webserver/main.go
   Listening on http://localhost:3000/index.html
   ```

6. Open `http://localhost:3000/index.html` in a browser and test that whatever lowercase letters you type in the text box are converted to uppercase:

Figure 9.2 – Go used to create Wasm

This concludes the introduction to Wasm, and I hope you have acquired a basic understanding of Wasm after reading this section. In the next section, we will learn about how Wasm helps to extend the Istio data plane.

Extending the Istio data plane using Wasm

The main goal for Wasm was to enable high-performance applications on web pages, and hence Wasm was originally designed for execution in web browsers. There is a **World Wide Web Consortium** (**W3C**) working group for Wasm, whose details are available at `https://www.w3.org/Wasm/`. The working group manages the Wasm specification available at `https://www.w3.org/TR/Wasm-core-1/` and `https://www.w3.org/TR/Wasm-core-2/`. Most internet browsers have implemented the specification, and you can find details for Google Chrome at `https://chromestatus.com/feature/5453022515691520`. Mozilla Foundation also maintains browser compatibility at `https://developer.mozilla.org/en-US/docs/WebAssembly#browser_compatibility`. When it comes to supporting the execution of Wasm on layer 4 and 7 proxies, most of the effort is recent. When executing Wasm on proxies, we need a way to communicate with the host environment. Similar to how web browsers are developed, Wasm should be written once, after which it should be able to run on any proxy.

Introducing Proxy-Wasm

For Wasm to communicate with the host environment and the development of Wasm to be agnostic of the underlying host environment, there is a `Proxy-Wasm` specification, also known as Wasm for proxies. The specification is made up of `Proxy-Wasm` ABIs, which are low-level. The specification is then abstracted in high-level languages, called `Proxy-Wasm` **software development kits** (**SDKs**), which are developer friendly and easy to understand and integrate with high-level language implementations. Every proxy also then implements a `Proxy-Wasm` ABI specification in the form of the `Proxy-Wasm` modules.

The concepts of `Proxy-Wasm` can be difficult to understand. To make it easy to digest them, let's break them down into the following sections and go through them one by one.

Proxy-Wasm ABI

ABI is a low-level interface specification that describes how Wasm communicates with the VM and host. The specification details are available at `https://github.com/proxy-Wasm/spec/blob/master/abi-versions/vNEXT/README.md`, and the specification itself is available at `https://github.com/proxy-Wasm/spec`. To understand the API, it is best to go through some of the most commonly used methods of the ABI specification to appreciate what it does:

- `_start`: This function needs to be implemented on Wasm and will be called when Wasm is loaded and initialized.
- `proxy_on_vm_start`: This is called when the host machine starts the Wasm VM. Wasm can use this method to retrieve any configuration details of the VM.
- `proxy_on_configure`: This is called when the host environment starts the plugin, which loads Wasm. Using this method, Wasm can retrieve any plugin-related configuration.

- `proxy_on_new_connection`: This is a level 4 extension that is called when a TCP connection is established between the proxy and the client.
- `proxy_on_downstream_data`: This is a level 4 extension that is called for each data chunk received from the client.
- `proxy_on_downstream_close`: This is a level 4 extension that is called when the connection with downstream is closed.
- `proxy_on_upstream_data`: This is a level 4 extension that is called for each data chunk received from upstream.
- `proxy_on_upstream_close`: This is a level 4 extension that is called when the connection with upstream is closed.
- `proxy_on_http_request_headers`: This is a level 7 extension that is called when HTTP request headers are received from the client.
- `proxy_on_http_request_body`: This is a level 7 extension that is called when the HTTP request body is received from the client.
- `proxy_on_http_response_headers`: This is a level 7 extension that is called when HTTP response headers are received from upstream.
- `proxy_on_http_response_body`: This is a level 7 extension that is called when the HTTP response body is received from upstream.
- `proxy_send_http_response`: This is also a level 7 extension that is implemented in the host environment, Envoy. Using this method, Wasm can instruct Envoy to send an HTTP response without actually calling the upstream services.

This list doesn't cover all methods in the ABI, but we hope it gave you a good understanding of what the ABI is used for. The following diagram illustrates what we covered in this section:

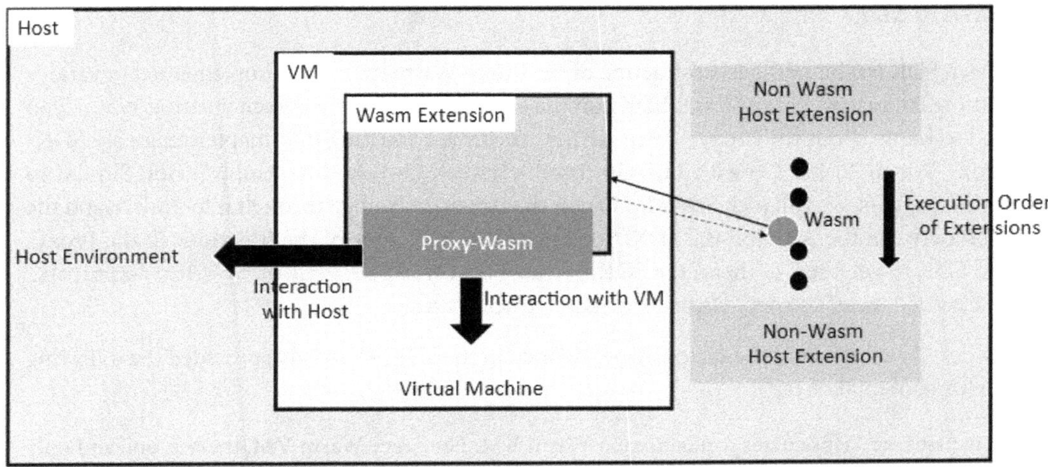

Figure 9.3 – Proxy-wasm ABI

If we analyze this diagram in the context of Envoy, we arrive at the following interpretation:

- Native extensions execute in the order specified in the configuration.
- There is also a native extension in Envoy for loading Wasm, specified at `https://www.envoyproxy.io/docs/envoy/latest/api-v3/extensions/Wasm/v3/Wasm.proto`. The extension is responsible for loading and asking Envoy to execute Wasm.
- Envoy executes Wasm on a VM.
- During execution, Wasm can interact with the request, VM, and Envoy via the `Proxy-Wasm` ABI, and we saw some of those interaction points earlier in the section.
- Once Wasm completes execution, the execution flows back to other native extensions defined in the configuration file.

While ABIs are elaborate, they are also very low-level and not programmer-friendly, who usually prefer writing code in high-level programming languages. In the following section, we will read about how the Proxy-Wasm SDK can solve this problem.

Proxy-Wasm SDK

Proxy-Wasm SDK is a higher-level abstraction of the Proxy-Wasm ABI and is implemented in various programming languages. Proxy-Wasm SDK complies with the ABI so that when creating Wasm, you don't need to know about the Proxy-Wasm ABI. At the time of writing this chapter, there are SDKs of the Proxy-Wasm API in Go with TinyGo compiler, Rust, C++, and AssemblyScript. Similar to what we did for ABIs, we will pick SDKs for one of the languages and go through it to understand the correlation between the ABI and the SDK. So, let's go through some of the functions in the Proxy-Wasm Go SDK to get a feel of them; the SDK is available at `https://pkg.go.dev/github.com/tetratelabs/proxy-Wasm-go-SDK/proxyWasm`.

First, you need to understand the various types defined in the SDK, so we have provided the following list of the fundamental ones:

- `VMContext`: This corresponds to each Wasm VM. For every Wasm VM, there is one and only one `VMContext`. `VMContext` has the following methods:
  - `OnVMStart(vmConfigurationSize int) OnVMStartStatus`: This method is called when the VM is created. From within this method, Wasm can retrieve the VM configuration.
  - `NewPluginContext(contextID uint32) PluginContext`: This creates a plugin context for each plugin configuration.
- `PluginContext`: This corresponds to each plugin configuration in the host. Plugins are configured at HTTP or network filters for listeners. Some of the methods in `PluginContext` are as follows:
  - `OnPluginStart(pluginConfigurationSize int) OnPluginStartStatus`: This is called for all plugins configured. Once the VM has been created, Wasm can retrieve the plugin configuration using this method.
  - `OnPluginDone() bool`: This is called when the host deletes `PluginContext`. If this method returns `true`, it signals to the host that `PluginContext` can be deleted, and `false` means that the plugin is in a pending state and cannot yet be deleted.
  - `NewTcpContext(contextID uint32) TcpContext`: This method creates `TCPContext`, corresponding to every TCP request.
  - `NewHttpContext(contextID uint32) HttpContext`: This method creates `HTTPContext`, corresponding to every HTTP request.
- `HTTPContext`: This method is created by `PluginContext` for every HTTP stream. The following are some of the methods available in this interface:
  - `OnHttpRequestHeaders(numHeaders int, endOfStream bool) Action`: This method provides access to HTTP headers as part of the request stream.

- `OnHttpRequestBody(bodySize int, endOfStream bool) Action`: This method provides access to data frames of the request body. It is called multiple times for every individual data frame in the request body.
- `OnHttpResponseHeaders(numHeaders int, endOfStream bool) Action`: This method provides access to response headers.
- `OnHttpResponseBody(bodySize int, endOfStream bool) Action`: This method provides access to response body frames.
- `OnHttpStreamDone()`: This method is called before the deletion of `HTTPContext`. From this method, Wasm can access all information about the request and response phases of the HTTP connection.

Among other types to read about is `TCPContext`. We have not covered all methods and types available in the SDK; you can find the complete list along with details at `https://pkg.go.dev/github.com/tetratelabs/proxy-Wasm-go-SDK@v0.20.0/proxyWasm/types#pkg-types`.

With this overview in mind, let's write a Wasm to inject a custom header in the response of the `envoydummy` Pod. Please note that in the *Customizing the data plane using the Envoy filter* section, we used EnvoyFilter to patch Istio and applied a Lua filter with inline code to inject headers to requests bound for the `httpbin` Pod.

Create the `chapter09-temp` namespace with `istio-injection` disabled:

```
% kubectl create ns chapter09-temp
namespace/chapter09-temp created
```

Run `envoydummy` to check that it is working as expected:

```
% kubectl apply -f Chapter09/01-envoy-dummy.yaml
namespace/chapter09-temp created
service/envoydummy created
configmap/envoy-dummy-2 created
deployment.apps/envoydummy-2 created
```

Forward the ports so that you can test locally:

```
% kubectl port-forward svc/envoydummy 18000:80 -n chapter09-t
emp
Forwarding from 127.0.0.1:18000 -> 10000
```

Then, test the endpoint:

```
% curl  localhost:18000
V2----------Bootstrap Service Mesh Implementation with Istio--
--------V2%
```

So, we have verified that envoydummy is working. The next step is to create Wasm to inject headers into the response. You will find the source code at Chapter09/go_Wasm_example_for_envoy.

There is only one main.go file in the Go module, and the following are the key parts of the code:

The entry point in the Go module is the main method. In the main method, we are setting up the Wasm VM by calling SetVMContext. The method is described in the Entrypoint.go file at https://github.com/tetratelabs/proxy-wasm-go-sdk/tree/main/proxywasm. The following code snippet shows the main method:

```
func main() {
    proxyWasm.SetVMContext(&vmContext{})
}
```

The following method injects a header into the response headers:

```
func (ctx *httpHeaders) OnHttpResponseHeaders(numHeaders int,
endOfStream bool) types.Action {
    if err := proxyWasm.AddHttpResponseHeader("X-ChapterName",
"ExtendingEnvoy"); err != nil {
        proxyWasm.LogCritical("failed to set response header:
X-ChapterName")
    }
    return types.ActionContinue
}
```

Also, notice AddHttpResponseHeader, which is defined at https://github.com/tetratelabs/proxy-Wasm-go-SDK/blob/v0.20.0/proxyWasm/hostcall.go#L395.

The next step is to compile the Go module for Wasm, for which we will need to use TinyGo. Please note that we cannot use the standard Go compiler due to a lack of support for the Proxy-Wasm Go SDK.

Install TinyGo for your host OS by following the instructions at https://tinygo.org/getting-started/install/macos/.

Using TinyGo, compile the Go module with Wasm using the following command:

```
% tinygo build -o main.Wasm -scheduler=none -target=wasi main.
go
```

Once the Wasm file is created, we need to load the Wasm file into `configmap`:

```
% kubectl create configmap 01-Wasm --from-file=main.Wasm -n chapter09-temp
configmap/01-Wasm created
```

Modify the `envoy.yaml` file to apply Wasm filters and load Wasm from `configmap`:

```
http_filters:
              - name: envoy.filters.http.Wasm
                typed_config:
                  "@type": type.googleapis.com/udpa.type.v1.TypedStruct
                  type_url: type.googleapis.com/envoy.extensions.filters.http.Wasm.v3.Wasm
                  value:
                    config:
                      vm_config:
                        runtime: "envoy.Wasm.runtime.v8"
                        code:
                          local:
                            filename: "/Wasm2/main.Wasm"
              - name: envoy.filters.http.router
                typed_config:
                  "@type": type.googleapis.com/envoy.extensions.filters.http.router.v3.Router
```

We specify `envoy` in the config to use the `v8` runtime for running Wasm. The changes are also available at `Chapter09/02-envoy-dummy.yaml`. Apply the changes, as shown here:

```
% kubectl apply -f Chapter09/02-envoy-dummy.yaml
service/envoydummy created
configmap/envoy-dummy-2 created
deployment.apps/envoydummy-2 created
```

Forward the port `80` to `18000`:

```
% kubectl port-forward svc/envoydummy 18000:80 -n chapter09-temp
```

Test the endpoint to check whether Wasm injected the response header:

```
% curl -v localhost:18000
* Mark bundle as not supporting multiuse
< HTTP/1.1 200 OK
< content-length: 72
< content-type: text/plain
< x-chaptername: ExtendingEnvoy
* Connection #0 to host localhost left intact
V2----------Bootstrap Service Mesh Implementation with Istio--
--------V2%
```

Hopefully, this section gave you confidence on how to create Wasm that is compliant with Proxy-Wasm and how to apply it to Envoy. We suggest you do more hands-on exercises by looking at examples available at `https://github.com/tetratelabs/proxy-Wasm-go-SDK/tree/main/examples`.

Before we conclude this section, let's also check how Wasm is compliant with the Proxy-Wasm ABI. For that, we will install the **Wasm Binary Toolkit** (**WABT**) available at `https://github.com/WebAssembly/wabt`. On MacOS, it is simple to install using `brew`:

```
% brew install wabt
```

WABT provides various methods to manipulate and introspect Wasm. One such tool, `Wasm-objdump`, prints information about a Wasm binary. Using the following command, you can print a list of all functions that become accessible to the host environment once Wasm has been instantiated:

```
% Wasm-objdump main.Wasm --section=export -x.
```

You will notice the output is a list of functions that are defined in the Proxy-Wasm ABI.

Important note

To do the cleanup, you can use the following command:

```
% kubectl delete ns chapter09-temp
```

That completes the section on Proxy-Wasm, and we hope you now understand how to create Proxy-Wasm-compliant Wasm using the Go SDK. In the next section, we will deploy Wasm in Istio.

Wasm with Istio

In this section, we will extend the Istio data plane using Wasm that we built in the previous section. We will be using Istio's **WasmPlugin API**, and we will go into the details of this plugin once we have configured it for the `httpbin` application:

1. The first step is to upload `main.Wasm` created in the Go module available at `Chapter09/go_Wasm_example_for_envoy` to an HTTPS location. You can make use of AWS S3 or something similar for that purpose; another option is to use an OCI registry such as **Docker Hub**. To complete this exercise, I uploaded `main.Wasm` to AWS S3. The HTTPS location of the S3 bucket hosting the file is `https://anand-temp.s3.amazonaws.com/main.Wasm`. Please note that for security reasons, the link might not be accessible to you while reading this book, but I am sure you can manage to create your own S3 buckets or Docker registry.
2. The second step is to deploy the `httpbin` application, which is already available at `Chapter09/01-httpbin-deployment.yaml`:

   ```
   % kubectl apply -f Chapter09/01-httpbin-deployment.yaml
   ```

   Check the response of the following commands and observe the headers added during the request:

   ```
   % curl -H "Host:httpbin.org" http://
   a816bb2638a5e4a8c990ce790b47d429-1565783620.us-east-1.
   elb.amazonaws.com/get
   ```

3. After this, we will apply the following changes using `WasmPlugin`:

   ```
   apiVersion: extensions.istio.io/v1alpha1
   kind: WasmPlugin
   metadata:
     name: addheaders
     namespace: chapter09
   spec:
     selector:
       matchLabels:
         app: httpbin
     url: https://anand-temp.s3.amazonaws.com/main.Wasm
     imagePullPolicy: Always
     phase: AUTHZ
   ```

Apply `WasmPlugin` using the following command:

```
% kubectl apply -f Chapter09/01-Wasmplugin.yaml
Wasmplugin.extensions.istio.io/addheaders configured
```

We will read more about `WasmPlugin` after *step 5*. For now, let's check the response headers from `httpbin`:

```
% curl --head -H "Host:httpbin.org" http://
a816bb2638a5e4a8c990ce790b47d429-1565783620.us-east-1.
elb.amazonaws.com/get
```

You will notice that, as expected, we have `x-chaptername: ExtendingEnvoy` in the response.

4. Let's create another Wasm to add a custom header to `request` so that we can see it in the response of the `httpbin` payload. There is a Wasm already created in `Chapter09/go_Wasm_example_for_istio` for this purpose. Notice the `OnHTTPRequestHeaders` function in `main.go`:

```
func (ctx *httpHeaders) OnHttpRequestHeaders(numHeaders
int, endOfStream bool) types.Action {
    if err := proxyWasm.AddHttpRequestHeader("X-Chapter",
"Chapter09"); err != nil {
        proxyWasm.LogCritical("failed to set request
header: X-ChapterName")
    }
    proxyWasm.LogInfof("added custom header to request")
    return types.ActionContinue
}
```

Compile that into Wasm and copy it to the S3 location. There is also another Istio config file available at `Chapter09/02-Wasmplugin.yaml`, which deploys this Wasm:

```
apiVersion: extensions.istio.io/v1alpha1
kind: WasmPlugin
metadata:
  name: addheaderstorequest
  namespace: chapter09
spec:
  selector:
    matchLabels:
      app: httpbin
  url: https://anand-temp.s3.amazonaws.com/
```

```
AddRequestHeader.Wasm
  imagePullPolicy: Always
  phase: AUTHZ
```

5. After applying the changes, test the endpoints, and you will find that both Wasm have executed, adding a header in the response as well as one in the request, which is reflected in the `httpbin` response. The following is a shortened version of the response:

```
% curl -v -H "Host:httpbin.org" http://
a816bb2638a5e4a8c990ce790b47d429-1565783620.us-east-1.
elb.amazonaws.com/get
< HTTP/1.1 200 OK
......
< x-chaptername: ExtendingEnvoy
<
{
  "args": {},
  "headers": {
    "Accept": "*/*",
    "Host": "httpbin.org",
    "User-Agent": "curl/7.79.1",
.....,
    "X-Chapter": "Chapter09",
...
  },
  "origin": "10.10.10.216",
  "url": "http://httpbin.org/get"
}
```

In *steps 3* and *4*, we used `WasmPlugin` to apply Wasm on the Istio data plane. The following are the parameters we configured in `WasmPlugin`:

- `selector`: Specify the resource on which the Wasm should be applied in the `selector` field. It can be the Istio gateway and Kubernetes Pods. You provide labels that must match the workload on whose Envoy sidecar the Wasm configuration will be applied. In the examples we implemented, we applied the `app:httpbin` label, which corresponds to the `httpbin` Pod.
- `url`: This is the location where the Wasm file is available to download. We provided the HTTP location, but OCI locations are also supported. The default value is `oci://`, used for referencing OCI images. To reference file-based locations, use `file://`, which is used for referencing Wasm files present locally within the proxy container.

- `imagePullPolicy`: The possible values for this are the following:
  - `UNSPECIFIED_POLICY`: This is the same as `IfNotPresent` unless the URL points to an OCI image with the latest tag. In that case, this field will default to `Always`.
  - `Always`: We will always pull the latest version of an image from the location specified in the URL.
  - `IfNotPresent`: Use this to pull Wasm only if the requested version is unavailable locally.
- `phase`: The possible values for this are the following:
  - `UNSPECIFIED_PHASE`: This means the Wasm filter will be inserted at the end of the filter chain.
  - `AUTHN`: This inserts the plugin before the Istio authentication filters.
  - `AUTHZ`: This inserts the plugin between the authentication and authorization filters.
  - `STATS`: This inserts the plugin after the authorization filter but before the stats filter.

We have described the values we used in the example, but various fields can be configured in `WasmPlugin`; you can find the detailed list at `https://istio.io/latest/docs/reference/config/proxy_extensions/Wasm-plugin/#WasmPlugin`.

For production deployment, we definitely suggest you use the `sha256` field to ensure the integrity of the Wasm modules.

Istio provides a reliable, out-of-the-box distribution mechanism for Wasm by leveraging the xDS proxy inside istio-agent and Envoy's **Extension Configuration Discovery Service** (**ECDS**). Details about ECDS are available at `https://www.envoyproxy.io/docs/envoy/latest/configuration/overview/extension`.

After applying `WasmPlugin`, you can check the `istiod` logs for ECDS entries:

```
% kubectl logs istiod-56fd889679-ltxg5 -n istio-system
```

You will find log entries similar to the following:

```
10-18T12:02:03.075545Z      info ads  ECDS: PUSH for
node:httpbin-7bffdcffd-4zrhj.chapter09 resources:1 size:305B
```

Istio makes an ECDS call to istio-proxy about applying the `WasmPlugin`. The following diagram describes the process of applying Wasm via the ECDS API:

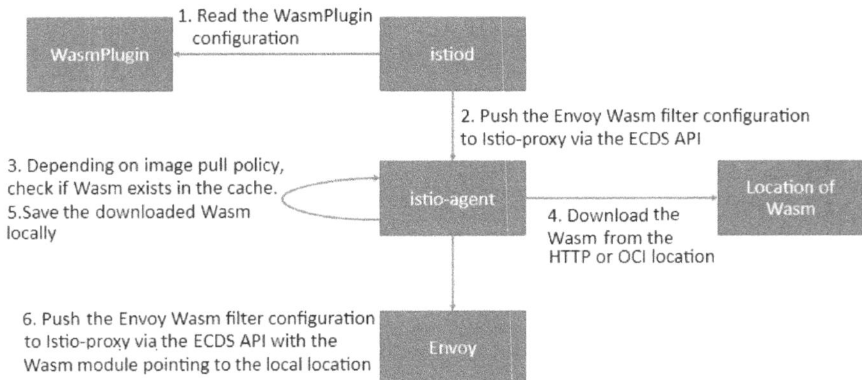

Figure 9.4 – Distributing Wasm to the Istio data plane

The istio-agent deployed alongside Envoy intercepts the ECDS call from `istiod`. It then downloads the Wasm module, saves it locally, and updates the ECDS configuration with the path of the downloaded Wasm module. If the WASM modules are not accessible to Istio-agent, it will reject the ECDS update. You will be able to see ECDS update failure in the `istiod` logs.

This concludes this section, and I hope it arms you with enough knowledge to start applying Wasm to your production workload.

Summary

In this chapter, we read about Wasm and its use. We learned about how Wasm is used on the web due to its high performance, and we also familiarized ourselves with how to build Wasm using Go and use it from a web browser using JavaScript. Wasm is also becoming a popular choice on the server side, especially among network proxies such as Envoy.

To get a standardized interface for implementing Wasm for proxies, there are the Proxy-Wasm ABI specifications which are low-level specifications describing the interface between Wasm and the proxy hosting the Wasm. Wasm for Envoy needs to be Proxy-Wasm compliant, but the Proxy-Wasm ABIs are difficult to work with; the Proxy-Wasm SDKs are much easier to work with. At the time of writing this chapter, there are many programming languages in which Proxy-Wasm SDK implementations are available, of which Rust, Go, C++, and AssemblyScript are among the most popular. We made use of the Envoy Wasm filter to configure a Wasm on an Envoy HTTP filter chain. We then built a few simple Wasm examples to manipulate request and response headers and deployed them on Istio using `WasmPlugin`. Wasm is not the only option to extend the Istio data plane, and there is another filter called EnvoyFilter, which can be used to apply the Envoy configuration as a patch on top of the Envoy configuration created by `Istiod`.

The next chapter is very interesting as we will learn about how to deploy an Istio Service Mesh for non-Kubernetes workloads.

10

Deploying Istio Service Mesh for Non-Kubernetes Workloads

Istio and Kubernetes are technologies that complement each other. Kubernetes solves the problem of managing distributed applications packaged as containers isolated from each other and deployed in a consistent environment with dedicated resources. Although Kubernetes solves container deployment, scheduling, and management, it doesn't solve traffic management between containers. Istio complements Kubernetes by providing traffic management capabilities, adding observability, and enforcing a zero-trust security model.

Istio is like a sidecar to Kubernetes; having said that, Kubernetes is a fairly new technology that got mainstream adoption approximately around 2017. From 2017 onward, most enterprises have used Kubernetes when building microservices and other cloud-native applications, but there are still many applications that are not built on Kubernetes and/or not migrated to Kubernetes; such applications are traditionally deployed on **virtual machines** (**VMs**). VMs are not just limited to traditional data centers but are also a mainstream offering from cloud providers. Organizations end up having this parallel universe of Kubernetes-based applications and VM-based applications deployed across the cloud and on-premises.

In this chapter, we will read about how Istio helps to marry these two worlds of legacy and modern technologies and how can you extend Service Mesh beyond Kubernetes. In this chapter, we will cover the following topics:

- Examining hybrid architecture
- Setting up a Service Mesh for hybrid architecture

Technical requirements

Using the following commands, we will set up the infrastructure in Google Cloud that will be used for hands-on exercises:

1. Create a Kubernetes cluster:

```
% gcloud container clusters create cluster1 --cluster-version latest --machine-type "e2-medium" --num-nodes "3" --network "default" --zone "australia-southeast1-a" --disk-type "pd-standard" --disk-size "30"
kubeconfig entry generated for cluster1.
NAME      LOCATION               MASTER_VERSION   MASTER_IP     MACHINE_TYPE  NODE_VERSION     NUM_NODES  STATUS
cluster1  australia-southeast1-a  1.23.12-gke.100  34.116.79.135  e2-medium     1.23.12-gke.100  3          RUNNING
```

2. Create a VM:

```
% gcloud compute instances create chapter10-instance --tags=chapter10-meshvm \
  --machine-type=e2-medium --zone=australia-southeast1-b
  --network=default --subnet=default \
  --image-project=ubuntu-os-cloud \
  --image=ubuntu-1804-bionic-v20221201, mode=rw, size=10
Created [https://www.googleapis.com/compute/v1/projects/istio-book-370122/zones/australia-southeast1-b/instances/chapter10-instance].
NAME        ZONE                   MACHINE_TYPE  PREEMPTIBLE  INTERNAL_IP  EXTERNAL_IP   STATUS
chapter10-instance australia-southeast1-b  e2-medium              10.152.0.13  34.87.233.38  RUNNING
```

3. Check your kubectl file to find the cluster name and set context appropriately:

```
% kubectl config view -o json | jq .contexts
[
  {
    "name": "gke_istio-book-370122_australia-southeast1-a_cluster1",
    "context": {
      "cluster": "gke_istio-book-370122_australia-southeast1-a_cluster1",
      "user": "gke_istio-book-370122_australia-southeast1-a_cluster1"
```

```
        }
      }
    ]
    % export CTX_CLUSTER1=gke_istio-book-370122_australia-
    southeast1-a_cluster1
```

4. Access the created server using **SSH** from the Google Cloud dashboard – you will find the **SSH** option in the bottom-right corner, as shown in the following screenshot:

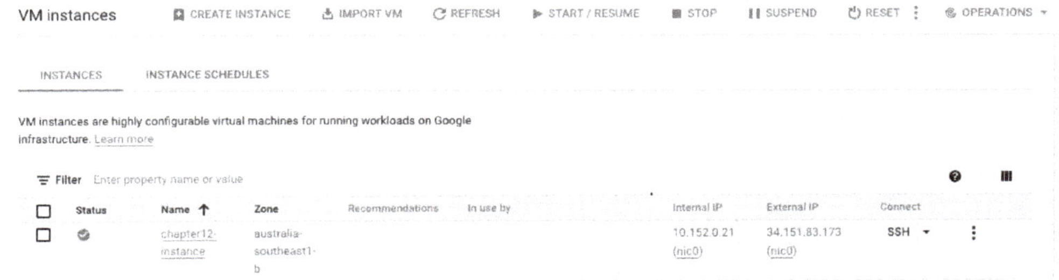

Figure 10.1 – Google Cloud dashboard

5. Click on **SSH**, which will open **SSH-in-browser**, as shown in the following figure:

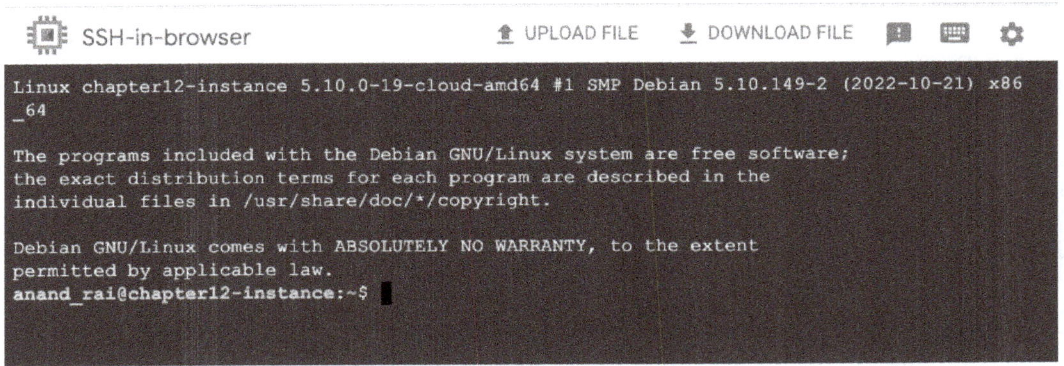

Figure 10.2 – SSH-in-browser

6. Find the username and then SSH from your terminal:

```
% gcloud compute ssh anand_rai@chapter10-instance
```

7. Set up the firewall to allow traffic between the Kubernetes cluster and VM using the following steps:

   A. Find the **Classless Inter-Domain Routing** (**CIDR**) of the cluster:

```
% CLUSTER_POD_CIDR=$(gcloud container clusters describe
cluster1 --format=json --zone=australia-southeast1-a | jq
-r '.clusterIpv4Cidr')
% echo $CLUSTER_POD_CIDR
10.52.0.0/14
```

   B. Create firewall rules:

```
% gcloud compute firewall-rules create "cluster1-pods-to-
chapter10vm" \
  --source-ranges=$CLUSTER_POD_CIDR \
  --target-tags=chapter10-meshvm  \
  --action=allow \
  --rules=tcp:10000
Creating firewall...⠹Created [https://www.googleapis.com/
compute/v1/projects/istio-book-370122/global/firewalls/
cluster1-pods-to-chapter10vm].
Creating firewall...done.
NAME  NETWORK  DIRECTION  PRIORITY  ALLOW      DENY  DISA-
BLED
cluster1-pods-to-chapter10vm
 default  INGRESS     1000      tcp:10000        False
```

That's all you need for the upcoming sections. We will first explore some fundamentals and then go through the actual setup.

Examining hybrid architecture

As mentioned in the introduction of this chapter, organizations have adopted Kubernetes and they run microservices and various other workloads as containers, but not all workloads are suitable for containers. So, organizations have to live with following a hybrid architecture:

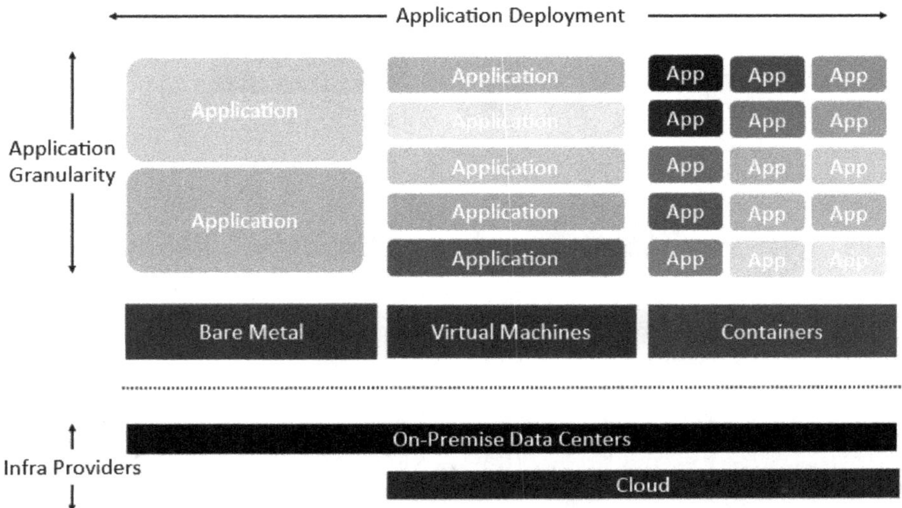

Figure 10.3 – Hybrid architecture

Appliances and legacy applications are usually deployed on bare metal servers. Monolithic applications, as well as several **commercial off-the-shelf (COTS)** applications, are deployed on VMs. Modern applications, as well as self-developed applications based on microservices architecture, are deployed as containers that are managed and orchestrated by platforms such as Kubernetes. All three deployment models – that is, bare metal, VMs, and containers – are spread across traditional data centers and various cloud providers. This intermingling of various application architectures and deployment patterns causes various problems:

- Management of traffic flows between the Service Mesh and VM is challenging because neither of them has any idea of the other's existence
- No operational visibility of application traffic between VM applications and applications within the Service Mesh
- Inconsistent governance of VMs and applications within the Service Mesh because there is no consistent way of defining and applying security policies for VM apps and apps within the mesh

The following is an example of how traffic flows in an environment with a VM and Service Mesh:

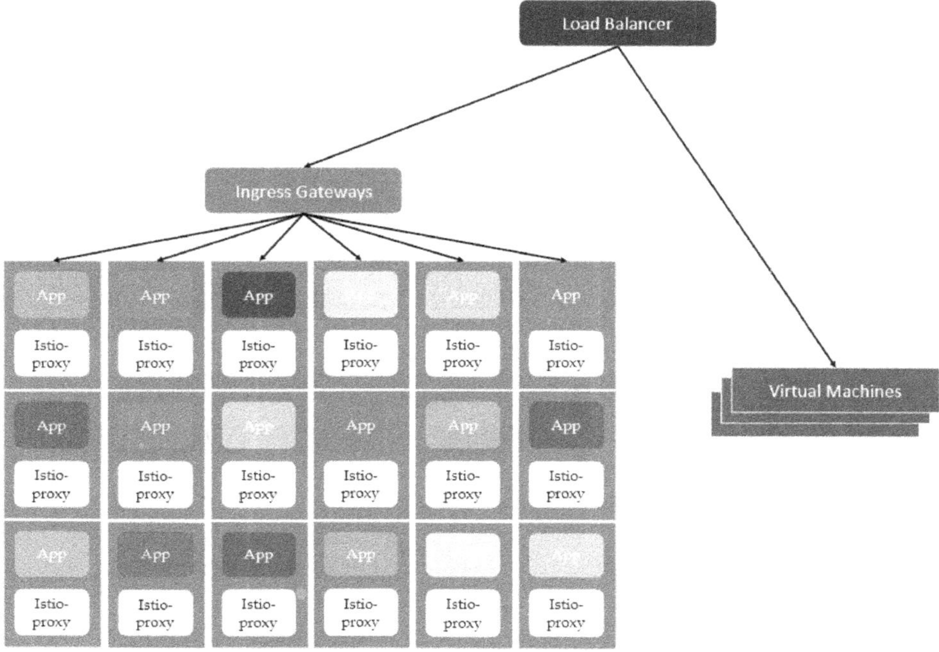

Figure 10.4 – Traffic across Service Mesh and VM is managed separately

A VM is treated as a separate universe. Developers have to choose one of the deployment patterns because they can't have system components spread across VMs and containers. This is okay for legacy systems but when building applications based on microservice architecture, it is constraining to choose between VMs and containers. For example, your system may need a database that might be best suited for VM-based deployment, whereas the rest of the application might be well suited for container-based deployment. Although there are many traditional solutions to route traffic between the Service Mesh and VM, doing so results in disparate networking solutions. Luckily, Istio provides options to establish a Service Mesh for VMs. The solution is to abstract VMs under Istio constructs so that the Service Mesh operator can operate the network between containers and VMs consistently.

In the next section, we will learn how to configure a Service Mesh for VMs.

Setting up a Service Mesh for hybrid architecture

In this section, we will set up the Service Mesh. But first, let's look at the steps at a high level in the *Overview of the setup* section and then perform implementation in the *Setting up a demo app on a virtual machine* section.

Overview of the setup

Envoy is a great networking software and an excellent reverse proxy; it is also widely adopted as a standalone reverse proxy. **Solo.io** and a few others have built API gateway solutions using Envoy, and **Kong Inc.** has both Kong Mesh and Kuma Service Mesh technologies, which make use of Envoy as sidecars for the data plane.

Envoy, when deployed as a sidecar, has no idea about being a sidecar; it communicates to `istiod` via the xDS protocols. Istio `init` bootstraps Envoy with the right configuration and details about `istiod`, and sidecar injection mounts the right certificates, which are then used by Envoy to authenticate itself with `istiod`; once bootstrapped, it keeps fetching correct configurations via xDS APIs.

Based on the same concepts, Istio packages Envoy as a sidecar for VMs. An Istio operator will need to perform the steps shown in the following figure to include a VM into the Service Mesh:

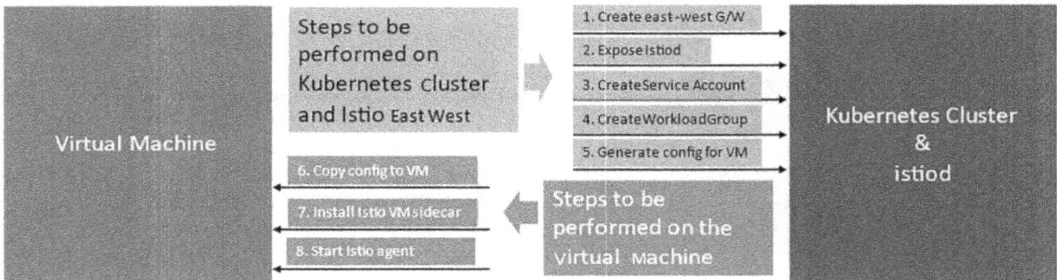

Figure 10.5 – Steps to include VM into the mesh

The following is a brief overview of the steps we will be implementing in the next section:

1. For the VM sidecar to access the Istio control plane, we need to expose `istiod` via the east-west gateway. So, we install another Ingress gateway for east-west traffic purposes.
2. We expose `istiod` services via the east-west gateway. This and the previous step are similar to the steps required for a multi-cluster Service Mesh setup, as discussed in *Chapter 8*.
3. The sidecar in the VM needs to access the Kubernetes API server but because the VM isn't part of the cluster, it does not have access to **Kubernetes credentials**. To solve that problem, we will manually create a service account in Kubernetes for the VM sidecar to access the API server. We are doing it manually here, but it can be automated using an external credential management service such as **HashiCorp Vault**.
4. The next step is the creation of the Istio **workload group**. `WorkloadGroup` provides specifications that are used by sidecars to bootstrap themselves. It can be shared by collections of VMs that are running similar types of workloads. In `WorkloadGroup`, you define labels through which the workload will be identified in Kubernetes as well as other nuances, such as what ports are exposed, the service account to be used, and various health check probes. To some extent, `WorkloadGroup` is similar to deployment descriptors in Kubernetes. We will look at it in more detail in the next section during the setup.

5. The operator needs to manually generate the configuration that will be used to configure the VM and the sidecar. This step has some challenges when it comes to generating configuration for auto-scalable VMs.
6. In this step, we need to copy the configuration from previous step to the VM at set locations.
7. The Istio sidecar needs to be installed.
8. Finally, the Istio sidecar needs to be started and some checks performed to ensure that it has picked up the configuration created in *step 5*.

Once the Istio sidecar is started, it will intercept the outgoing traffic and route it according to the Service Mesh rules as long as the target service endpoints, which can be VMs or Kubernetes Pods, are on the same network. The Ingress gateway is fully aware of the VM workload and can route traffic, and the same applies to any traffic inside the mesh, as shown in the following figure:

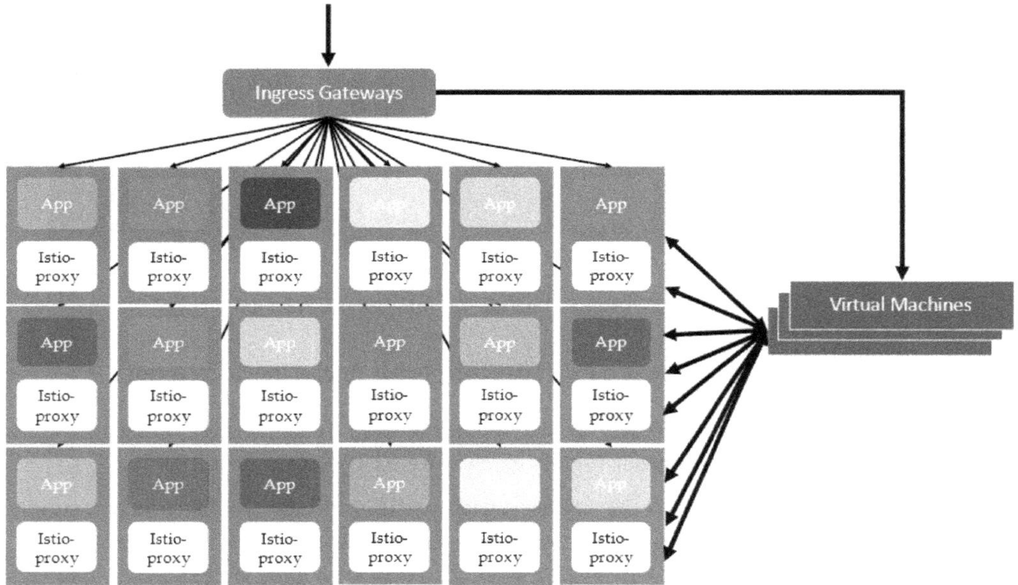

Figure 10.6 – VM workload treated similarly to other workloads in the mesh

Pod connectivity assumes that the cluster network uses the same address space as standalone machines. For cloud-managed Kubernetes (Google GKE and Amazon EKS), it is the default networking mode, but for self-managed clusters, you need a networking subsystem such as Calico to implement a flat routable network address space. In the next section, we will perform the setup of Istio on VMs, so roll up your sleeves and make sure that you have completed the tasks described in the *Technical requirements* section.

Setting up a demo app on a VM

We will first install an application on the VM to mimic a VM workload/application that we can use for testing the overall setup. To do this, we need to perform the following steps:

1. Set up Envoy on the VM. Follow the instructions as provided by Envoy at `https://www.envoyproxy.io/docs/envoy/latest/start/install` for the operating system you selected for creating the VM. This can be done as follows:

   I. Install `envoy`, as shown in the following code block:

   ```
   $ sudo apt update
   $ sudo apt install debian-keyring debian-archive-keyring
   apt-transport-https curl lsb-release
   $ curl -sL 'https://deb.dl.getenvoy.io/public/
   gpg.8115BA8E629CC074.key' | sudo gpg --dearmor -o /usr/
   share/keyrings/getenvoy-keyring.gpg
   # Verify the keyring - this should yield "OK"
   $ echo
   a077cb587a1b622e03aa4bf2f3689de14658a9497a9af2c427b-
   ba5f4cc3c4723 /usr/share/keyrings/getenvoy-keyring.gpg |
   sha256sum --check
   $ echo "deb [arch=amd64 signed-by=/usr/share/keyrings/
   getenvoy-keyring.gpg] https://deb.dl.getenvoy.io/public/
   deb/debian $(lsb_release -cs) main" | sudo tee /etc/apt/
   sources.list.d/getenvoy.list
   $ sudo apt update
   $ sudo apt install getenvoy-envoy
   ```

   II. Check that `envoy` is properly installed:

   ```
   $ envoy --version
   envoy  version:
   d362e791eb9e4efa8d87f6d878740e72dc8330ac/1.18.2/clean-
   getenvoy-76c310e-envoy/RELEASE/BoringSSL
   ```

2. Configure Envoy to run a dummy application:

   I. Using **vi** or any other editor, create `envoy-demo.yaml` and copy the contents of `Chapter4/envoy-config-2.yaml`.

   II. Check that the contents of `envoy-demo.yaml` match what you have copied or created.

3. Run envoy with the config provided in envoy-demo.yaml:

```
$ envoy -c envoy-demo.yaml &
[2022-12-06 03:46:31.679][55335][info][main] [external/
envoy/source/server/server.cc:330] initializing epoch 0
(base id=0, hot restart version=11.104)
```

4. Test that the application is running on the VM:

```
$ curl localhost:10000
V2----------Bootstrap Service Mesh Implementation with
Istio----------V2
```

With an application running on the VM, we can now proceed with the rest of the setup. Please note that it is not mandatory to set up the application before installing Istio on the VM. You can install the demo application on the VM at any time before or after setting up the Istio sidecar on the VM.

Setting up Istio in the cluster

We are assuming that you don't have Istio running in the cluster, but if you have, then you can skip this section. Use the following steps to set it up:

1. Configure the IstioOperator config file for installation, providing the cluster and network names. The file is also available at Chapter10/01-Cluster1.yaml:

```
apiVersion: install.istio.io/v1alpha1
kind: IstioOperator
spec:
  values:
    global:
      meshID: mesh1
      multiCluster:
        clusterName: cluster1
      network: network1
```

2. Install Istio, as shown in the following code block:

```
% istioctl install -f Chapter10/01-Cluster1.yaml
--set values.pilot.env.PILOT_ENABLE_WORKLOAD_ENTRY_
AUTOREGISTRATION=true --set values.pilot.env.PILOT_
ENABLE_WORKLOAD_ENTRY_HEALTHCHECKS=true --context="${CTX_
CLUSTER1}"
```

```
This will install the Istio 1.16.0 default profile with
["Istio core" "Istiod" "Ingress gateways"] components
into the cluster. Proceed? (y/N) y
 Istio core installed
 Istiod installed
 Ingress gateways installed
 Installation complete
                              Making this installation
the default for injection and validation.
```

This concludes the basic installation of Istio in the cluster. Next, we will configure Istio to make it ready for integration with Istio on the VM.

Configuring the Kubernetes cluster

In this section, we will prepare the mesh for integration with Istio on VMs:

1. Install the east-west gateway to expose the `istiod` validation webhook and services:

```
% samples/multicluster/gen-eastwest-gateway.sh \
--mesh mesh1 --cluster cluster1 --network network1  | \
istioctl install -y --context="${CTX_CLUSTER1}" -f -

✓ Ingress gateways installed
✓ Installation complete
```

2. Expose the `istiod` services:

```
% kubectl apply -n istio-system -f samples/multicluster/
expose-istiod.yaml --context="${CTX_CLUSTER1}"
gateway.networking.istio.io/istiod-gateway created
virtualservice.networking.istio.io/istiod-vs created
```

3. Create a service account by following these steps:

   A. Create a namespace to host `WorkloadGroup` and `ServiceAccount`:

```
% kubectl create ns chapter10vm --context="${CTX_
CLUSTER1}"
namespace/chapter10vm created
```

B. Create a service account to be used by `istiod` on the VM to connect with the Kubernetes API server:

```
% kubectl create serviceaccount chapter10-sa -n
chapter10vm --context="${CTX_CLUSTER1}"
serviceaccount/chapter10-sa created
```

4. Set up `WorkloadGroup` as follows:

A. Use the following configuration to create a workload template; the file is also available at `Chapter10/01-WorkloadGroup.yaml`:

```
apiVersion: networking.istio.io/v1alpha3
kind: WorkloadGroup
metadata:
  name: "envoydummy"
  namespace: "chapter10vm"
spec:
  metadata:
    labels:
      app: "envoydummy"
  template:
    serviceAccount: "chapter10-sa"
    network: "network1"
  probe:
    periodSeconds: 5
    initialDelaySeconds: 1
    httpGet:
      port: 10000
      path: /
```

B. Apply the configuration. This template will be used by Istio to create workload entries representing the workload running on the VM:

```
% kubectl --namespace chapter10vm apply -f "Chapter10/01-
WorkloadGroup.yaml" --context="${CTX_CLUSTER1}"
workloadgroup.networking.istio.io/envoyv2 created
```

Before moving to the next section, let's inspect the contents of `WorkloadGroup.yaml`.

`WorkloadGroup` is a way to define the characteristics of the workload hosted in the VM and is similar to deployments in Kubernetes. `WorkloadGroup` has the following configuration:

- `metadata`: This is primarily used for defining Kubernetes labels to identify the workloads. We have set up an app label with the value of envoydummy, which we can use in the Kubernetes service description to identify the endpoints to be abstracted by service definitions.
- `template`: This defines the values that will be copied over to the `WorkloadEntry` configuration generated by the Istio agent. The two most important values are the service account name and network name. `ServiceAccount` specifies the account name whose token will be used to generate workload identities. The network name is used to group endpoints based on their network location and to understand which endpoints are directly reachable from each other and which ones need to be connected via the east-west gateway, like the ones we set up for a multi-cluster environment in *Chapter 8*. In this instance, we have allocated the value of `network1`, which is the same as what we configure in `01-cluster1.yaml` (that is `cluster1`) and the VM are on the same network and directly reachable to each other, so we don't need any special provisions for connecting them.
- `probe`: This is the configuration to be used to understand the health and readiness of the VM workload. Traffic is not routed to unhealthy workloads, providing a resilient architecture. In this instance, we are configuring to perform an HTTP `Get` probe with a delay of 1 second after the creation of `WorkloadEntry` and then at regular intervals of 5 seconds. You can also define success and failure thresholds, for which the default values are `1` and `3` seconds, respectively. We have configured that the endpoint on the VM is exposed on port `10000` on the `root` path and it should be used to determine the health of the application.

Next, let's get started with setting up Istio on a VM.

Setting up Istio on a VM

To configure and set up Istio on a VM, we will need to perform the following steps:

1. Generate a configuration for the Istio sidecar:

```
% istioctl x workload entry configure -f "Chapter10/01-
WorkloadGroup.yaml" -o . --clusterID "cluster1"
--autoregister --context="${CTX_CLUSTER1}"
Warning: a security token for namespace "chapter10vm" and
service account "chapter10-vm-sa" has been generated and
stored at "istio-token"
Configuration generation into directory . was successful
```

This will generate the following five files in the current directory:

```
% ls
hosts root-cert.pem istio-token cluster.env mesh.yaml
```

2. Copy all files to the VM home directory first, and then copy them to various folders, as shown in the following code block:

```
% sudo mkdir -p /etc/certs
% sudo cp "${HOME}"/root-cert.pem /etc/certs/root-cert.
pem
% sudo  mkdir -p /var/run/secrets/tokens
% sudo cp "${HOME}"/istio-token /var/run/secrets/tokens/
istio-token
% sudo cp "${HOME}"/cluster.env /var/lib/istio/envoy/
cluster.env
% sudo cp "${HOME}"/mesh.yaml /etc/istio/config/mesh
% sudo sh -c 'cat $(eval echo ~$SUDO_USER)/hosts >> /etc/
hosts'
% sudo mkdir -p /etc/istio/proxy
% sudo chown -R istio-proxy /var/lib/istio /etc/certs /
etc/istio/proxy /etc/istio/config /var/run/secrets /etc/
certs/root-cert.pem
```

3. Install the Istio VM integration runtime. Download and install the package from `https://storage.googleapis.com/istio-release/releases/`:

```
$ curl -LO https://storage.googleapis.com/istio-release/
releases/1.16.0/deb/istio-sidecar.deb
$ sudo dpkg -i istio-sidecar.deb
Selecting previously unselected package istio-sidecar.
(Reading database ... 54269 files and directories
currently installed.)
Preparing to unpack istio-sidecar.deb ...
Unpacking istio-sidecar (1.16.0) ...
Setting up istio-sidecar (1.16.0) ...
```

4. Start the Istio agent on the VM and then check the status:

```
$ sudo systemctl start istio
$ sudo systemctl status istio
● istio.service - istio-sidecar: The Istio sidecar
```

```
     Loaded: loaded (/lib/systemd/system/istio.service;
disabled; vendor preset: enable>
     Active: active (running) since Tue 20XX-XX-06
07:56:03 UTC; 15s ago
       Docs: http://istio.io/
   Main PID: 56880 (sudo)
      Tasks: 19 (limit: 4693)
     Memory: 39.4M
        CPU: 1.664s
     CGroup: /system.slice/istio.service
             ├─56880 sudo -E -u istio-proxy -s /bin/bash
-c ulimit -n 1024; INSTANCE_IP>
             ├─56982 /usr/local/bin/pilot-agent proxy
             └─56992 /usr/local/bin/envoy -c etc/istio/
proxy/envoy-rev.json --drain-tim>
```

This completes the installation and configuration of the Istio sidecar on the VM.

Next, verify that `WorkloadEntry` is created in the `chapter10vm` namespace:

```
% kubectl get WorkloadEntry -n chapter10vm
NAME                                AGE    ADDRESS
envoydummy-10.152.0.21-network1    115s   10.152.0.21
```

`WorkloadEntry` is automatically created and is a sign that the VM has onboarded itself successfully into the mesh. It describes the properties of the application running on the VM and inherits the template from the `WorkloadGroup` configuration.

Inspect the contents of `WorkloadEntry` using the following code block:

```
% kubectl get WorkloadEntry/envoydummy-10.152.0.21-network1 -n
chapter10vm -o yaml
```

`WorkloadEntry` contains the following values:

- `address`: This is the network address at which the application is running on the VM. This can also be DNS names. In this instance, the VM's private IP is `10.152.0.21`.
- `labels`: These are inherited from the `WorkloadGroup` definition and are used to identify endpoints selected by service definitions.
- `locality`: In a multi-data center, this field is used to identify the location of the workload at the rack level. This field is used for locality/proximity-based load balancing.

- `network`: This value is inherited from the `WorkloadGroup` entry.
- `serviceAccount`: This value is inherited from the `WorkloadGroup` entry.
- `status`: This value specifies the health of the application.

By now, we have configured the Istio agent on a VM and have verified that the agent can communicate with Istiod. In the next section, we will integrate the workload on the VM with the mesh.

Integrating the VM workload with the mesh

Let's get started with performing configurations so that the mesh can route traffic to workloads running on the VM:

1. Expose the `envoydummy` app on the VM as a Kubernetes service using the following code block:

   ```
   apiVersion: v1
   kind: Service
   metadata:
     name: envoydummy
     labels:
       app: envoydummy
     namespace: chapter10vm
   spec:
     ports:
     - port: 80
       targetPort: 10000
       name: tcp
     selector:
       app: envoydummy
   ---
   ```

   The configuration is standard, and it treats the VM as a Pod that needs to be exposed by a service. Notice the labels, which match the metadata values in `WorkloadGroup` definitions. When defining the service, just assume that the VM is nothing other than a Kubernetes Pod as defined in the `WorkloadGroup` configuration file. The service description file is available at `Chapter10/01-istio-gateway.yaml`. Apply the configuration using the following command:

   ```
   % kubectl apply -f Chapter10/02-envoy-proxy.yaml -n
   chapter10vm
   ```

2. Next, we will deploy version v1 of the `envoydummy` application in the Kubernetes cluster:

```
$ kubectl create ns chapter10 --context="${CTX_CLUSTER1}"
$ kubectl label namespace chapter10 istio-
injection=enabled --context="${CTX_CLUSTER1}"
$ kubectl create configmap envoy-dummy --from-
file=Chapter3/envoy-config-1.yaml -n chapter10
--context="${CTX_CLUSTER1}"
$ kubectl create -f "Chapter10/01-envoy-proxy.yaml"
--namespace=chapter10 --context="${CTX_CLUSTER1}"
$ kubectl apply -f Chapter10/01-istio-gateway.yaml" -n
chapter10 --context="${CTX_CLUSTER2}"
```

Notice the `route` configuration in `01-istio-gateway.yaml`:

```
    route:
    - destination:
        host: envoydummy.chapter10.svc.cluster.local
      weight: 50
    - destination:
        host: envoydummy.chapter10vm.svc.cluster.local
      weight: 50
```

We are routing half of the traffic to `envoydummy.chapter10vm.svc.cluster.local`, which represents the application running in the VM, and the other half to `envoydummy.chapter10.svc.cluster.local`, which represents the application running in the Kubernetes cluster.

We have configured all the steps for the integration of the VM workload with the mesh. To test the connectivity and DNS resolution of the VM, run the following command from the VM:

```
$ curl envoydummy.chapter10.svc:80
Bootstrap Service Mesh Implementation with Istio
```

This shows that the VM is cognizant of endpoints exposed in the Service Mesh. You can do it the other way round, from the Kubernetes cluster of the `curl` Pod described in the `utilities` folders in the GitHub repository of this book, please make sure that it is part of the mesh and not just a Pod running on Kubernetes.

Now, test it from the Istio Ingress gateway:

```
% for i in {1..10}; do curl -Hhost:mockshop.com -s
"http://34.87.194.86:80";echo '\n'; done
```

```
V2----------Bootstrap Service Mesh Implementation with Istio--
--------V2
Bootstrap Service Mesh Implementation with Istio
V2----------Bootstrap Service Mesh Implementation with Istio--
--------V2
Bootstrap Service Mesh Implementation with Istio
V2----------Bootstrap Service Mesh Implementation with Istio--
--------V2
Bootstrap Service Mesh m bbhgc  Mesh Implementation with Istio-
---------V2
Bootstrap Service Mesh Implementation with Istio
V2----------Bootstrap Service Mesh Implementation with Istio--
--------V2
Bootstrap Service Mesh Implementation with Istio
```

Now, let's also take a peek into the Kiali dashboard and see what the graph looks like. Please install Kiali using the instructions in *Chapter 7*.

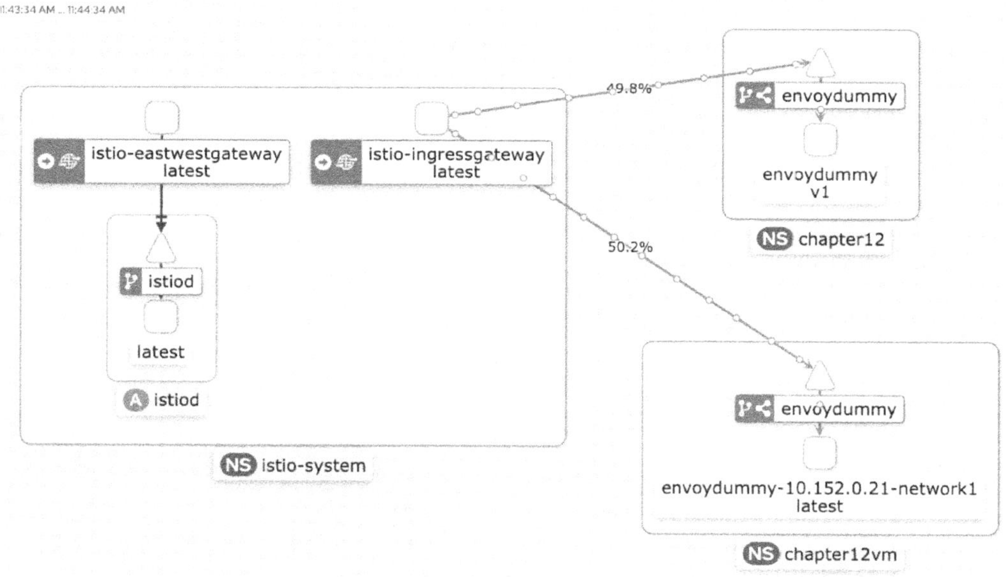

Figure 10.7 – Kiali dashboard showing the traffic distribution to VM workload

From the preceding figure, you can see `WorkloadEntry` in the `chapter10vm` namespace represented like another Pod, just like `envoydummyv1` in the `chapter10` namespace.

Now that the setup is complete, let's summarize what we've learned in this chapter.

Summary

VMs are an important piece of the puzzle in modern architecture and are here to stay for the foreseeable future along with containers. With Istio, you can integrate traditional workloads running on VMs into Istio Service Mesh and leverage all the benefits of traffic management and security provided by Istio. Istio support for VMs enables the inclusion of legacy applications, as well as those applications that cannot run on a container due to certain constraints in the mesh.

After reading this chapter, you should be able to create a mesh for hybrid architectures. You can now install Istio on a VM and integrate workloads with the mesh along with Kubernetes-based workloads. To get yourself hardened with concepts in this chapter, practice creating multiple VMs with different versions of the `envoydummy` application and then implement traffic management via virtual services and destination rules.

In the next chapter, we will read about various troubleshooting strategies and techniques to manage Istio.

11
Troubleshooting and Operating Istio

Deploying microservices involves many moving parts, including the application, the underlying Kubernetes platform, and the application network provided by Istio. It is not uncommon for the mesh to be operating in an unintended way. Istio, in its early days, was infamous for being complex and too difficult to troubleshoot. The istio community took that perception very seriously and has been working toward simplifying its installation and day-2 operations to make it easier and more reliable to use in production-scale deployments.

In this chapter, we will read about the common problems you will encounter when operating istio and how to distinguish and isolate them from other issues. We will then learn how to **troubleshoot** these problems once they are identified. We will also explore various **best practices** for deploying and operating istio and how to automate the enforcement of best practices.

In a nutshell, this chapter will cover the following topics:

- Understanding interactions between istio components
- Inspecting and analyzing the istio configuration
- Troubleshooting errors using access logs
- Troubleshooting errors using debug logs
- Debugging istio agents
- Understanding istio's best practices
- Automating best practices using OPA Gatekeeper

Understanding interactions between Istio components

When troubleshooting problems with the Service Mesh, the unexpected behavior of the mesh is likely caused by one of the following underlying issues:

- Invalid control plane configuration
- Invalid data plane configuration
- Unexpected data plane

In the upcoming sections, we will explore how to diagnose the underlying reason for any such unexpected behavior with the help of various diagnostic tools provided by istio. But first, let's look at various interactions that happen inside the mesh between istiod, data planes, and other components.

Exploring Istiod ports

Istiod exposes various ports, some of which can be used for troubleshooting. In this section, we will go through those ports, and understand what they do and how they can help with troubleshooting.

Let's get started by looking at those ports:

- **Port 15017**: This port is exposed on istiod for sidecar injection and validation. Whenever a Pod is created in the Kubernetes cluster with the Istio injection enabled, the mutating admission controller sends a request on this port to istiod to fetch the sidecar injection template and later validate the configuration. The port is originally exposed on port `443` but is forwarded to port `15017`, which we saw in action when setting up primary remote clusters.
- **Port 15014**: This port is used by Prometheus to scrape control plane metrics. You can check the metric using the following command:

```
% kubectl -n istio-system port-forward deploy/istiod
15014 &
[1] 68745
Forwarding from 127.0.0.1:15014 -> 15014
```

Then, you can fetch the metrics from port `15014`:

```
% curl http://localhost:15014/metrics
Handling connection for 15014
# HELP citadel_server_csr_count The number of CSRs
received by Citadel server.
# TYPE citadel_server_csr_count counter
citadel_server_csr_count 6
# HELP citadel_server_root_cert_expiry_timestamp The
unix timestamp, in seconds, when Citadel root cert will
```

```
expire. A negative time indicates the cert is expired.
# TYPE citadel_server_root_cert_expiry_timestamp gauge
citadel_server_root_cert_expiry_timestamp 1.986355163e+09
.........
```

- **Ports 15010 and 15012**: These two ports serve the xDS and CA APIs. The difference is that port `15010` is insecure and port `15012` is secure so can be used for production environments.
- **Port 9876**: This port exposes the **ControlZ** interface, which is an istiod introspection framework to inspect and manipulate the internal state of an `istiod` instance. This port is used to access the ControlZ interface either via REST API calls from within the mesh or via a dashboard, which can be accessed using the following command:

```
% istioctl dashboard controlz deployment/istiod.istio-system
```

The following is a screenshot of the ControlZ interface:

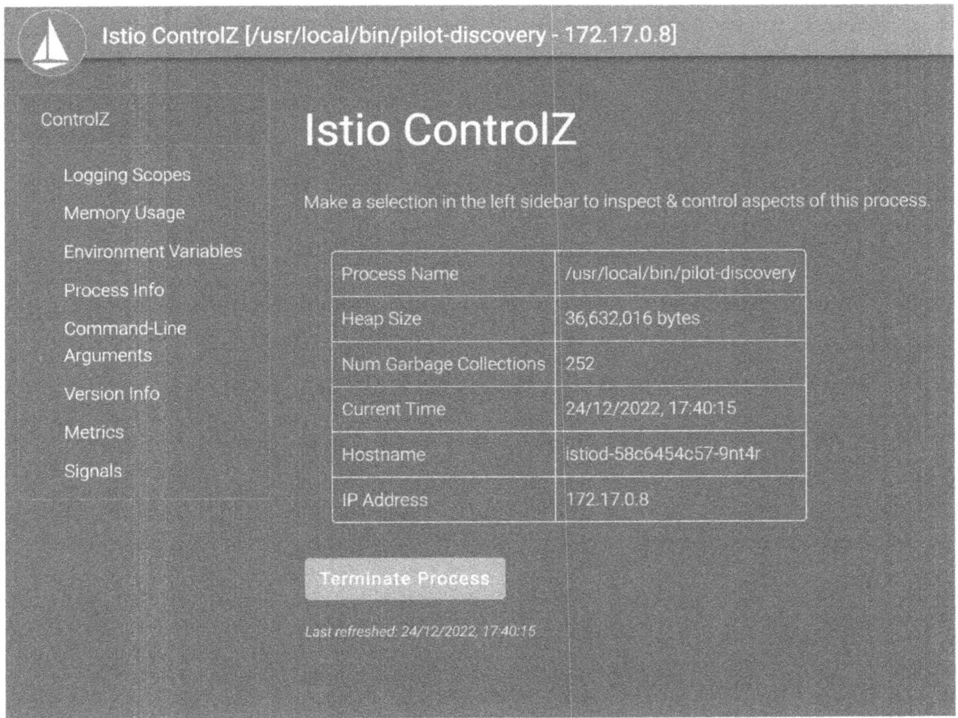

Figure 11.1 – Istio ControlZ interface

The ControlZ interface can be used to inspect logging scopes, environment variables, and so on. The interface can also be used to change logging levels.

In this section, we read about various ports exposed by istiod. Let's move on to read about the ports exposed by the Istio data plane.

Exploring Envoy ports

Envoy, which is the data plane for Istio, exposes various ports for interaction with the Istio control plane and observability tools. Let's look at those ports, what they do, and how they can help with troubleshooting:

- **Port 15000**: This is the Envoy admin interface, which can be used to inspect and query envoy configuration. We will read about this port in detail in the next section, *Inspecting and analyzing the Istio configuration*.
- **Port 15001**: This port is used for receiving all outbound traffic from the application Pods.
- **Port 15004**: This port can be used to debug the data plane configuration.
- **Port 15006**: All inbound application traffic from within the mesh is routed to this port.
- **Port 15020**: This port provides merged metrics information from Envoy, `istio-agent`, and the application, which is then scraped by Prometheus.
- **Port 15021**: This port is exposed for performing health checks of the data plane.
- **Port 15053**: This port is used to serve the DNS proxy.
- **Port 15090**: This port provides Envoy telemetry information.

In the next section, we will explore how to analyze and inspect the Istio configuration.

Inspecting and analyzing the Istio configuration

When debugging the Istio data plane, it is useful to check whether there is any configuration mismatch between the Istio control plane and the data plane. When working with multi-cluster mesh, it is a good idea to first check the connectivity between the control plane and data plane; if your Pod supports `curl`, then you can use the following command to check the connectivity between the two:

```
$ kubectl exec -it curl -c curl -n chapter11 -- curl  istiod.
istio-system.svc.cluster.local:15014/version
1.16.0-8f2e2dc5d57f6f1f7a453e03ec96ca72b2205783-Clean
```

To inspect the configuration, the first checkpoint can be to use the `istioctl proxy-status` command to find the synchronization state of the cluster, listener, routes, and endpoints configuration between istiod and istio-proxy.

You can use the following command to check the configuration status of the whole cluster:

```
% istioctl proxy-status
NAME           CLUSTER            CDS              LDS              EDS
       RDS           ECDS            ISTIOD                       VERSION
curl.chapter11
Kubernetes      SYNCED       SYNCED       SYNCED       SYNCED
       NOT SENT      istiod-58c6454c57-9nt4r      1.16.0
envoydummy.chapter11
Kubernetes      SYNCED       SYNCED       SYNCED       SYNCED
       NOT SENT      istiod-58c6454c57-9nt4r      1.16.0
istio-egressgateway-5bdd756dfd-bjqrg.istio-
system      Kubernetes      SYNCED      SYNCED      SYNCED      NOT
SENT      NOT SENT      istiod-58c6454c57-9nt4r      1.16.0
istio-ingressgateway-67f7b5f88d-xx5fb.istio-system
Kubernetes      SYNCED       SYNCED       SYNCED       SYNCED
       NOT SENT      istiod-58c6454c57-9nt4r      1.16.0
```

The following are the possible values of the synchronization status:

- `SYNCED`: Envoy has the latest config.
- `NOT SENT`: istiod has not sent any config to Envoy; in most cases, the reason is that istiod has no config to send. In this example, the status is `NOT SENT` for the Istio Egress Gateway because there is no route information to be synced.
- `STALE`: Envoy doesn't have the latest config, which is an indication of a networking issue between Envoy and istiod.

You can check the status of a workload using the following command:

```
% istioctl proxy-status envoydummy.chapter11
Clusters Match
Listeners Match
Routes Match (RDS last loaded at Sat, 17 Dec 2022 11:53:31
AEDT)Access Logs
```

If the Pod being investigated supports `curl`, then you can perform a dump of the `istio-proxy` configuration by fetching config from Envoy's admin interface exposed at port `15000` using the following command:

```
% kubectl exec curl -c curl -n chapter11 -- curl
localhost:15000/config_dump > config_dump.json
```

You can selectively dump the configuration of listeners, clusters, routes, and endpoints by using the following `istioctl proxy-config` command:

```
% istioctl proxy-config endpoints envoydummy.chapter11 -o json
> endpoints-envoydummy.json
% istioctl proxy-config routes envoydummy.chapter11 -o json >
routes-envoydummy.json
% istioctl proxy-config listener envoydummy.chapter11 -o json >
listener-envoydummy.json
% istioctl proxy-config cluster envoydummy.chapter11 -o json >
cluster-envoydummy.json
```

You can fetch the Envoy configuration from the istiod perspective and compare it with the config fetched from Envoy's admin interface:

```
% kubectl exec istiod-58c6454c57-gj6cw -n istio-system -- curl
'localhost:8080/debug/config_dump?proxyID=curl.chapter11' | jq
. > Chapter11/config-from-istiod-for-curl.json
```

You can inspect the `istio-proxy` configuration in a Pod from a web browser using the following command:

```
$ kubectl port-forward envoydummy -n chapter11 15000
```

From the browser, you can now access the Envoy dashboard, as shown in the following figure:

Command	Description
certs	print certs on machine
clusters	upstream cluster status
config_dump	dump current Envoy configs (experimental)
	The resource to dump
	The mask to apply. When both resource and mask are specified, the mask is applied to every element in the desired repeated field so that only a subset of fields are returned. The mask is parsed as a ProtobufWkt::FieldMask
	Dump only the currently loaded configurations whose names match the specified regex. Can be used with both resource and mask query parameters.
	Dump currently loaded configuration including EDS. See the response definition for more information
contention	dump current Envoy mutex contention stats (if enabled)
cpuprofiler	enable/disable the CPU profiler
y	enables the CPU profiler
drain_listeners	drain listeners
	When draining listeners, enter a graceful drain period prior to closing listeners. This behaviour and duration is configurable via server options or CLI
	Drains all inbound listeners. traffic_direction field in envoy_v3_api_msg_config.listener.v3.Listener is used to determine whether a listener is inbound or outbound.
healthcheck/fail	cause the server to fail health checks
healthcheck/ok	cause the server to pass health checks
heap_dump	dump current Envoy heap (if supported)
heapprofiler	enable/disable the heap profiler
y	enable/disable the heap profiler
help	print out list of admin commands
hot_restart_version	print the hot restart compatibility version
init_dump	dump current Envoy init manager information (experimental)
	The desired component to dump unready targets. The mask is parsed as a ProtobufWkt::FieldMask. For example, get the unready targets of all listeners with /init_dump?mask=listener`
listeners	print listener info
text	File format to use
logging	query/change logging levels
	Change multiple logging levels by setting to <logger_name1>:<desired_level1>,<logger_name2>:<desired_level2>.

Figure 11.2 – Envoy dashboard

The Envoy dashboard is a good option to inspect values but proceed with caution when changing configuration parameters because, from this dashboard, you will be changing data plane configuration outside the purview of istiod.

At the time of writing this book, `istioctl describe` is an experimental feature. It is used to describe a Pod, and if the Pod meets all the requirements to be part of the mesh, this command (when run against a Pod) can also tell whether istio-proxy in the Pod has been started or whether the Pod is part of the mesh or not. It will emit any warnings and suggestions to enable better integration of the Pod into the mesh.

The following runs the `istioctl describe` command against the `envoydummy` Pod:

```
% istioctl x describe pod envoydummy -n chapter11
Pod: envoydummy
   Pod Revision: default
   Pod Ports: 10000 (envoyproxy), 15090 (istio-proxy)
Suggestion: add 'app' label to pod for Istio telemetry.
--------------------
Service: envoydummy
   Port:  80/auto-detect targets pod port 10000
--------------------
Effective PeerAuthentication:
   Workload mTLS mode: PERMISSIVE
```

In the output, you can see that it is suggesting to apply the `app` label to the Pod for Istio telemetry. Istio recommends adding `app` and `version` labels explicitly to workloads in the mesh. These labels add contextual information to the metrics and telemetry collected by Istio. The output also describes some other important pieces of information, such as the service is exposed on port `80` and the endpoint is on port `10000`, the `istio-proxy` Pod is exposed at port `15090`, and `mTLS mode` is permissive. It also describes and warns about any issues with destination rules and virtual services.

The following command is another example of when an incorrect virtual service configuration is applied to `envoydummy`. The correct configuration is available at `Chapter11/04-istio-gateway-chaos.yaml`:

```
% kubectl apply -f Chapter11/04-istio-gateway-chaos.yaml
gateway.networking.istio.io/chapter11-gateway configured
virtualservice.networking.istio.io/mockshop configured
destinationrule.networking.istio.io/envoydummy configured
% istioctl x describe pod envoydummy -n chapter11
Pod: envoydummy
```

```
   Pod Revision: default
   Pod Ports: 10000 (envoyproxy), 15090 (istio-proxy)
--------------------
Service: envoydummy
   Port:  80/auto-detect targets pod port 10000
DestinationRule: envoydummy for "envoydummy"
   Matching subsets: v1
      (Non-matching subsets v2)
   No Traffic Policy
--------------------
Effective PeerAuthentication:
   Workload mTLS mode: PERMISSIVE
Exposed on Ingress Gateway http://192.168.49.2
VirtualService: mockshop
   WARNING: No destinations match pod subsets (checked 1 HTTP
routes)
      Warning: Route to UNKNOWN subset v3; check
DestinationRule envoydummy
```

The warning at the bottom of the preceding snippet clearly describes that the virtual service is routing traffic to `UNKNOWN subset v3`. To fix this problem, you need to either configure the virtual service to correct the subset defined in the destination rule or add a `v3` subset in the destination rules. The correct configuration is available at `Chapter11/04-istio-gateway.yaml`.

Another diagnostic tool to inspect and detect any misconfiguration in the mesh is `istioctl analyze`. It can be run against the whole cluster as well as against any configuration before applying it to the mesh:

```
% istioctl analyze Chapter11/04-istio-gateway-chaos.yaml -n
chapter11
Error [IST0101] (VirtualService chapter11/mockshop
Chapter11/04-istio-gateway-chaos.yaml:35) Referenced
host+subset in destinationrule not found: "envoydummy+v3"
Info [IST0118] (Service chapter11/envoydummy) Port name  (port:
80, targetPort: 10000) doesn't follow the naming convention of
Istio port.
Error: Analyzers found issues when analyzing namespace:
chapter11.
See https://istio.io/v1.16/docs/reference/config/analysis for
more information about causes and resolutions.
```

In the preceding example, `istioctl analyze` pointed out the error by analyzing the configuration file. This is very handy to validate any erroneous configuration before applying it to the mesh. In the following section, we will read about how to troubleshoot errors using envoy access logs.

Troubleshooting errors using access logs

Access logs are produced by the Envoy proxy and can be directed toward the standard output of Envoy. If Istio is installed in demo mode, then access logs are enabled by default and printed to the standard output of the `istio-proxy` container. Access logs are records of traffic flow to Envoy and can be intertwined together with the access of the Ingress and Egress gateway along with all other upstream and downstream workloads in the mesh, to follow the journey of the request.

By default, access logs are turned off unless you have installed Istio using the demo profile. You can check whether access logs are enabled by inspecting the Istio config map:

```
% kubectl get cm/istio -n istio-system -o json | jq .data.mesh
"accessLogFile: \"\"\ndefaultConfig:\n  discoveryAddress:
istiod.istio-system.svc:15012\n  proxyMetadata:
{}\n  tracing:\n    zipkin:\n      address: zipkin.
istio-system:9411\nenablePrometheusMerge: true\
nextensionProviders:\n- envoyOtelAls:\n    port:
4317\n    service: opentelemetry-collector.istio-system.svc.
cluster.local\n  name: otel\nrootNamespace: istio-system\
ntrustDomain: cluster.local"
```

Access logs can be enabled via the following command:

```
% istioctl install --set profile=demo --set meshConfig.
accessLogFile="/dev/stdout"
```

Depending on your needs and system performance requirements, you can decide to turn access logging on or off. If it is turned on and you want to disable it, you can do so by providing a blank value for the `accessLogFile` parameter:

```
% istioctl install --set profile=demo --set meshConfig.
accessLogFile=""
```

The access log can also be enabled at the workload or namespace level. Assuming you have turned off access logs globally, we can turn them on selectively for the `envoydummy` workload using the `Telemetry` resource:

```
apiVersion: telemetry.istio.io/v1alpha1
kind: Telemetry
```

```
metadata:
  name: envoy-dummy-accesslog-overwrite
  namespace: chapter11
spec:
  selector:
    matchLabels:
      service.istio.io/canonical-name: envoydummy
  accessLogging:
  - providers:
    - name: envoy
  - disabled: false
```

The configuration will turn on the access logging for the `envoydummy` Pod. The configuration is available at `Chapter11/03-telemetry-01.yaml` on GitHub:

1. Apply the configuration using the following commands:

   ```
   $ kubectl apply -f Chapter11/03-telemetry-01.yaml
   telemetry.telemetry.istio.io/envoy-dummy-accesslog-
   overwrite configured
   ```

2. Now, you can make a `curl` request to `envoydummy` from the `curl` Pod, as shown in the following command:

   ```
   % kubectl exec -it curl -n chapter11 -c curl -- curl
   envoydummy.chapter11
   Bootstrap Service Mesh Implementation with Istio
   ```

3. Next, if you check the access logs of `istio-proxy` in the `curl` and `envoydummy` Pods, you will find that there are no access logs in the `curl` Pod but there are access logs in the `envoydummy` Pod:

   ```
   [2022-12-15T00:48:54.972Z] "GET / HTTP/1.1" 200 -
   via_upstream - "-" 0 48 0 0 "-" "curl/7.86.0-DEV"
   "2d4eec8a-5c17-9e2c-8699-27a341c21b8b" "envoydummy.
   chapter11" "172.17.0.9:10000" inbound|10000||
   127.0.0.6:49977 172.17.0.9:10000 172.17.0.8:56294
   outbound_.80_._.envoydummy.chapter11.svc.cluster.local
   default
   ```

The reason for not having any access logs in the `curl` Pod is that we turned them off globally but selectively turned them on for the `envoydummy` Pod using the `Telemetry` resource.

You can read more about how to configure `accessLogging` using `Telemetry` at `https://istio.io/latest/docs/reference/config/telemetry/#AccessLogging`.

The default encoding of access logs is a string formatted using the following specification:

```
[%START_TIME%] \"%REQ(:METHOD)% %REQ(X-ENVOY-ORIGINAL-
PATH?:PATH)% %PROTOCOL%\" %RESPONSE_CODE% %RESPONSE_FLAGS%
%RESPONSE_CODE_DETAILS% %CONNECTION_TERMINATION_DETAILS%
\"%UPSTREAM_TRANSPORT_FAILURE_REASON%\" %BYTES_RECEIVED%
%BYTES_SENT% %DURATION% %RESP(X-ENVOY-UPSTREAM-SERVICE-TIME)%
\"%REQ(X-FORWARDED-FOR)%\" \"%REQ(USER-AGENT)%\" \"%REQ(X-
REQUEST-ID)%\" \"%REQ(:AUTHORITY)%\" \"%UPSTREAM_HOST%\"
%UPSTREAM_CLUSTER% %UPSTREAM_LOCAL_ADDRESS% %DOWNSTREAM_LOCAL_
ADDRESS% %DOWNSTREAM_REMOTE_ADDRESS% %REQUESTED_SERVER_NAME%
%ROUTE_NAME%\n
```

A detailed definition of each of these fields is available at `https://www.envoyproxy.io/docs/envoy/latest/configuration/observability/access_log/usage`.

Access logs can also be configured to be displayed in JSON format, which can be achieved by setting `accessLogEncoding` to `JSON`:

```
% istioctl install --set profile=demo --set meshConfig.
accessLogFile="" --set meshConfig.accessLogFormat="" --set
meshConfig.accessLogEncoding="JSON"
```

Once set, the access log will be displayed in JSON format, which I have simplified for ease of reading:

```
{
   "duration":0,
   "start_time":"2022-12-15T01:03:02.725Z",
   "bytes_received":0,
   "authority":"envoydummy.chapter11",
   "upstream_transport_failure_reason":null,
   "upstream_cluster":"inbound|10000||",
   "x_forwarded_for":null,
   "response_code_details":"via_upstream",
   "upstream_host":"172.17.0.9:10000",
   "user_agent":"curl/7.86.0-DEV",
   "request_id":"a56200f2-da0c-9396-a168-8dfddf8b623f",
   "response_code":200,
   "route_name":"default",
```

```
    "method":"GET",
    "downstream_remote_address":"172.17.0.8:45378",
    "upstream_service_time":"0",
    "requested_server_name":"outbound_.80_._.envoydummy.
chapter11.svc.cluster.local",
    "protocol":"HTTP/1.1",
    "path":"/",
    "bytes_sent":48,
    "downstream_local_address":"172.17.0.9:10000",
    "connection_termination_details":null,
    "response_flags":"-",
    "upstream_local_address":"127.0.0.6:42313"
}
```

In the access logs, there is a field called `response_flags` (as seen in the preceding code snippet), which is a very useful piece of information when troubleshooting via access logs.

Next, we will learn about response flags by injecting some errors in the `envoydummy` Pod:

1. Let's first turn on access logs for the `curl` Pod using the following command:

   ```
   % kubectl apply -f Chapter11/03-telemetry-02.yaml
   telemetry.telemetry.istio.io/curl-accesslog-overwrite
   created
   ```

2. Then, delete the `envoydummy` Pod but keep the `envoydummy` service:

   ```
   % kubectl delete po envoydummy -n chapter11
   pod "envoydummy" deleted
   ```

3. Now, the service is broken, and you will not be able to use `curl`:

   ```
   % kubectl exec -it curl -n chapter11 -c curl -- curl
   envoydummy.chapter11
   no healthy upstream
   ```

4. Check the access logs for the `curl` Pod:

   ```
   % kubectl logs -f curl -n chapter11 -c istio-proxy | grep
   response_flag
   ```

```
{"path":"/","response_code":503,"method":"GET","upstream_
cluster":"outbound|80||envoydummy.chapter11.svc.cluster.
local","user_agent":"curl/7.86.0-DEV","connection_
termination_details":null,"authority":"envoydummy.
chapter11","x_forwarded_for":null,"upstream_
transport_failure_reason":null,"downstream_local_
address":"10.98.203.175:80","bytes_received":0,"requested_
server_name":null,"response_code_details":"no_healthy_
upstream","upstream_service_time":null,"request_
id":"4b39f4ca-ffe3-9c6a-a202-0650b0eea8ef","route_
name":"default","upstream_local_address":null,"response_
flags":"UH","protocol":"HTTP/1.1","start_
time":"2022-12-15T03:49:38.504Z","duration"
:0,"upstream_host":null,"downstream_remote_
address":"172.17.0.8:52180","bytes_sent":19}
```

The value of `response_flags` is `UH`, which means there is no healthy upstream host in the upstream cluster. Possible values of `response_flag` are shown in the following table and referenced from `https://www.envoyproxy.io/docs/envoy/latest/configuration/observability/access_log/usage`:

Name	Description
UH	No healthy upstream hosts in the upstream cluster in addition to a 503 response code
UF	Upstream connection failure in addition to a 503 response code
UO	Upstream overflow (circuit breaking) in addition to a 503 response code
NR	No route configured for a given request in addition to a 404 response code, or no matching filter chain for a downstream connection
URX	The request was rejected because the upstream retry limit (HTTP) or maximum connect attempts (TCP) was reached
NC	Upstream cluster not found
DT	A request or connection exceeded `max_connection_duration` or `max_downstream_connection_duration`

Table 11.1 – Response flag values for HTTP and TCP connections

The following table describes the value of the response flag for HTTP connections:

Name	Description
DC	Downstream connection termination
LH	Local service failed health check request in addition to a 503 response code
UT	Upstream request timeout in addition to a 504 response code
LR	Connection local reset in addition to a 503 response code
UR	Upstream remote reset in addition to a 503 response code
UC	Upstream connection termination in addition to a 503 response code
DI	The request processing was delayed for a period specified via fault injection
FI	The request was aborted with a response code specified via fault injection
RL	The request was rate-limited locally by the HTTP rate limit filter in addition to a 429 response code
UAEX	The request was denied by the external authorization service
RLSE	The request was rejected because there was an error in the rate limit service
IH	The request was rejected because it set an invalid value for a strictly-checked header, in addition to a 400 response code
SI	Stream idle timeout in addition to a 408 or 504 response code
DPE	The downstream request had an HTTP protocol error
UPE	The upstream response had an HTTP protocol error
UMSDR	The upstream request reached max stream duration
OM	The overload manager terminated the request
DF	The request was terminated due to a DNS resolution failure

Table 11.2 – Response flag values for HTTP connections

Response flags are very useful in troubleshooting access logs and good indicators of what might have gone wrong with the upstream systems. With that knowledge, let's move our focus to how Istio debug logs can be used for troubleshooting.

Troubleshooting errors using debug logs

Istio components support a flexible logging scheme for **debug** logs. The debug log levels can be changed from a high level to a very verbose level to get details of what's happening in the Istio control and data planes. The following two sections will describe the process for changing the log level for the Istio data and control planes. Let's dive right in!

Changing debug logs for the Istio data plane

The following are various log levels for the Istio data plane – that is, the Envoy sidecar:

- `trace`: Highest verbose log messages
- `debug`: Very verbose log messages
- `info`: Informative messages to know about Envoy's execution state
- `warning/warn`: Events that indicate problems and may lead to error events
- `error`: Error events that are important and may impair some of Envoy's capability but will not make Envoy completely non-functional
- `critical`: Severe error events that may cause Envoy to stop functioning
- `off`: Produces no logs

The log levels can be changed using the following command format:

```
istioctl proxy-config log [<type>/]<name>[.<namespace>] [flags]
```

An example of such a command is as follows:

```
% istioctl proxy-config log envoydummy.chapter11 -n chapter11
--level debug
```

This way of changing log levels doesn't require a restart of the Pods. As discussed in the *Exploring istiod ports* section earlier in this chapter, the log levels can also be changed using the ControlZ interface.

Changing log levels for the Istio control plane

The Istio control plane supports the following log levels:

- `none`: Produces no logs
- `error`: Produces only errors
- `warn`: Produces warning messages
- `info`: Produces detailed information for normal conditions
- `debug`: Produces the maximum amount of log messages

Each component inside istiod categorizes the logs based on the type of message being logged. These categories are called **scopes**. Using scopes, we can control the logging of messages across all components in istiod. Messages that cannot be categorized into a scope are logged under a default scope. In Istio 1.16.0, there are 25 scopes: `ads`, `adsc`, `all`, `analysis`, `authn`, `authorization`, `ca`, `cache`, `cli`, `default`, `installer`, `klog`, `mcp`, `model`, `patch`, `processing`, `resource`, `source`, `spiffe`, `tpath`, `translator`, `util`, `validation`, `validationController`, and `wle`.

The following is an example of a command used to change log levels for various scopes:

```
% istioctl analyze --log_output_level
validation:debug,validationController:info,ads:debug
```

In this example, we are changing the log levels for the `validation` scope to `debug`, the `validationController` scope to `info`, and `ads` to `debug`:

```
% istioctl admin log | grep -E 'ads|validation'
ads                      ads
debugging                                  debug
adsc                     adsc
debugging                                 info
validation               CRD validation
debugging                     debug
validationController     validation webhook
controller                  info
validationServer         validation webhook
server                      info
```

You can use `istioctl admin log` to retrieve the log level for all Istio components:

```
% istioctl admin log
ACTIVE    SCOPE      DESCRIPTION                   LOG LEVEL
ads                      ads
debugging                                  debug
adsc                     adsc
debugging                                 info
analysis                 Scope for configuration analysis
runtime      info
authn                    authn
debugging                                info
authorization            Istio Authorization
Policy                     info
ca                       ca
client                                     info
controllers              common controller
logic                        info
default                  Unscoped logging
messages.                     info
delta                    delta xds
```

```
debugging                           info
file                   File client messa
ges                         info
gateway                gateway-api
controller                       info
grpcgen                xDS Generator for Proxyless
gRPC              info
......
```

We just looked at Envoy, which is on the `request` path of the request, but there is another critical component that is not on the `request` path but is important for smooth Envoy operations. In the next section, we will read about how to debug any issues in istio-agent.

Debugging the Istio agent

In this section, we will read how to troubleshoot any issues in the data plane caused by the misconfiguration of the Istio agent. An Istio agent may not act as expected due to a multitude of reasons; in this section, we will discuss various options to debug and troubleshoot such issues.

The following command can be used to inspect the initial bootstrap configuration file that is used by Envoy to start itself and connect with istiod:

```
$ istioctl proxy-config bootstrap envoydummy -n chapter11 -o
json >bootstrap-envoydummy.json
```

The bootstrap configuration is composed of the information provided by the Istiod controller during sidecar injection via validation and the sidecar injection webhook.

We can check the certificate and `secret` configured for Envoy by `istio-agent` using the following command:

```
% istioctl proxy-config secret envoydummy -n chapter11
RESOURCE NAME     TYPE           STATUS     VALID
CERT     SERIAL NUMBER                              NOT
AFTER                  NOT BEFORE
default           Cert
Chain      ACTIVE      true      151990293406794295074718429679
77531899      20XX-12-26T01:02:53Z      20XX-12-25T01:00:53Z
ROOTCA         CA         ACTIVE         true
177195801324177165655021729164749485784      20XX-12-
11T05:19:23Z      20XX-12-14T05:19:23Z
```

You can also display the detailed information in JSON format by adding `-o json` to the command. `ROOTCA` is the root certificate, and `default` is the workload certificate. When performing a multi-cluster setup, ROOTCA must match in different clusters.

You can inspect the certificate values using the following command:

```
% istioctl proxy-config secret envoydummy -n chapter11 -o
json | jq '.dynamicActiveSecrets[0].secret.tlsCertificate.
certificateChain.inlineBytes' -r | base64 -d | openssl x509
-noout -text
```

There might be other Secrets also depending on configured gateways and destination rules. In the logs, if you find that Envoy is stuck in a warning state, then that means that the correct Secret has not been loaded in Envoy. Issues related to the `default` certificate and `ROOTCA` are usually caused by connectivity issues between istio-proxy and istiod.

To check that Envoy has successfully started, you can log into the `istio-proxy` container using the following command:

```
% kubectl exec -it envoydummy -n chapter11 -c istio-proxy
--  pilot-agent wait
2022-12-25T05:29:44.310696Z info Waiting for Envoy proxy to be
ready (timeout: 60 seconds)...
2022-12-25T05:29:44.818220Z info Envoy is ready!
```

During the bootstrap of the sidecar proxy, `istio-agent` checks the readiness of Envoy by pinging `http://localhost:15021/healthz/ready`. It also uses the same endpoint for determining the readiness of Envoy during the lifetime of the Pod. An HTTP status code of 200 means that Envoy is ready and the `istio-proxy` container is marked as initialized. If the `istio-proxy` container is in a pending state and not initializing, then that means that Envoy has not received the configuration from istiod, which can be either because of connectivity issues with istiod or a config rejected by Envoy.

Errors such as the following in `istio-proxy` logs denote connectivity issues either because of the network or unavailability of istiod:

```
20XX-12-25T05:58:02.225208Z     warning envoy
config    StreamAggregatedResources gRPC config stream to
xds-grpc closed since 49s ago: 14, connection error: desc =
"transport: Error while dialing dial tcp 10.107.188.192:15012:
connect: connection refused"
```

If Envoy can connect with istiod, then you will find a message similar to the following in the log files:

```
20XX-12-
25T06:00:08.082523Z     info    xdsproxy        connected to
upstream XDS server: istiod.istio-system.svc:15012
```

In this section, we did a deep dive into Istio debug logs, and we looked at how to troubleshoot using the debug logs of Envoy, `istio-agent`, and the Istio control plane. In the next section, we will read about Istio's best practices for securely managing and efficiently operating Istio.

Understanding Istio's best practices

When operating the Service Mesh, it is advised to assume that security threats will not just originate from outside of the organization's security boundaries but also from within the security perimeter. You should always assume that networks are not impregnable and create security controls that can secure assets, even if network boundaries are breached. In this section, we will discuss some of the various attack vectors to be mindful of when implementing Service Mesh.

Examining attack vectors for the control plane

The following list shows common strategies for initiating attacks on the control plane:

- Causing configuration to deliberately make the control plane malfunction so that the Service Mesh becomes inoperable, thus impacting business-critical applications being managed by the mesh. This can also be a precursor to forthcoming attacks targeting Ingress or any other applications.
- Obtaining privileged access to be able to perform control plane and data plane attacks. By gaining privileged access, an attacker can then modify security policies to allow the exploitation of the assets.
- Eavesdropping to exfiltrate sensitive data from the control plane, or tampering and spoofing communication between data and control planes.

Examining attack vectors for the data plane

The following are common strategies for initiating attacks on the data plane:

- Eavesdropping on service-to-service communication to exfiltrate sensitive data and send it to an attacker.
- Masquerading as a trusted service in the mesh; the attacker can then perform man-in-the-middle attacks between service-to-service communication. By using a man-in-the-middle attack, the attacker can steal sensitive data or tamper with the communication between the services to produce a favorable outcome for the attacker.
- Manipulating the applications to perform botnet attacks.

The other component susceptible to attacks is the infrastructure hosting istio, which can be the Kubernetes cluster, virtual machines, or any other components of the underlying stack hosting istio. We will not dive into how to protect Kubernetes, as there are various books on how to secure Kubernetes clusters; one such book is *Learn Kubernetes Security*, written by Kaizhe Huang and Pranjal Jumde and published by Packt. Other best practices include protecting the Service Mesh, which we will discuss in the following section.

Securing the Service Mesh

Some best practices on how to secure the Service Mesh are as follows:

- Deploy a **web application firewall** (**WAF**) to protect Ingress traffic. WAF implements security controls, including threats identified by **Open Web Application Security Project** (**OWASP**); you can read about OWASP at `https://owasp.org/`. Most cloud providers provide WAF as part of their cloud offering; some examples are **AWS WAF** by AWS, **Cloud Armor** by Google Cloud, and **Azure Web Application Firewall** from Azure. There are other vendors, such as Cloudflare, Akamai, Imperva, and AppTrana, who provide WAF as a SaaS offering, whereas vendors such as Fortinet and Citrix also provide self-hosted WAF offerings. WAFs are one of your first lines of defense and will take care of many attack vectors bound for Ingress to the mesh.
- Define policies to control access from outside the mesh to services inside the mesh. Ingress access control policies are important to prohibit unauthorized access to services. Every Ingress should be well-defined and have associated authentication and authorization policies to verify whether external requests are authorized to access services exposed by the Ingress gateway. Nevertheless, all Ingress should happen via Ingress gateways, and every Ingress should be routed via a virtual service and the destination rules associated with it.
- All Egress systems should be known and defined, and traffic to unknown Egress points should not be allowed. Security policies should enforce TLS origination for Egress traffic and for all Egress to happen via Egress gateways. Authorization policies should be used to control what workloads are allowed to send Egress traffic and, if allowed, all Egress endpoints should be known and approved by security administrators. Egress security policies also help prevent data exfiltration; with Egress policies, you can control traffic to known Egresses only and thus stop an attacker who has infiltrated your system from sending data to the attacker's systems. This also stops applications within the mesh from participating in any botnet attacks.
- All services in the mesh should communicate over mTLS and should have associated authentication and authorization policies. By default, all service-to-service communication should be denied unless authorized via authorization policies, and any service-to-service communication should be explicitly enabled via well-defined service identities.

- Where a service-to-service communication is happening on behalf of an end user or system, all such communication (apart from mTLS) should also make use of JWTs. A JWT acts as a credential to prove that a service request is explicitly being made on behalf of the end user; the caller service needs to present a JWT as a credential identifying an end user, combined with authentication and authorization policies that you can enforce after determining what services can be accessed and what level of access is granted. This helps to stop any compromised application from performing data exfiltration or service exploitation.
- If any external long-lived authentication token is used for authenticating any subject, be it an end user or a system, then such a token should be replaced by a short-lived token. Token replacement should happen at Ingress, and then the short-lived token should be used throughout the mesh. Doing so helps prevent attacks where an attacker steals tokens and uses them for unauthorized access. Also, whereas the long-lived external attack might have many broader scopes attached to it that might be misused by a compromised application, having a short-lived token with restricted scope helps to avoid misuse of tokens.
- When applying exceptions to mesh or security rules, then you should proceed with caution when defining exception policies. For example, if you want to enable workload A in the mesh to allow HTTP traffic from another workload, B, then instead of explicitly allowing all HTTP traffic, you should explicitly define an exception to allow HTTP traffic from workload B, whereas all other traffic should be on HTTPS only.
- Access to istiod must be restricted and controlled. Firewall rules should restrict access to the control plane to known sources. The rule should cater to human operators as well the data plane's access to the control plane in single and multi-cluster setups.
- All workloads being managed by the Service Mesh should be managed by the Service Mesh only via **role-based access control** (**RBAC**) policies for the Kubernetes environment and user groups for non-Kubernetes workloads. Kubernetes administrators should carefully define RBAC policies for application users and mesh administrators and allow only the latter to make any changes to the mesh. Mesh operators should be further classified according to the operations they are authorized to perform. For example, mesh users with permission to deploy applications in a namespace should not have access to other namespaces.
- Restrict what repositories are accessible to users from where images can be pulled for deployment.

In this section, we read about istio's best practices; as well as this section, you should also read about best practices on the istio website at `https://istio.io/latest/docs/ops/best-practices/`. The website is frequently updated based on feedback from the istio community.

Even after all controls, mesh operators may accidentally misconfigure the Service Mesh, which can result in unexpected outcomes and even security breaches. One option is to enforce a stringent review and governance process for making changes, but doing so manually is expensive, time-consuming, error-prone, and often annoying. In the following section, we will read about **OPA Gatekeeper** and how to use it for the automation of best practices policies.

Automating best practices using OPA Gatekeeper

To avoid human errors, you can define the best practices and constraints in the form of policies that can then be enforced automatically whenever a resource is created, deleted, or updated in the cluster. Automated policy enforcement ensures consistency and adherence to best practices without compromising agility and deployment velocity. One such software is **Open Policy Agent** (**OPA**) Gatekeeper, which is an admission controller that enforces policies based on the **custom resource definition** (**CRD**), executed by OPA. OPA Gatekeeper enables the enforcement of guard rails; any istio configuration not within the guard rails is automatically rejected. It also allows **cluster administrators** to audit the resources in breach of best practices. Using the following steps, we will set up OPA Gatekeeper, followed by the configuration to enforce some of the best practices for istio. Let's get started!

1. Install Gatekeeper using the following command:

```
% kubectl apply -f https://raw.githubusercontent.com/
open-policy-agent/gatekeeper/master/deploy/gatekeeper.
yaml
```

2. Configure Gatekeeper to sync namespaces, Pods, Services, istio CRD gateways, virtual services, destination rules, policy, and service role bindings into its cache. We have defined that in the following file available at `Chapter11/05-GatekeeperConfig.yaml`:

```
apiVersion: config.gatekeeper.sh/v1alpha1
kind: Config
metadata:
  name: config
  namespace: gatekeeper-system
spec:
  sync:
    syncOnly:
      - group: ""
        version: "v1"
        kind: "Namespace"
      - group: ""
        version: "v1"
        kind: "Pod"
      - group: ""
        version: "v1"
        kind: "Service"
      - group: "networking.istio.io"
```

```
    version: "v1alpha3"
    kind: "Gateway"
  - group: "networking.istio.io"
    version: "v1alpha3"
    kind: "VirtualService"
  - group: "networking.istio.io"
    version: "v1alpha3"
    kind: "DestinationRule"
  - group: "authentication.istio.io"
    version: "v1alpha1"
    kind: "Policy"
  - group: "rbac.istio.io"
    version: "v1alpha1"
    kind: "ServiceRoleBinding"
```

3. Now, apply the configuration using the following command:

```
% kubectl apply -f Chapter11/05-GatekeeperConfig.yaml
config.config.gatekeeper.sh/config created
```

This completes the installation of Gatekeeper; next, we will configure constraints.

We will start with a simple policy for ensuring Pod naming conventions as per istio best practices. As discussed in the *Inspecting and analyzing istio configuration* section, istio recommends adding an explicit `app` and `version` label to every Pod deployment. The `app` and `version` labels add contextual information to the metrics and telemetry that istio collects.

Perform the following steps to enforce this governance rule so that any deployment not adhering to this rule is automatically rejected:

1. First, we must define `ConstraintTemplate`. In the constraint template, we do the following:
   - Describe the policy that will be used to enforce the constraints
   - Describe the schema of the constraint

> **Important note**
> Gatekeeper constraints are defined using a purpose-built, high-level declarative language called **Rego**. Rego is particularly used for writing OPA policies; you can read more about Rego at `https://www.openpolicyagent.org/docs/latest/#rego`.

The following steps defines the constraint template and the schema:

I. We will declare a constraint template using the OPA CRD:

```
apiVersion: templates.gatekeeper.sh/v1beta1
kind: ConstraintTemplate
metadata:
  name: istiorequiredlabels
  annotations:
    description: Requires all resources to contain a
specified label with a value
      matching a provided regular expression.
```

II. Next, we will define the schema:

```
spec:
  crd:
    spec:
      names:
        kind: istiorequiredlabels
      validation:
        # Schema for the `parameters` field
        openAPIV3Schema:
          properties:
            message:
              type: string
            labels:
              type: array
              items:
                type: object
                properties:
                  key:
                    type: string
```

III. Finally, we will define `rego` to check the labels and detect any violation:

```
 targets:
    - target: admission.k8s.gatekeeper.sh
      rego: |
        package istiorequiredlabels
```

```
        get_message(parameters, _default) = msg {
          not parameters.message
          msg := _default
        }
        get_message(parameters, _default) = msg {
          msg := parameters.message
        }
        violation[{"msg": msg, "details": {"missing_
labels": missing}}] {
          provided := {label | input.review.object.
metadata.labels[label]}
          required := {label | label := input.parameters.
labels[_].key}
          missing := required - provided
          count(missing) > 0
          def_msg := sprintf("you must provide labels:
%v", [missing])
          msg := get_message(input.parameters, def_msg)
        }
```

The configuration is available at `Chapter11/gatekeeper/01-istiopodlabelconstraint_template.yaml`. Apply the configuration using the following command:

```
% kubectl apply -f Chapter11/gatekeeper/01-
istiopodlabelconstraint_template.yaml
constrainttemplate.templates.gatekeeper.sh/
istiorequiredlabels created
```

2. Next, we will define `constraints` that are used to inform Gatekeeper that we want to enforce the constraint template named `istiorequiredlabels`, as per the configuration defined in the `mesh-pods-must-have-app-and-version` constraint. The sample file is available at `Chapter11/gatekeeper/01-istiopodlabelconstraint.yaml`:

```
apiVersion: constraints.gatekeeper.sh/v1beta1
kind: istiorequiredlabels
metadata:
  name: mesh-pods-must-have-app-and-version
spec:
  enforcementAction: deny
```

```
    match:
      kinds:
        - apiGroups: [""]
          kinds: ["Pod"]
      namespaceSelector:
        matchExpressions:
          - key: istio-injection
            operator: In
            values: ["enabled"]
    parameters:
      message: "All pods must have an `app and version`
  label"
      labels:
        - key: app
        - key: version
```

For the constraint configuration, we have defined the following fields:

- `enforcementAction`: This field defines the action for handling constraint violations. The field is set to `deny`, which is also the default behavior; any resource creation or update that is in violation of this constraint will be handled as per enforcement action. Other supported `enforcementAction` values include `dryrun` and `warn`. When rolling out new constraints to running clusters, the `dryrun` functionality can be helpful to test them in a running cluster without enforcing them. The `warn` enforcement action offers the same benefits as `dryrun`, such as testing constraints without enforcing them. In addition to this, it also provides immediate feedback on why that constraint would have been denied.
- `match`: This field defines the selection criteria for identifying the objects to which the constraints will be applied. In the configuration, we have defined that the constraints should be applied to `pod` resources that are deployed in a namespace with the label of `istio-injection` and a value of `enabled`. By doing this, we can selectively apply the constraints to namespaces that are part of the mesh data plane.

3. Finally, we have defined the message to be displayed when constraints are violated. Apply the following constraints:

```
% kubectl apply -f Chapter11/gatekeeper/01-
istiopodlabelconstraint.yaml
istiorequiredlabels.constraints.gatekeeper.sh/mesh-pods-
must-have-app-and-version created
```

4. As a test, we will deploy the envoydummy Pod with missing labels:

```
% kubectl apply -f Chapter11/06-envoy-proxy-chaos.yaml -n
chapter11
service/envoydummy created
Error from server (Forbidden): error when creating
"Chapter11/01-envoy-proxy.yaml": admission webhook
"validation.gatekeeper.sh" denied the request: [all-must-
have-owner] All pods must have an `app` label
```

The deployment was verified by Gatekeeper and rejected because it is violating constraints by not having app and version labels. Please fix the labels and redeploy to check that you can successfully deploy with the correct labels.

I hope this gave you some idea about how to use Gatekeeper for automating some of the best practices. We will practice one more example to give you some further confidence on how to use Gatekeeper. At first, it might appear a daunting task to define the constraints, but once you build some experience using rego, then you will find it simple and easy to use.

As another example, let's write another constraint to enforce port naming conventions. As per istio best practices described at https://istio.io/v1.0/docs/setup/kubernetes/spec-requirements/, Service ports must be named. The port names must be of the <protocol>[-<suffix>] form, with http, http2, grpc, mongo, or redis as <protocol>. This is important if you want to take advantage of istio's routing features. For example, name: http2-envoy and name: http are valid port names, but name: http2envoy is an invalid name.

The rego in the constraint template will be as follows:

```
package istio.allowedistioserviceportname
        get_message(parameters, _default) = msg {
          not parameters.message
          msg := _default
        }
        get_message(parameters, _default) = msg {
          msg := parameters.message
        }
        violation[{"msg": msg, "details": {"missing_prefixes":
prefixes}}] {
          service := input.review.object
          port := service.spec.ports[_]
          prefixes := input.parameters.prefixes
          not is_prefixed(port, prefixes)
```

```
            def_msg := sprintf("service %v.%v port name missing
prefix",
              [service.metadata.name, service.metadata.
namespace])
            msg := get_message(input.parameters, def_msg)
          }
          is_prefixed(port, prefixes) {
            prefix := prefixes[_]
            startswith(port.name, prefix)
          }
```

In the `rego`, we are defining that port names should start with a prefix, and if the prefix is missing, then it should be considered a violation.

The constraint is defined as follows:

```
apiVersion: constraints.gatekeeper.sh/v1beta1
kind: AllowedIstioServicePortName
metadata:
  name: port-name-constraint
spec:
  enforcementAction: deny
  match:
    kinds:
      - apiGroups: [""]
        kinds: ["Service"]
    namespaceSelector:
      matchExpressions:
        - key: istio-injection
          operator: In
          values: ["enabled"]
  parameters:
    message: "All services declaration must have port name will
one of following  prefix http-, http2-, grpc-, mongo-,redis-"
    prefixes: ["http-", "http2-","grpc-","mongo-","redis-"]
```

In the constraint, we are defining the various prefixes that are allowed and the corresponding error message to be displayed when the constraints are violated.

Now, let's apply the constraint template and configuration as follows:

```
% kubectl apply -f Chapter11/gatekeeper/02-
istioportconstraints_template.yaml
constrainttemplate.templates.gatekeeper.sh/
allowedistioserviceportname created
% kubectl apply -f Chapter11/gatekeeper/02-
istioportconstraints.yaml
allowedistioserviceportname.constraints.gatekeeper.sh/port-
name-constraint configured
```

After applying the config, let's deploy envoydummy with incorrect names for the service ports. We will create an envoydummy service without specifying any port names, as described in the following code snippet:

```
spec:
  ports:
  - port: 80
    targetPort: 10000
  selector:
    name: envoydummy
```

The file is available at Chapter11/07-envoy-proxy-chaos.yaml. Apply the configuration using the following code and observe the error message to see OPA Gatekeeper in action:

```
% kubectl apply -f Chapter11/07-envoy-proxy-chaos.yaml
pod/envoydummy created
Error from server (Forbidden): error when creating
"Chapter11/07-envoy-proxy-chaos.yaml": admission webhook
"validation.gatekeeper.sh" denied the request: [port-name-
constraint] All services declaration must have port name with
one of following prefix http-, http2-, grpc-, mongo-, redis-
```

In the response, you can see the error message caused by the incorrect naming of the port in the service configuration. The issue can be resolved by adding a name to the port declaration as follows:

```
spec:
  ports:
  - port: 80
    targetPort: 10000
    name: http-envoy
```

```
  selector:
    name: envoydummy
```

OPA Gatekeeper is a powerful tool to automate the enforcement of best practices for the Service Mesh. It compensates for any misconfiguration caused by human operators and reduces the cost and time required to keep your mesh aligned with the best practices guard rails. You can read more about OPA Gatekeeper at `https://open-policy-agent.github.io/gatekeeper/website/docs/`, and there are also some good examples of Gatekeeper available at `https://github.com/crcsmnky/gatekeeper-istio`.

Uninstalling OPA Gatekeeper

To uninstall OPA Gatekeeper, you may use the following command:

```
% kubectl delete -f https://raw.githubusercontent.com/open-policy-agent/gatekeeper/master/deploy/gatekeeper.yaml
```

Summary

In this chapter, we read about various troubleshooting techniques as well as best practices for configuring and operating istio. By now, you should have a good understanding of various ports exposed by istio and how they can help diagnose any errors in the mesh. You also read about debugs and access logs produced by Envoy and istiod and how they can help you pinpoint the root cause of errors. istio provides various tools in its diagnostic toolkit that are very helpful for troubleshooting and analyzing issues and errors in the Service Mesh.

Security is of utmost importance when running the Service Mesh, which is why we discussed various attack vectors for the control and data planes. You should now have a good understanding of the list of controls you can put in place to secure the Service Mesh. Finally, we read about how to automate best practices using OPA Gatekeeper to catch most, if not all, non-compliant configurations. You learned how to set up OPA Gatekeeper, how to define constraint templates using Rego and constraint schema, and how to use them to catch poor configurations. I hope this chapter provides you with the confidence to troubleshoot and operate istio and use automation tools such as OPA Gatekeeper to enforce configuration hygiene in your istio Service Mesh implementation.

In the next chapter, we will put the learning from this book into practice by deploying an application on the mesh.

12

Summarizing What We Have Learned and the Next Steps

Throughout the book, you learned about and practiced various concepts of Service Mesh and how to apply them using Istio. It is strongly recommended that you practice the hands-on examples in each chapter. Don't just limit yourself to the scenarios presented in this book but rather explore, tweak, and extend the examples and apply them to real-world problems you are facing in your organizations.

In this chapter, we will revise the concepts discussed in this book by implementing Istio for an Online Boutique application. It will be a good idea to look at scenarios presented in this chapter and try to implement them yourself before looking at code examples. I hope reading this last chapter provides you with more confidence in using Istio. We will go through the following topics in this chapter:

- Enforcing best practices using OPA Gatekeeper
- Applying the learnings of this book to a sample Online Boutique application
- Istio roadmap, vision, and documentation, and how to engage with the community
- Certification, learning resources, and various pathways to learning
- The Extended Berkeley Packet Filter

Technical requirements

The technical requirements in this chapter are similar to *Chapter 4*. We will be using AWS EKS to deploy a website for an online boutique store, which is an open source application available under Apache License 2.0 at `https://github.com/GoogleCloudPlatform/microservices-demo`.

Please check *Chapter 4*'s *Technical requirements* section to set up the infrastructure in AWS using Terraform, set up kubectl, and install Istio including observability add-ons. To deploy the Online Boutique store application, please use the deployment artifacts in the `Chapter12/online-boutique-orig` file on GitHub.

You can deploy the Online Boutique store application using the following commands:

```
$ kubectl apply -f Chapter12/online-boutique-orig/00-online-
boutique-shop-ns.yaml
namespace/online-boutique created
$ kubectl apply -f Chapter12/online-boutique-orig
```

The last command should deploy the Online Boutique application. After some time, you should be able to see all the Pods running:

```
$ kubectl get po -n online-boutique
NAME                           READY  STATUS    RESTARTS   AGE
adservice-8587b48c5f-
v7nzq                1/1     Running   0           48s
cartservice-5c65c67f5d-
ghpq2              1/1     Running   0           60s
checkoutservice-54c9f7f49f-
9qgv5          1/1     Running   0           73s
currencyservice-5877b8dbcc-
jtgcg          1/1     Running   0           57s
emailservice-5c5448b7bc-
kpgsh             1/1     Running   0           76s
frontend-67f6fdc769-
r8c5n                 1/1     Running   0           68s
paymentservice-7bc7f76c67-
r7njd            1/1     Running   0           65s
productcatalogservice-67fff7c687-
jrwcp    1/1     Running   0           62s
recommendationservice-b49f757f-
9b78s      1/1     Running   0           70s
redis-cart-58648d854-
jc2nv                1/1     Running   0           51s
shippingservice-76b9bc7465-
qwnvz          1/1     Running   0           55s
```

The name of the workloads also reflects their role in the Online Boutique application, but you can find more about this freely available open source application at `https://github.com/GoogleCloudPlatform/microservices-demo`.

For now, you can access the application via the following command:

```
$ kubectl port-forward svc/frontend 8080:80 -n online-boutique
Forwarding from 127.0.0.1:8080 -> 8079
Forwarding from [::1]:8080 -> 8079
```

You can then open it on the browser using `http://localhost:8080`. You should see something like the following:

Figure 12.1 – Online Boutique application by Google

This completes the technical setup required for code examples in this chapter. Let's get into the main topics of the chapter. We will begin with setting up the OPA Gatekeeper to enforce Istio deployment best practices.

Enforcing workload deployment best practices using OPA Gatekeeper

In this section, we will deploy OPA Gatekeeper using our knowledge from *Chapter 11*. We will then configure OPA policies to enforce that every deployment has `app` and `version` as labels, and all port names have protocol names as a prefix:

1. Install OPA Gatekeeper. Deploy it by following the instructions in *Chapter 11*, in the *Automating best practices using OPA Gatekeeper* section:

```
% kubectl apply -f https://raw.githubusercontent.com/
open-policy-agent/gatekeeper/master/deploy/gatekeeper.
yaml
```

2. After deploying OPA Gatekeeper, you need to configure it to sync namespaces, Pods, services and Istio CRD gateways, virtual services, destination rules, and policy and service role bindings into its cache. We will make use of the configuration file we created in *Chapter 11*:

```
$ kubectl apply -f Chapter11/05-GatekeeperConfig.yaml
config.config.gatekeeper.sh/config created
```

3. Configure OPA Gatekeeper to apply the constraints. In *Chapter 11*, we configured constraints to enforce that Pods should have `app` and `version` numbers as labels (defined in `Chapter11/gatekeeper/01-istiopodlabelconstraint_template.yaml` and `Chapter11/gatekeeper/01-istiopodlabelconstraint.yaml`), and all port names should have a protocol name as a prefix (defined in `Chapter11/gatekeeper/02-istioportconstraints_template.yaml` and `Chapter11/gatekeeper/02-istioportconstraints.yaml`). Apply the constraints using the following commands:

```
$ kubectl apply -f Chapter11/gatekeeper/01-
istiopodlabelconstraint_template.yaml
constrainttemplate.templates.gatekeeper.sh/
istiorequiredlabels created
$ kubectl apply -f Chapter11/gatekeeper/01-
istiopodlabelconstraint.yaml
istiorequiredlabels.constraints.gatekeeper.sh/mesh-pods-
must-have-app-and-version created
$ kubectl apply -f Chapter11/gatekeeper/02-
istioportconstraints_template.yaml
constrainttemplate.templates.gatekeeper.sh/
allowedistioserviceportname created
$ kubectl apply -f Chapter11/gatekeeper/02-
istioportconstraints.yaml
allowedistioserviceportname.constraints.gatekeeper.sh/
port-name-constraint created
```

This completes the deployment and configuration of OPA Gatekeeper. You should extend the constraints with anything else you might like to be included to ensure good hygiene of deployment descriptors of the workloads.

In the next section, we will redeploy the Online Boutique application and enable istio sidecar injection and then discover the configurations that are in violation of OPA constraints and resolve them one by one.

Applying our learnings to a sample application

In this section, we will apply the learnings of the book – specifically, the knowledge from *Chapters 4* to *6* – to our Online Boutique application. Let's dive right in!

Enabling Service Mesh for the sample application

Now that OPA Gatekeeper is in place with all the constraints we want it to enforce on deployments, it's time to deploy a sample application. We will first start with un-deploying the `online-boutique` application and redeploying with istio-injection enabled at the namespace level.

Undeploy the Online Boutique application by deleting the `online-boutique` namespace:

```
% kubectl delete ns online-boutique
namespace " online-boutique " deleted
```

Once undeployed, let's modify the namespace and add an `istio-injection:enabled` label and redeploy the application. The updated namespace configuration will be as follows:

```
apiVersion: v1
kind: Namespace
metadata:
  name: online-boutique
  labels:
    istio-injection: enabled
```

The sample file is available at `Chapter12/OPAGatekeeper/automaticsidecarinjection/00-online-boutique-shop-ns.yaml` on GitHub.

With automatic sidecar injection enabled, let's try to deploy the application using the following commands:

```
$ kubectl apply -f Chapter12/OPAGatekeeper/
automaticsidecarinjection
namespace/online-boutiquecreated
$ kubectl apply -f Chapter12/OPAGatekeeper/
automaticsidecarinjection
Error from server (Forbidden): error when creating "Chapter12/
OPAGatekeeper/automaticsidecarinjection/02-carts-svc.yml":
admission webhook "validation.gatekeeper.sh" denied the
request: [port-name-constraint] All services declaration must
have port name with one of following  prefix http-, http2-
,https-,grpc-,grpc-web-,mongo-,redis-,mysql-,tcp-,tls-
```

There will be errors caused by constraint violations imposed by OPA Gatekeeper. The output in the preceding example is truncated to avoid repetitions but from the output in your terminal, you must notice that all deployments are in violation and hence no resource is deployed to the online-boutique namespace.

Try to fix the constraint violation by applying the correct labels and naming ports correctly as suggested by Istio best practices.

You need to apply `app` and `version` labels to all deployments. The following is an example for a `frontend` deployment:

```
apiVersion: apps/v1
kind: Deployment
metadata:
  name: frontend
  namespace: online-boutique
spec:
  selector:
    matchLabels:
      app: frontend
  template:
    metadata:
      labels:
        app: frontend
        version: v1
```

Similarly, you need to add `name` to all port definitions in the service declaration. The following is an example of a `carts` service:

```
apiVersion: v1
kind: Service
metadata:
  name: frontend
  namespace: online-boutique
spec:
  type: ClusterIP
  selector:
    app: frontend
  ports:
  - name: http-frontend
```

```
    port: 80
    targetPort: 8080
```

For your convenience, the updated files are available in `Chapter12/OPAGatekeeper/automaticsidecarinjection`. Deploy the Online Boutique application using the following command:

```
% kubectl apply -f Chapter12/OPAGatekeeper/
automaticsidecarinjection
```

With that, we have practiced the deployment of the Online Boutique application in the Service Mesh. You should have the Online Boutique application along with automatic sidecar injection deployed in your cluster. The Online Boutique application is part of the Service Mesh but not yet completely ready for it. In the next section, we will apply the learning from *Chapter 5* on managing application traffic.

Configuring Istio to manage application traffic

In this section, using the learnings from *Chapter 4*, we will configure the Service Mesh to manage application traffic for the Online Boutique application. We will first start with configuring the Istio Ingress gateway to allow traffic inside the mesh.

Configuring Istio Ingress Gateway

In *Chapter 4*, we read that a gateway is like a load balancer on the edge of the mesh that accepts incoming traffic that is then routed to underlying workloads.

In the following source code block, we have defined the gateway configuration:

```
apiVersion: networking.istio.io/v1alpha3
kind: Gateway
metadata:
  name: online-boutique-ingress-gateway
  namespace: online-boutique
spec:
  selector:
    istio: ingressgateway
  servers:
  - port:
      number: 80
      name: http
      protocol: HTTP
```

```
    hosts:
    - "onlineboutique.com"
```

The file is also available in `Chapter12/trafficmanagement/01-gateway.yaml` on GitHub. Apply the configuration using the following command:

```
$ kubectl apply -f Chapter12/trafficmanagement/01-gateway.yaml
gateway.networking.istio.io/online-boutique-ingress-gateway
created
```

Next, we need to configure `VirtualService` to route traffic for the `onlineboutique.com` host to the corresponding `frontend` service.

Configuring VirtualService

`VirtualService` is used to define route rules for every host as specified in the gateway configuration. `VirtualService` is associated with the gateway and the hostname is managed by that gateway. In `VirtualService`, you can define rules on how a traffic/route can be matched and, if matched, then where it should be routed to.

The following source code block defines `VirtualService` that matches any traffic handled by `online-boutique-ingress-gateway` with a hostname of `onlineboutique.com`. If matched, the traffic is routed to subset `v1` of the destination service named `frontend`:

```
apiVersion: networking.istio.io/v1alpha3
kind: VirtualService
metadata:
  name: onlineboutique-frontend-vs
  namespace: online-boutique
spec:
  hosts:
  - "onlineboutique.com"
  gateways:
  - online-boutique-ingress-gateway
  http:
  - route:
    - destination:
        host: frontend
        subset: v1
```

The configuration is available in `Chapter12/trafficmanagement/02-virtualservice-frontend.yaml` on GitHub.

Next, we will configure `DestinationRule`, which defines how the request will be handled by the destination.

Configuring DestinationRule

Though they might appear unnecessary, when you have more than one version of the workload, then `DestinationRule` is used for defining traffic policies such as a load balancing policy, connection pool policy, outlier detection policy, and so on. The following code block configures `DestinationRule` for the `frontend` service:

```
apiVersion: networking.istio.io/v1alpha3
kind: DestinationRule
metadata:
  name: frontend
  namespace: online-boutique
spec:
  host: frontend
  subsets:
  - name: v1
    labels:
      app: frontend
```

The configuration is available along with the `VirtualService` configuration in `Chapter12/trafficmanagement/02-virtualservice-frontend.yaml` on GitHub.

Next, let's create `VirtualService` and `DestinationRule` by using the following commands:

```
$ kubectl apply -f Chapter12/trafficmanagement/02-
virtualservice-frontend.yaml
virtualservice.networking.istio.io/onlineboutique-frontend-vs
created
destinationrule.networking.istio.io/frontend created
```

You should now be able to access the Online Boutique store site from the web browser. You need to find the public IP of the AWS load balancer exposing the Ingress gateway service – do not forget to add a **Host** header using the **ModHeader** extension to Chrome, as discussed in *Chapter 4* and as seen in the following screenshot:

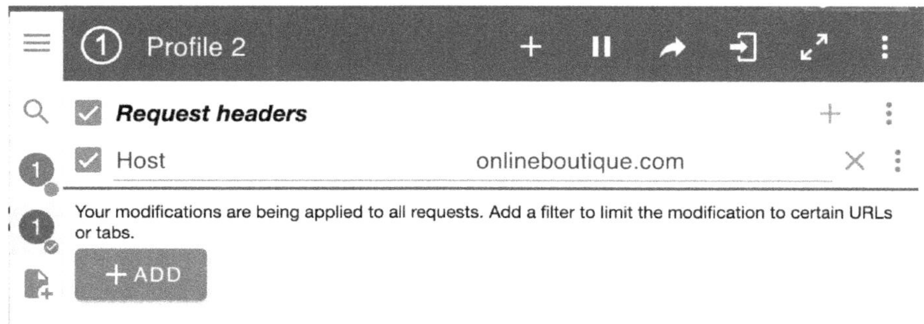

Figure 12.2 – ModHeader extension with Host header

Once the correct Host header is added, you can access the Online Boutique from Chrome using the AWS load balancer public DNS:

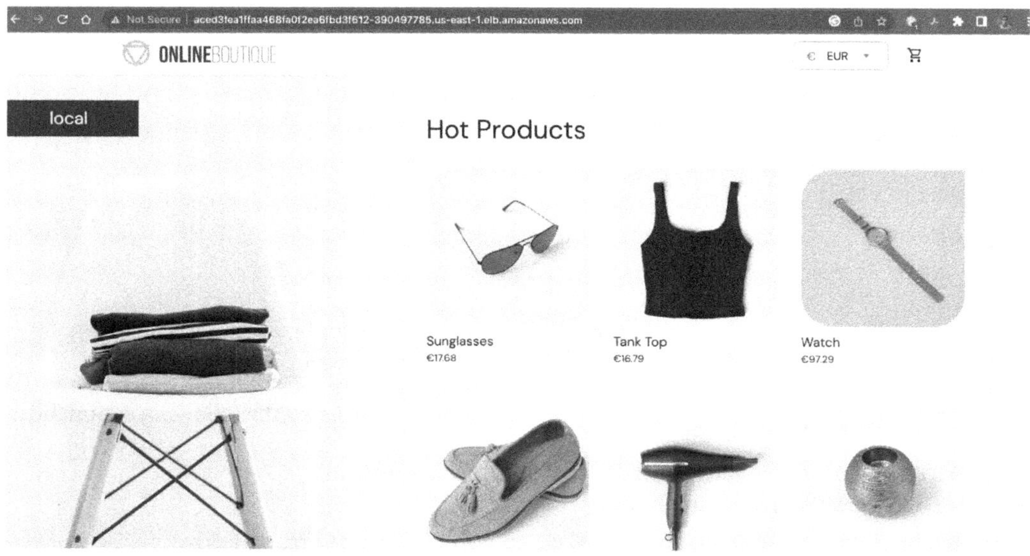

Figure 12.3 – Online Boutique landing page

So far, we have created only one virtual service to route traffic from the Ingress gateway to the `frontend` service in the mesh. By default, Istio will send traffic to all respective microservices in the mesh, but as we discussed In the previous chapter, the best practice is to define routes via `VirtualService` and how the request should be routed via destination rules. Following the best practice, we need to define `VirtualService` and `DestinationRule` for the remaining microservices. Having `VirtualService` and `DestinationRule` helps you manage traffic when there is more than one version of underlying workloads.

For your convenience, `VirtualService` and `DestinationRule` are already defined in the `Chapter12/trafficmanagement/03-virtualservicesanddr-otherservices.yaml` file on GitHub. You can apply the configuration using the following command:

```
$ kubectl apply -f Chapter12/trafficmanagement/03-virtualservicesanddr-otherservices.yaml
```

After applying the configuration and generating some traffic, check out the Kiali dashboard:

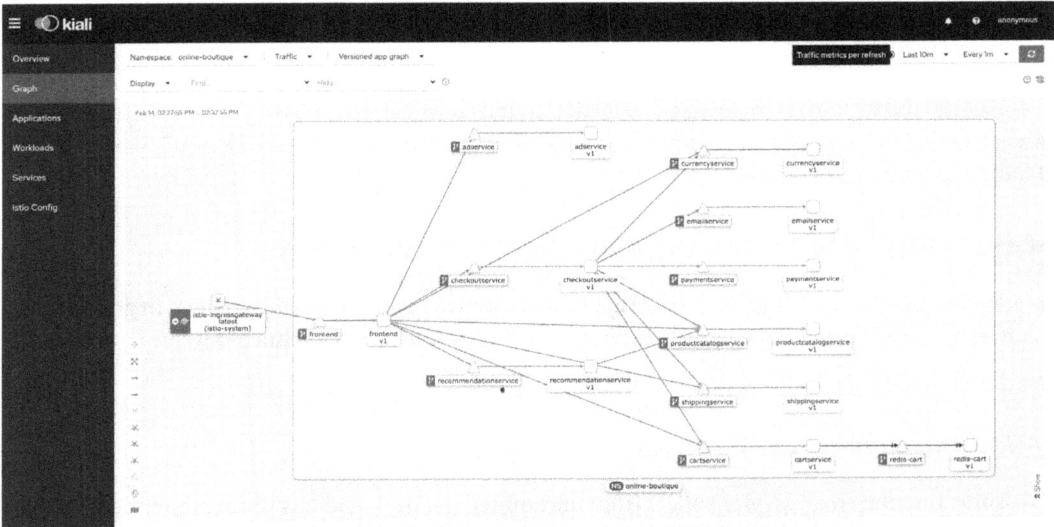

Figure 12.4 – Versioned app graph for the Online Boutique shop

In the Kiali dashboard, you can observe the Ingress gateway, all virtual services, and underlying workloads.

Configuring access to external services

Next, we quickly revise concepts on routing traffic to destinations outside the mesh. In *Chapter 4*, we learned about `ServiceEntry`, which enables us to add additional entries to Istio's internal service registry so that service in the mesh can route traffic to these endpoints that are not part of the Istio service registry. The following is an example of `ServiceRegistry` adding `xyz.com` to the Istio service registry:

```
apiVersion: networking.istio.io/v1alpha3
kind: ServiceEntry
metadata:
  name: allow-egress-to-xyv.com
spec:
```

```
  hosts:
  - "xyz.com"
  ports:
  - number: 80
    protocol: HTTP
    name: http
  - number: 443
    protocol: HTTPS
    name: https
```

This concludes the section on managing application traffic, in which we exposed `onlineboutique.com` via Istio Ingress Gateway and defined `VirtualService` and `DestinationRule` for routing and handling traffic in the mesh.

Configuring Istio to manage application resiliency

Istio provides various capabilities to manage application resiliency, and we discussed them in great detail in *Chapter 5*. We will apply some of the concepts from that chapter to the Online Boutique application.

Let's start with timeouts and retries!

Configuring timeouts and retries

Let's assume that the email service suffers from intermittent failures, and it is prudent to timeout after 5 seconds if a response is not received from the email service, and then retry sending the email a few times rather than aborting it. We will configure retries and timeout for the email service to revise application resiliency concepts.

Istio provides a provision to configure timeouts, which is the amount of time that an Istio-proxy sidecar should wait for replies from a given service. In the following configuration, we have applied a timeout of 5 seconds for the email service:

```
apiVersion: networking.istio.io/v1alpha3
kind: VirtualService
metadata:
  namespace: online-boutique
  name: emailvirtualservice
spec:
  hosts:
  - emailservice
```

```
  http:
  - timeout: 5s
    route:
    - destination:
        host: emailservice
        subset: v1
```

Istio also provides provision for automated retries that are implemented as part of the `VirtualService` configuration. In the following source code block, we have configured Istio to retry the request to the email service twice, with each retry to timeout after 2 seconds and a retry to happen only if `5xx,gateway-error,reset,connect-failure,refused-stream,retriable-4xx` errors are returned from downstream:

```
apiVersion: networking.istio.io/v1alpha3
kind: VirtualService
metadata:
  namespace: online-boutique
  name: emailvirtualservice
spec:
  hosts:
  - emailservice
  http:
  - timeout: 5s
    route:
    - destination:
        host: emailservice
        subset: v1
    retries:
      attempts: 2
      perTryTimeout: 2s
      retryOn: 5xx,gateway-error,reset,connect-failure,refused-
stream,retriable-4xx
```

We have configured `timeout` and `retries` via the `VirtualService` configuration. With the assumption that the email service is fragile and suffers interim failure, let's try to alleviate this issue by mitigating any potential issue caused by a traffic surge or spike.

Configuring rate limiting

Istio provides controls to handle a surge of traffic from consumers, as well as to control the traffic to match consumers' capability to handle the traffic.

In the following destination rule, we are defining rate-limiting controls for the email service. We have defined that the number of active requests to the email service will be `1` (as per `http2MaxRequests`), there will be only 1 request per connection (as defined in `maxRequestsPerConnection`), and there will be 0 requests queued while waiting for connection from the connection pool (as defined in `http1MaxPendingRequests`):

```
apiVersion: networking.istio.io/v1alpha3
kind: DestinationRule
metadata:
  namespace: online-boutique
  name: emaildr
spec:
  host: emailservice
  trafficPolicy:
      connectionPool:
        http:
          http2MaxRequests: 1
          maxRequestsPerConnection: 1
          http1MaxPendingRequests: 0
  subsets:
  - name: v1
    labels:
      version: v1
      app: emailservice
```

Let's make some more assumptions and assume that there are two versions of the email service, with `v1` being more rogue than the other, `v2`. In such scenarios, we need to apply outlier detection policies to perform circuit breakers. Istio provides good control for outlier detection. The following code block describes the config you need to add to `trafficPolicy` in the corresponding destination rule for the email service:

```
      outlierDetection:
        baseEjectionTime: 5m
        consecutive5xxErrors: 1
        interval: 90s
        maxEjectionPercent: 50
```

In the outlier detection, we have defined `baseEjectionTime` with a value of `5` minutes, which is the minimum duration per ejection. It is then also multiplied by the number of times an email service is found to be unhealthy. For example, if the `v1` email service is found to be an outlier 5 times, then it will be ejected from the connection pool for `baseEjectionTime*5`. Next, we have defined `consecutive5xxErrors` with a value of `1`, which is the number of `5x` errors that need to occur to qualify the upstream to be an outlier. Then, we have defined `interval` with a value of `90s`, which is the time between the checks when Istio scans the upstream for the health status. Finally, we have defined `maxEjectionPercent` with a value of `50%`, which is the maximum number of hosts in the connection pool that can be ejected.

With that, we revised and applied various controls for managing application resiliency for the Online Boutique application. Istio provides various controls for managing application resiliency without needing to modify or build anything specific in your application. In the next section, we will apply the learning of *Chapter 6* to our Online Boutique application.

Configuring Istio to manage application security

Now that we have created Ingress via Istio Gateway, routing rules via Istio `VirtualService`, and `DestinationRules` to handle how traffic will be routed to the end destination, we can move on to the next step of securing the traffic in the mesh. The following policy enforces that all traffic in the mesh should strictly happen over mTLS:

```
apiVersion: security.istio.io/v1beta1
kind: PeerAuthentication
metadata:
  name: strictmtls-online-boutique
  namespace: online-boutique
spec:
  mtls:
    mode: STRICT
```

The configuration is available in the `Chapter12/security/strictMTLS.yaml` file on GitHub. Without this configuration, all the traffic in the mesh is happening in *PERMISSIVE* mode, which means that the traffic can happen over mTLS as well as plain text. You can validate that by deploying a `curl` Pod and making an HTTP call to any of the microservices in the mesh. But once you apply the policy, Istio will enforce *STRICT* mode, which means mTLS will be strictly enforced for all traffic. Apply the configuration using the following:

```
$ kubectl apply -f Chapter12/security/strictMTLS.yaml
peerauthentication.security.istio.io/strictmtls-online-boutique
created
```

You can check in Kiali that all traffic in the mesh is happening over mTLS:

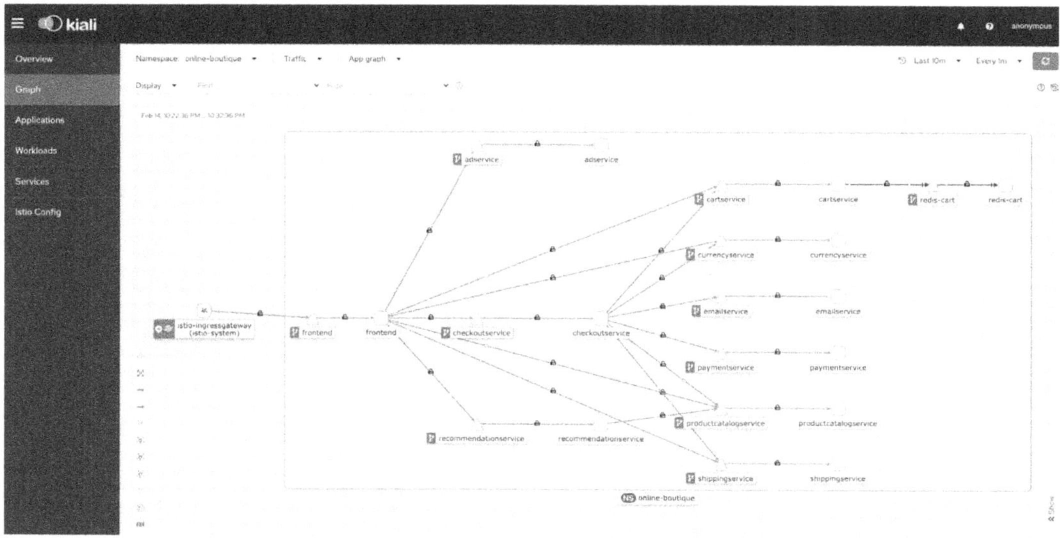

Figure 12.5 – App graph showing mTLS communication between services

Next, we will be securing Ingress traffic using `https`. This step is important to revise but the outcome of it creates a problem in accessing the application, so we will perform the steps to revise the concepts and then revert them back so that we can continue accessing the application.

We will use the learning from *Chapter 4*'s *Exposing Ingress over HTTPS* section. The steps are much easier if you have a **Certificate Authority** (**CA**) and registered DNS name, but if not, simply follow these steps to create a certificate to be used for the `onlineboutique.com` domain:

1. Create a CA. Here, we are creating a CA with a **Common Name** (**CN**) as `onlineboutique.inc`:

```
$ openssl req -x509 -sha256 -nodes -days 365 -newkey
rsa:2048 -subj '/O=Online Boutique./CN=onlineboutique.
inc' -keyout onlineboutique.inc.key -out onlineboutique.
inc.crt
Generating a 2048 bit RSA private key
writing new private key to 'onlineboutique.inc.key'
```

2. Generate a **Certificate Signing Request** (**CSR**) for the Online Boutique. Here, we are generating a CSR for `onlineboutique.com`, which also generates a private key:

```
$ openssl req -out onlineboutique.com.csr -newkey
rsa:2048 -nodes -keyout onlineboutique.com.key -subj "/
CN=onlineboutique.com/O=onlineboutique.inc"
Generating a 2048 bit RSA private key
.........................................................
..................+++
.........+++
writing new private key to 'onlineboutique.com.key'
```

3. Sign the CSR using the CA using the following command:

```
$ openssl x509 -req -sha256 -days 365 -CA onlineboutique.
inc.crt -CAkey onlineboutique.inc.key -set_serial 0 -in
onlineboutique.com.csr -out onlineboutique.com.crt
Signature ok
subject=/CN= onlineboutique.com/O= onlineboutique.inc
```

4. Load the certificate and private key as a Kubernetes Secret:

```
$ kubectl create -n istio-system secret tls
onlineboutique-credential --key=onlineboutique.com.key
--cert=onlineboutique.com.crt
secret/onlineboutique-credential created
```

We have created the certificate and stored them as Kubernetes Secret. In the next steps, we will modify the Istio Gateway configuration to expose the traffic over HTTPS using the certificates.

5. Create the Gateway configuration as described in the following command:

```
apiVersion: networking.istio.io/v1alpha3
kind: Gateway
metadata:
  name: online-boutique-ingress-gateway
  namespace: online-boutique
spec:
  selector:
    istio: ingressgateway
  servers:
  - port:
```

```
      number: 443
      name: https
      protocol: HTTPS
    tls:
      mode: SIMPLE
      credentialName: onlineboutique-credential
    hosts:
    - "onlineboutique.com"
```

Apply the following configuration:

```
$ kubectl apply -f Chapter12/security/01-istio-gateway.
yaml
```

You can access and check the certificate using the following commands. Please note that the output is truncated to highlight relevant sections only:

```
$ curl -v -HHost:onlineboutique.
com --connect-to "onlineboutique.
com:443:aced3fea1ffaa468fa0f2ea6fbd3f612-390497785.
us-east-1.elb.amazonaws.com" --cacert onlineboutique.inc.crt
--head  https://onlineboutique.com:443/
..
* Connected to aced3fea1ffaa468fa0f2ea6fbd3f612-390497785.
us-east-1.elb.amazonaws.com (52.207.198.166) port 443 (#0)
--
* Server certificate:
*  subject: CN=onlineboutique.com; O=onlineboutique.inc
*  start date: Feb 14 23:21:40 2023 GMT
*  expire date: Feb 14 23:21:40 2024 GMT
*  common name: onlineboutique.com (matched)
*  issuer: O=Online Boutique.; CN=onlineboutique.inc
*  SSL certificate verify ok.
..
```

The configuration will secure the Ingress traffic to the `online-boutique` store, but it also means that you will not be able to access it from the browser because of a mismatch of the FQDN being used in the browser and the CN configured in the certificates. You can alternatively register DNS names against the AWS load balancer but for now, you might find it easier to remove the HTTPS configuration and revert to using the `Chapter12/trafficmanagement/01-gateway.yaml` file on GitHub.

Let's dive deeper into security and perform `RequestAuthentication` and authorization for the Online Boutique store. In *Chapter 6*, we did an elaborate exercise of building authentication and authorization using Auth0. Along the same lines, we will be building an authentication and authorization policy for the `frontend` service but this time, we will use a dummy JWKS endpoint, which is shipped in with Istio.

We will start with creating a `RequestAuthentication` policy to define the authentication method supported by the `frontend` service:

```
apiVersion: security.istio.io/v1beta1
kind: RequestAuthentication
metadata:
 name: frontend
 namespace: online-boutique
spec:
  selector:
    matchLabels:
      app: frontend
  jwtRules:
  - issuer: "testing@secure.istio.io"
    jwksUri: "https://raw.githubusercontent.com/istio/istio/
release-1.17/security/tools/jwt/samples/jwks.json"
```

We are making use of dummy `jwksUri`, which comes along with Istio for testing purposes. Apply the `RequestAuthentication` policy using the following:

```
$ kubectl apply -f Chapter12/security/requestAuthentication.
yaml
requestauthentication.security.istio.io/frontend created
```

After applying the `RequestAuthentication` policy, you can test that by providing a dummy token to the `frontend` service:

1. Fetch the dummy token and set it as an environment variable to be used in requests later:

   ```
   TOKEN=$(curl -k https://raw.githubusercontent.com/istio/
   istio/release-1.17/security/tools/jwt/samples/demo.jwt
   -s); echo $TOKEN
   ```

```
eyJhbGciOiJSUzI1NiIsImtpZCI6IkRIRmJwb0lVcXJZOHQyen
BBMnFYZkNtcjVWTzVaRXI0UnpIVV8tZW52dlEiLCJ0eXAiOiJKV-
1QifQ.eyJleHAiOjQ2ODU5ODk3MDAsImZvbyI6ImJhciIsImlh-
dCI6MTUzMjM4OTcwMCwiaXNzIjoidGVzdGluZ0BzZWN1cmUuaX-
N0aW8uaW8iLCJzdWIiOiJ0ZXN0aW5nQHNlY3VyZS5pc3Rpby5pbyJ9.
CfNnxWP2tcnR9q0vxyxweaF3ovQYHYZl82hAUsn21bwQd9zP7c-LS9qd_
vpdLG4Tn1A15NxfCjp5f7QNBUo-KC9PJqYpgGbaXhaGx7bEdFWjcwv3n-
Zzvc7M__ZpaCERdwU7igUmJqYGBYQ51vr2njU9ZimyKkfDe3axcyiB-
Zde7G6dabliUosJvvKOPcKIWPccCgefSj_GNfwIip3-SsFdlR7BtbVUc-
qR-yv-XOxJ3Uc1MI0tz3uMiiZcyPV7sNCU4KRnemRIMHVOfuvHsU60_
GhGbiSFzgPTAa9WTltbnarTbxudb_YEOx12JiwYToeX0DCPb43W1tzI-
Bxgm8NxUg
```


2. Test using `curl`:

```
$ curl -HHost:onlineboutique.com http://
aced3fea1ffaa468fa0f2ea6fbd3f612-390497785.us-east-1.
elb.amazonaws.com/ -o /dev/null --header "Authorization:
Bearer $TOKEN" -s -w '%{http_code}\n'
200
```

   Notice that you received a `200` response.

3. Now try testing with an invalid token:

```
$ curl -HHost:onlineboutique.com http://
aced3fea1ffaa468fa0f2ea6fbd3f612-390497785.us-east-1.
elb.amazonaws.com/ -o /dev/null --header "Authorization:
Bearer BLABLAHTOKEN" -s -w '%{http_code}\n'
401
```

   The `RequestAuthentication` policy plunged into action and denied the request.

4. Test without any token:

```
% curl -HHost:onlineboutique.com http://
aced3fea1ffaa468fa0f2ea6fbd3f612-390497785.us-east-1.elb.
amazonaws.com/ -o /dev/null  -s -w '%{http_code}\n'
200
```

The outcome of the request is not desired but is expected because the `RequestAuthentication` policy is only responsible for validating a token if a token is passed. If there is no `Authorization` header in the request, then the `RequestAuthentication` policy will not be invoked. We can solve this problem using `AuthorizationPolicy`, which enforces an access control policy for workloads in the mesh.

Let's build `AuthorizationPolicy`, which enforces that a principal must be present in the request:

```
apiVersion: security.istio.io/v1beta1
kind: AuthorizationPolicy
metadata:
  name: require-jwt
  namespace: online-boutique
spec:
  selector:
    matchLabels:
      app: frontend
  action: ALLOW
  rules:
  - from:
    - source:
       requestPrincipals: ["testing@secure.istio.io/testing@
secure.istio.io"]
```

The configuration is available in the `Chapter12/security/requestAuthorizationPolicy.yaml` file in GitHub. Apply the configuration using the following command:

```
$ kubectl apply -f Chapter12/security/
requestAuthorizationPolicy.yaml
authorizationpolicy.security.istio.io/frontend created
```

After applying the configuration test using *Steps 1* to *4*, which we performed after applying the `RequestAuthentication` policy, you will notice that all steps work as expected, but for *Step 4*, we are getting the following:

```
$ curl -HHost:onlineboutique.com http://
aced3fea1ffaa468fa0f2ea6fbd3f612-390497785.us-east-1.elb.
amazonaws.com/ -o /dev/null  -s -w '%{http_code}\n'
403
```

That is because the authorization policy enforces the required presence of a JWT with the `["testing@secure.istio.io/testing@secure.istio.io"]` principal.

This concludes the security configuration for our Online Boutique application. In the next section, we will read about various resources that will help you become an expert and certified in using and operating Istio.

Certification and learning resources for Istio

The primary resource for learning Istio is the Istio website (`https://istio.io/latest/`). There is elaborative documentation on performing basic to multi-cluster setups. There are resources for beginners and advanced users, and various exercises on performing traffic management, security, observability, extensibility, and policy enforcement. Outside of the Istio documentation, the other organization providing lots of supportive content on Istio is Tetrate (`https://tetrate.io/`), which also provides labs and certification courses. One such certification provided by Tetrate Academy is **Certified Istio Administrator**. Details about the course and exam are available at `https://academy.tetrate.io/courses/certified-istio-administrator`. Tetrate Academy also provides a free course to learn about Istio fundamentals. You can find the details of the course at `https://academy.tetrate.io/courses/istio-fundamentals`. Similarly, there is a course from Solo.io named **Get Started with Istio**; you can find details of the course at `https://academy.solo.io/get-started-with-istio`. Another good course from The Linux Foundation is named **Introduction to Istio**, and you can find the details of the course at `https://training.linuxfoundation.org/training/introduction-to-istio-lfs144x/`.

I personally enjoy the learning resources available at `https://istiobyexample.dev/`; the site explains various use cases of Istio (such as canary deployment, managing Ingress, managing gRPC traffic, and so on) in great detail, along with configuration examples. For any technical questions, you can always head to StackOverflow at `https://stackoverflow.com/questions/tagged/istio`. There is an energetic and enthusiastic community of Istio users and builders who are discussing various topics about Istio at `https://discuss.istio.io/`; feel free to sign up for the discussion board.

Tetrate Academy also provides a free course on Envoy fundamentals; the course is very helpful to understand the fundamentals of Envoy and, in turn, the Istio data plane. You can find the details of this course at `https://academy.tetrate.io/courses/envoy-fundamentals`. The course is full of practical labs and quizzes that are very helpful in mastering your Envoy skills.

The Istio website has compiled a list of helpful resources to keep you updated with Istio and engage with the Istio community; you can find the list at `https://istio.io/latest/get-involved/`. The list also provides you with details on how to report bugs and issues.

To summarize, there are not many resources except a few books and websites, but you will find most of the answers to your questions at `https://istio.io/latest/docs/`. It is also a great idea to follow **IstioCon**, which is the Istio Community conference and happens on a yearly cadence. You can find a session of IstioCon 2022 at `https://events.istio.io/istiocon-2022/sessions/` and 2021 at `https://events.istio.io/istiocon-2021/sessions/`.

Understanding eBPF

As we are at the end of this book, it is important to also look at other technologies that are relevant to Service Mesh. One such technology is the **Extended Berkeley Packet Filter** (**eBPF**). In this section, we will read about eBPF and its role in Service Mesh evolution.

eBPF is a framework that allows users to run custom programs within the kernel of the operating system without needing to change kernel source code or load kernel modules. The custom programs are called **eBPF programs** and are used to add additional capabilities to the operating system at runtime. The eBPF programs are safe and efficient and, like the kernel modules, they are like lightweight sandbox virtual machines run in a privileged context by the operating system.

eBPF programs are triggered based on events happening at the kernel level, which is achieved by associating them to hook points. Hooks are predefined at kernel levels and include system calls, network events, function entry and exit, and so on. In scenarios where an appropriate hook doesn't exist, then users can make use of kernel probes, also called kprobes. The kprobes are inserted into the kernel routine; ebPF programs are defined as a handler to kprobes and are executed whenever a particular breakpoint is hit in the kernel. Like hooks and kprobes, eBPF programs can also be attached to uprobes, which are probes at user space levels and are tied to an event at the user application level, thus eBPF programs can be executed at any level from the kernel to the user application. When executing programs at the kernel level, the biggest concern is the security of the program. In eBPF, that is assured by BPF libraries. The BPF libraries handle the system call to load the eBPF programs in two steps. The first step is the verification step, during which the eBPF program is validated to ensure that it will run to completion and will not lock up the kernel, the process loading the eBPF program has correct privileges, and the eBPF program will not harm the kernel in any way. The second step is a **Just-In-Time** (**JIT**) compilation step, which translates the generic bytecode of the program into the machine-specific instruction and optimizes it to get the maximum execution speed of the program. This makes eBPF programs run as efficiently as natively compiled kernel code as if it was loaded as a kernel module. Once the two steps are complete, the eBPF program is loaded and compiled into the kernel waiting for the hook or kprobes to trigger the execution.

BPF has been widely used as an add-on to the kernel. Most of the applications have been at the network level and mostly in observability space. eBPF has been used to provide visibility into system calls at packet and socket levels, which are then used for building security solution systems that can operate with low-level context from the kernel. eBPF programs are also used for introspection of user applications along with the part of the kernel running the application, which provides a consolidated insight to troubleshoot application performance issues. You might be wondering why we are discussing eBPF in the context of Service Mesh. The programmability and plugin model of eBPF is particularly useful in networking. eBPF can be used to perform IP routing, packet filtering, monitoring, and so on at native speeds of kernel modules. One of the drawbacks of the Istio architecture is its model of deploying a sidecar with every workload, as we discussed in *Chapter 2* – the sidecar basically works by intercepting network traffic, making use of iptables to configure the kernel's netfilter packet filter functionality. The drawback of this approach is less optimal performance, as the data path created for service traffic is much longer than what it would have been if the workload was just by itself without any sidecar traffic interception. With eBPF socket-related program types, you can filter socket data, redirect socket data, and monitor socket events. These programs have the potential for replacing the iptables-based traffic interception; using eBPF, there are options to intercept and manage network traffic without incurring any negative impacts on the data path performance.

Isovalent (at `https://isovalent.com/`) is one such organization that is revolutionizing the architecture of API Gateway and Service Mesh. **Cilium** is a product from Isovalent, and it provides a variety of functionality, including API Gateway function, Service Mesh, observability, and networking. Cilium is built with eBPF as its core technology where it injects eBPF programs at various points in the Linux kernel to achieve application networking, security, and observability functions. Cilium is getting adopted in Kubernetes networking to solve performance degradation caused by packets needing to traverse the same network stack multiple times between the host and the Pod. Cilium is solving this problem by bypassing iptables in the networking stacking, avoiding net filters and other overheads caused by iptables, which has led to significant gains in network performance. You can read more about the Cilium product stack at `https://isovalent.com/blog/post/cilium-release-113/`; you will be amazed to see how eBPF is revolutionizing the application networking space.

Istio has also created an open source project called Merbridge, which replaces iptables with eBPF programs to allow the transporting of data directly between inbound and outbound sockets of sidecar containers and application containers to shorten the overall data path. Merbridge is in its early days but has produced some promising results; you can find the open source project at `https://github.com/merbridge/merbridge`.

With eBPF and products like Cilium, it is highly likely that there will be an advancement in how network proxy-based products will be designed and operated in the future. eBPF is being actively explored by various Service Mesh technologies, including Istio, on how it can be used to overcome drawbacks and improve the overall performance and experience of using Istio. eBPF is a very promising technology and is already being used for doing awesome things with products such as Cilium and Calico.

Summary

I hope this book has provided you with a good insight into Istio. *Chapters 1* to *3* set the context on why Service Mesh is needed and how Istio the control and data planes operate. The information in these three chapters is important to appreciate Istio and to build an understanding of Istio architecture. *Chapters 4* to *6* then provided details on how to use Istio for building the application network that we discussed in the earlier chapters.

Then, in *Chapter 7*, you learned about observability and how Istio provides integration into various observation tools, as the next steps you should explore integration with other observability and monitoring tools such as Datadog. Following that, *Chapter 8* showed practices on how to deploy Istio across multiple Kubernetes clusters, which should have given you confidence on how to install Istio in production environments. *Chapter 9* then provided details on how Istio can be extended using WebAssembly and its applications, while *Chapter 10* discussed how Istio helps bridge the old world of virtual machines with the new world of Kubernetes by discussing how the Service Mesh can be extended to include workloads deployed on virtual machines. Lastly, *Chapter 11* covered the best practices for operating Istio and how tools such as OPA Gatekeeper can be used to automate some of the best practices.

In this chapter, we managed to revise the concepts of *Chapters 4* to *6* by deploying and configuring another open source demo application, which should have provided you with the confidence and experience to take on the learnings from the book to real-life applications and to take advantage of application networking and security provided by Istio.

You also read about eBPF and what a game-changing technology it is, making it possible to write code at the kernel level without needing to understand or experience the horrors of the kernel. eBPF will possibly bring lots of changes to how Service Mesh, API Gateway, and networking solutions in general operate. In the Appendix of this book, you will find information about other Service Mesh technologies: Consul Connect, Kuma Mesh, Gloo Mesh, and Linkerd. The Appendix provides a good overview of these technologies and helps you appreciate their strength and limitations.

I hope you enjoyed learning about Istio. To establish your knowledge of Istio, you can also explore taking the Certified Istio Administrator exam provided by Tetrate. You can also explore the other learning avenues provided in this chapter. I hope reading this book was an endeavor that will take you to the next level in your career and experience of building scalable, resilient, and secure applications using Istio.

BEST OF LUCK!

Appendix – Other Service Mesh Technologies

In this appendix, we will learn about the following Service Mesh implementations:

- Consul Connect
- Gloo Mesh
- Kuma
- Linkerd

These Service Mesh technologies are popular and are gaining recognition and adoption by organizations. The information provided in this *Appendix* about these Service Mesh technologies is not exhaustive; rather, the goal here is to make you familiar with and aware of the alternatives to Istio. I hope reading this *Appendix* will provide some basic awareness of these alternative technologies and help you understand how these technologies fare in comparison to Istio. Let's dive in!

Consul Connect

Consul Connect is a Service Mesh solution offered by HashiCorp. It is also known as Consul Service Mesh. On the HashiCorp website, you will find that the terms Consul Connect and Consul Service Mesh are used interchangeably. It is built upon Consul, which is a service discovery solution and a key-value store. Consul is a very popular and long-established service discovery solution; it provides and manages service identities for every type of workload, which are then used by Service Mesh to manage traffic between Services in Kubernetes. It also supports using ACLs to implement zero-trust networking and provides granular control over traffic flow in the mesh.

Consul uses Envoy as its data plane and injects it into workload Pods as sidecars. The injection can be based on annotations as well as global configurations to automatically inject sidecar proxies into all workloads in specified namespaces. We will start by installing Consul Service Mesh on your workstation, followed by some exercises to practice the basics of using Consul Service Mesh.

Let's begin by installing Consul:

1. Clone the Consul repository:

```
% git clone https://github.com/hashicorp-education/learn-
consul-get-started-kubernetes.git
…..
Resolving deltas: 100% (313/313), done.
```

2. Install the Consul CLI:

   - For MacOS, follow these steps:

     i. Install the HashiCorp tap:

```
% brew tap hashicorp/tap
```

     ii. Install the Consul Kubernetes CLI:

```
% brew install hashicorp/tap/consul-k8s
Running `brew update --auto-update`...
==> Auto-updated Homebrew!
Updated 1 tap (homebrew/core).
You have 4 outdated formulae installed.
You can upgrade them with brew upgrade
or list them with brew outdated.
==> Fetching hashicorp/tap/consul-k8s
==> Downloading https://releases.hashicorp.com/consul-
k8s/1.0.2/consul-k8s_1.0.2_darwin_arm64.zip
##########################################################
############### 100.0%
==> Installing consul-k8s from hashicorp/tap
/opt/homebrew/Cellar/consul-k8s/1.0.2: 3 files, 64MB,
built in 2 seconds
```

     iii. Check the version of `consul-k8s` on the Consul CLI:

```
% consul-k8s version
    consul-k8s v1.0.2
```

   - For Linux Ubuntu/Debian, follow these steps:

     iv. Add the HashiCorp GPG key:

```
% curl -fsSL https://apt.releases.hashicorp.com/gpg |
sudo apt-key add -
```

     v. Add the HashiCorp apt repository:

```
% sudo apt-add-repository "deb [arch=amd64] https://apt.
releases.hashicorp.com $(lsb_release -cs) main"
```

vi. Run `apt-get install` to install the `consul-k8s` CLI:

```
% sudo apt-get update && sudo apt-get install consul-k8s
```

- For CentOS/RHEL, follow these steps:

vii. Install yum-config-manager to manage your repositories:

```
% sudo yum install -y yum-utils
```

viii. Use yum-config-manager to add the official HashiCorp Linux repository:

```
% sudo yum-config-manager --add-repo https://rpm.
releases.hashicorp.com/RHEL/hashicorp.repo
```

ix. Install the `consul-k8s` CLI:

```
% sudo yum -y install consul-k8s
```

3. Start minikube:

```
% minikube start --profile dc1 --memory 4096
--kubernetes-version=v1.24.0
```

4. Install Consul on minikube using the Consul CLI.

Run the following in learn-consul-get-started-kubernetes/local:

```
% consul-k8s install -config-file=helm/values-v1.yaml
-set global.image=hashicorp/consul:1.14.0
==> Checking if Consul can be installed
 ✓ No existing Consul installations found.
 ✓ No existing Consul persistent volume claims found
 ✓ No existing Consul secrets found.
==> Consul Installation Summary
    Name: consul
    Namespace: consul
….
--> Starting delete for "consul-server-acl-init-cleanup"
Job
 ✓ Consul installed in namespace "consul".
```

5. Check the Consul Pods in the namespace:

```
% kubectl get po -n consul
NAME                          READY   STATUS    RESTARTS     AGE
consul-connect-injector-57dcdd54b7-
hhxl4         1/1      Running   1 (21h ago)    21h
consul-server-0          1/1      Running    0              21h
consul-webhook-cert-manager-76bbf7d768-
2kfhx    1/1      Running    0             21h
```

6. Configure the Consul CLI to be able to communicate with Consul.

   We will set environment variables so that the Consul CLI can communicate with your Consul cluster.

   Set CONSUL_HTTP_TOKEN from `secrets/consul-bootstrap-acl-token` and set it as an environment variable:

```
% export CONSUL_HTTP_TOKEN=$(kubectl get --namespace
consul secrets/consul-bootstrap-acl-token --template={{.
data.token}} | base64 -d)
```

   Set the Consul destination address. By default, Consul runs on port 8500 for HTTP and 8501 for HTTPS:

```
% export CONSUL_HTTP_ADDR=https://127.0.0.1:8501
```

   Remove SSL verification checks to simplify communication with your Consul cluster:

```
% export CONSUL_HTTP_SSL_VERIFY=false
```

7. Access the Consul dashboard using the following command:

```
% kubectl port-forward pods/consul-server-0 8501:8501
--namespace consul
```

   Open `localhost:8501` in your browser to access the Consul dashboard, as shown in the following screenshot:

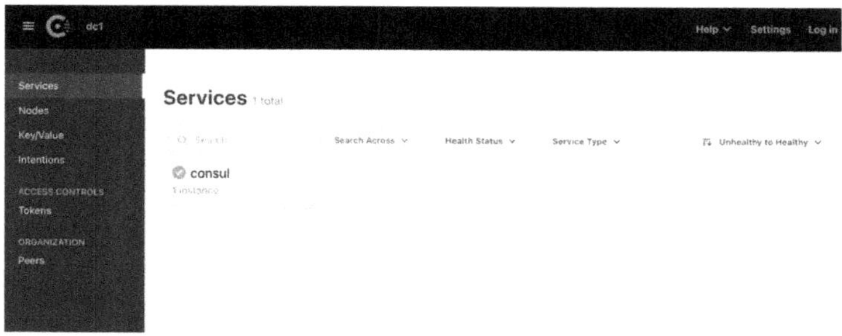

Figure A.1 – Consul Dashboard

Now that we have installed Consul Service Mesh on minikube, let's deploy an example application and go through the fundamentals of Consul Service Mesh.

Deploying an example application

In this section, we will deploy envoydummy along with a curl application. The `sampleconfiguration` file is available in `AppendixA/envoy-proxy-01.yaml`.

In the configuration file, you will notice the following annotation:

```
annotations:
  consul.hashicorp.com/connect-inject: "true"
```

This annotation allows Consul to automatically inject a proxy for each service. The proxies create a data plane to handle requests between services based on the configuration from Consul.

Apply the configuration to create `envoydummy` and the `curl` Pods:

```
% kubectl create -f AppendixA/Consul/envoy-proxy-01.yaml -n
appendix-consul
configmap/envoy-dummy created
service/envoydummy created
deployment.apps/envoydummy created
servicedefaults.consul.hashicorp.com/envoydummy created
serviceaccount/envoydummy created
servicedefaults.consul.hashicorp.com/curl created
serviceaccount/curl created
pod/curl created
service/curl created
```

In a few seconds, you will notice that Consul automatically injects a sidecar into the Pods:

```
% % kubectl get po -n appendix-consul
NAME                          READY   STATUS    RESTARTS   AGE
curl                          2/2     Running   0          16s
envoydummy-77dfb5d494-2dx5w   2/2     Running   0          17s
```

To find out more about the sidecar, please inspect the `envoydummy` Pod using the following commands:

```
% kubectl get po/envoydummy-77dfb5d494-pcqs7 -n appendix-consul
-o json | jq '.spec.containers[].image'
"envoyproxy/envoy:v1.22.2"
```

```
"hashicorp/consul-dataplane:1.0.0"
% kubectl get po/envoydummy-77dfb5d494-pcqs7 -n appendix-consul
-o json | jq '.spec.containers[].name'
"envoyproxy"
"consul-dataplane"
```

In the output, you can see a container named `consul-dataplane` created from an image called `hashicorp/consul-dataplane:1.0.0`. You can inspect the image at `https://hub.docker.com/layers/hashicorp/consul-dataplane/1.0.0-beta1/images/sha256-f933183f235d12cc526099ce90933cdf43c7281298b3cd34a4ab7d4ebeeabf84?context=explore` and you will notice that it is made up of envoy proxy.

Let's try to access `envoydummy` from the `curl` Pod:

```
% kubectl exec -it pod/curl -n appendix-consul -- curl http://
envoydummy:80
curl: (52) Empty reply from server
command terminated with exit code 52
```

So far, we have successfully deployed the `envoydummy` Pod along with `consul-dataplane` as a sidecar. We have observed Consul Service Mesh security in action by seeing that the `curl` Pod, while deployed in the same namespace, is unable to access the `envoydummy` Pod. In the next section, we will understand this behavior and learn how to configure Consul to perform zero-trust networking.

Zero-trust networking

Consul manages inter-service authorization with Consul constructs called intentions. Using Consul CRDs, you need to define intentions that prescribe what services are allowed to communicate with each other. Intentions are the cornerstones of zero-trust networking in Consul.

Intentions are enforced by the sidecar proxy on inbound connections. The sidecar proxy identifies the inbound service using its TLS client certificate. After identifying the inbound service, the sidecar proxy then checks if an intention exists to allow the client to communicate with the destination service.

In the following code block, we are defining an intention to allow traffic from the `curl` service to the `envoydummy` service:

```
apiVersion: consul.hashicorp.com/v1alpha1
kind: ServiceIntentions
metadata:
  name: curl-to-envoydummy-api
  namespace: appendix-consul
spec:
```

```
  destination:
    name: envoydummy
  sources:
    - name: curl
      action: allow
```

In the configuration, we have specified the names of the destination service and the source service. In `action`, we have specified `allow` to allow traffic from source to destination. Another possible value of `action` is `deny`, which denies traffic from source to destination. If you do not want to specify the name of a service, you will need to use `*`. For example, if the service name in `sources` is `*` then it will allow traffic from all services to `envoydummy`.

Let's apply intentions using the following command:

```
% kubectl create -f AppendixA/Consul/curl-to-envoy-dummy-intentions.yaml
serviceintentions.consul.hashicorp.com/curl-to-envoydummy-api created
```

You can verify the created intentions in the Consul dashboard:

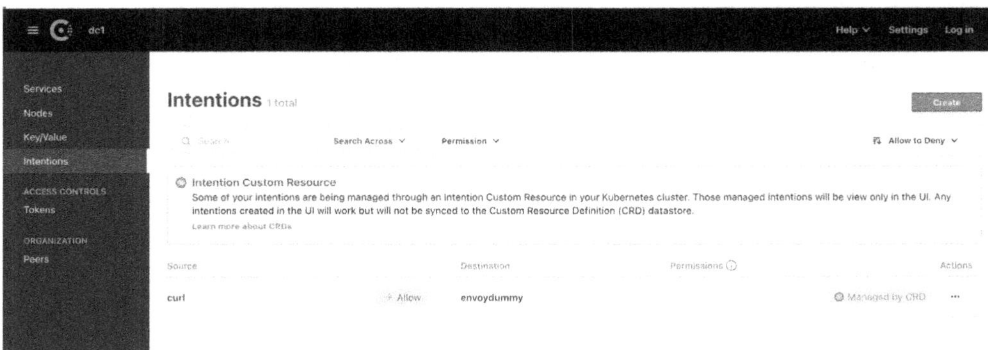

Figure A.2 – Consul intentions

Considering that we have created the intention to allow traffic from the `curl` service to the `envoydummy` service, let's test that the `curl` Pod is able to communicate with the `envoydummy` Pod using the following command:

```
% kubectl exec -it pod/curl -n appendix-consul -- curl http://envoydummy
V1----------Bootstrap Service Mesh Implementation with Istio----------V1%
```

Using intentions, we were able to define rules to control traffic between services without needing to configure a firewall or any changes in the cluster. Intentions are key building blocks of Consul for creating zero-trust networks.

Traffic management and routing

Consul provides a comprehensive set of service discovery and traffic management features. The service discovery comprises three stages: routing, splitting, and resolution. These three stages are also referred to as the service discovery chain, and it can be used to implement traffic controls based on HTTP headers, path, query strings, and workload version.

Let's go through each stage of the service discovery chain.

Routing

This is the first stage of the service discovery chain, and it is used to intercept traffic using Layer 7 constructs such as HTTP header and path. This is achieved via service-router config entry through which you can control the traffic routing using various criteria. For example, for `envoydummy`, let's say we want to enforce that any request send to `envoydummy` version v1 with `/latest` in the URI should be routed to `envoydummy` version v2 instead, and any request to version v2 of the `envoydummy` app but with `/old` in the path should be routed to version v1 of the `envoydummy` app. This can be achieved using the following `ServiceRouter` configuration:

```
apiVersion: consul.hashicorp.com/v1alpha1
kind: ServiceRouter
metadata:
  name: envoydummy
spec:
  routes:
    - match:
        http:
          pathPrefix: '/latest'
      destination:
        service: 'envoydummy2'
```

In the configuration, we are specifying that any request destined for the `envoydummy` service but with `pathPrefix` set to `'/latest'` will be routed to `envoydummy2`. And in the following configuration, we are specifying that any request destined for the `envoydummy2` service but with `pathPrefix` set to `'/old'` will be routed to `envoydummy`:

```
apiVersion: consul.hashicorp.com/v1alpha1
kind: ServiceRouter
```

```
metadata:
  name: envoydummy2
spec:
  routes:
    - match:
        http:
          pathPrefix: '/old'
      destination:
        service: 'envoydummy'
```

Both `ServiceRouter` configurations are saved in `AppendixA/Consul/routing-to-envoy-dummy.yaml`. A deployment descriptor for `envoydummy` version v2 and the intentions that allow traffic from the `curl` Pod are also available in `AppendixA/Consul/envoy-proxy-02.yaml` on GitHub.

Go ahead and deploy version v2 of `envoydummy` along with the ServiceRouter configuration using the following commands:

```
% kubectl apply -f AppendixA/Consul/envoy-proxy-02.yaml
% kubectl apply -f AppendixA/Consul/routing-to-envoy-dummy.yaml
-n appendix-consul
servicerouter.consul.hashicorp.com/envoydummy configured
servicerouter.consul.hashicorp.com/envoydummy2 configured
```

You can check the configuration using the Consul dashboard. The following two screenshots show the two `ServiceRouter` configurations we have applied:

- `ServiceRouter` configuration to send traffic with the prefix `/latest` to `envoydummy2`:

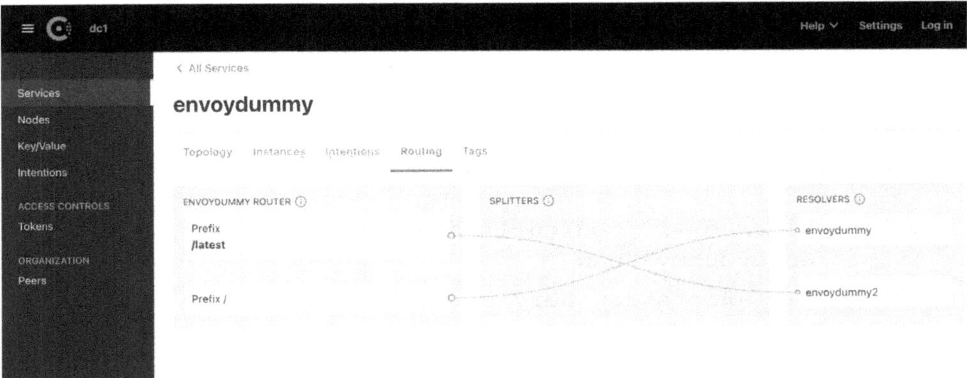

Figure A.3

- `ServiceRouter` configuration to send traffic with the prefix `/old` to `envoydummy`:

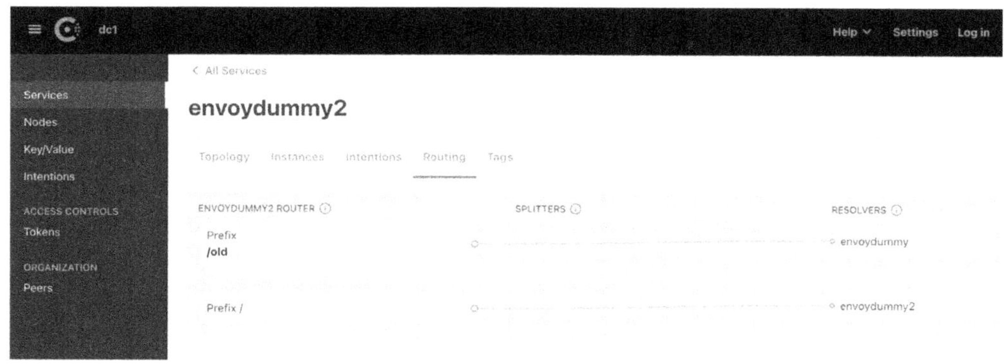

Figure A.4

Now that we have configured the service routes, let's test the routing behavior:

1. Make any request to version v1 of `envoydummy` with a URI that is not `/latest`:

```
% kubectl exec -it pod/curl -n appendix-consul -- curl
http://envoydummy/new
V1----------Bootstrap Service Mesh Implementation with
Istio----------V1%
```

   The output is as expected: the request should be routed to version v1 of `envoydummy`.

2. Make a request to version v1 of `envoydummy` with a URI that is `/latest`:

```
% kubectl exec -it pod/curl -n appendix-consul -- curl
http://envoydummy/latest
V2----------Bootstrap Service Mesh Implementation with
Istio----------V2%
```

   The output is as expected: the request, although addressed to version v1 of `envoydummy`, is routed to version v2 of `envoydummy`.

3. Make any request to version v2 of `envoydummy` with a URI that is not `/old`:

```
% kubectl exec -it pod/curl -n appendix-consul -- curl
http://envoydummy2/new
V2----------Bootstrap Service Mesh Implementation with
Istio----------V2%
```

   The output is as expected: the request should be routed to version v2 of `envoydummy`.

4. Make a request to version v2 of envoydummy with a URI that is /old:

```
% kubectl exec -it pod/curl -n appendix-consul -- curl
http://envoydummy2/old
V1----------Bootstrap Service Mesh Implementation with
Istio----------V1%
```

The output is as expected: the request although addressed to version v2 of envoydummy is routed to version v1 of envoydummy.

In these examples, we made use of path prefixes as criteria for routing. The other options are query parameters and HTTP headers. ServiceRouter also supports retry logic, which can be added to the destination configuration. Here is an example of retry logic added to the ServiceRouter config:

```
apiVersion: consul.hashicorp.com/v1alpha1
kind: ServiceRouter
metadata:
  name: envoydummy2
spec:
  routes:
    - match:
        http:
          pathPrefix: '/old'
      destination:
        service: 'envoydummy'
        requestTimeout = "20s"
        numRetries = 3
        retryOnConnectFailure = true
```

You can read more about ServiceRouter configuration on the HashiCorp website: https://developer.hashicorp.com/consul/docs/connect/config-entries/service-router.

Next in the service discovery chain is splitting, which we will learn about in the following section.

Splitting

Service splitting is the second stage in the Consul service discovery chain and is configured via the ServiceSplitter configuration. ServiceSplitter allows you to split a request to a service to multiple subset workloads. Using this configuration, you can also perform canary deployments. Here is an example where traffic for the envoydummy service is routed in a 20:80 ratio to version v1 and v2 of the envoydummy application:

```
apiVersion: consul.hashicorp.com/v1alpha1
kind: ServiceSplitter
```

```
metadata:
  name: envoydummy
spec:
  splits:
    - weight: 20
      service: envoydummy
    - weight: 80
      service: envoydummy2
```

In the `ServiceSplitter` configuration, we have configured 80% of the traffic to `envoydummy` to be routed to the `envoydummy2` service and the remaining 20% of the traffic to be routed to the `envoydummy` service. The configuration is available in `AppendixA/Consul/splitter.yaml`. Apply the configuration using the following command:

```
% kubectl apply -f AppendixA/Consul/splitter.yaml -n appendix-consul
servicesplitter.consul.hashicorp.com/envoydummy created
```

After applying the configuration, you can check out the routing config on the Consul dashboard. In the following screenshot, we can see that all traffic to `envoydummy` is routed to `envoydummy` and `envoydummy2`. The following screenshot doesn't show the percentage, but you can hover the mouse over the arrows connecting the splitters and resolvers and you should be able to see the percentage:

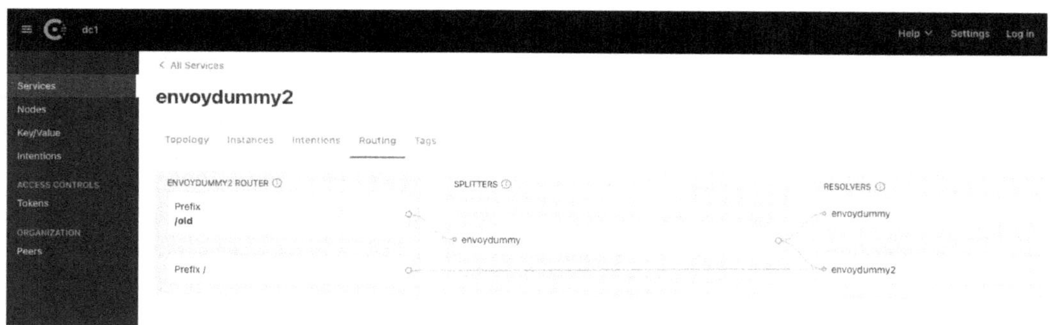

Figure A.5 – Split of traffic to envoydummy2

The following screenshot shows the split of traffic for `envoydummy`:

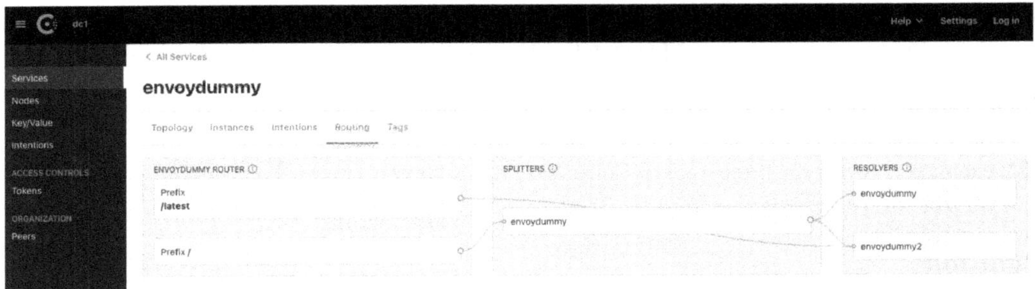

Figure A.6 – Split of traffic to envoydummy service

Now that the `ServiceSplitter` configuration is in place, test that traffic to our services is routed in the ratio specified in the config file:

```
% for ((i=0;i<10;i++)); do kubectl exec -it pod/curl -n
appendix-consul -- curl http://envoydummy/new ;done
V2----------Bootstrap Service Mesh Implementation with Istio--
--------V2
V2----------Bootstrap Service Mesh Implementation with Istio--
--------V2
V2----------Bootstrap Service Mesh Implementation with Istio--
--------V2
V1----------Bootstrap Service Mesh Implementation with Istio--
--------V1
V2----------Bootstrap Service Mesh Implementation with Istio--
--------V2
V1----------Bootstrap Service Mesh Implementation with Istio--
--------V1
V2----------Bootstrap Service Mesh Implementation with Istio--
--------V2
V2----------Bootstrap Service Mesh Implementation with Istio--
--------V2
V2----------Bootstrap Service Mesh Implementation with Istio--
--------V2
V2----------Bootstrap Service Mesh Implementation with Istio--
--------V2
```

You will observe that the traffic is routed in a 20:80 ratio between the two services. `ServiceSplitter` is a powerful feature that can be used for A/B testing, as well as canary and blue/green deployments. Using `ServiceSplitter`, you can also perform weight-based routing between subsets of the same service. It also allows you to add HTTP headers while routing the service. You can read more about `ServiceSplitter` at `https://developer.hashicorp.com/consul/docs/connect/config-entries/service-splitter`.

We have looked at two of the three steps in Consul's service discovery chain. The final stage is resolution, which we will cover in the next section.

Resolution

Consul has another config type called `ServiceResolver`, which is used to define which instances of a service map to the service name requested by the client. They control the service discovery and decide where the request is finally routed to. Using `ServiceResolver`, you can control the resilience of your system by routing the request to healthy upstreams. `ServiceResolver` distributes load to services when they are spread across more than one data center and provides failover when the services are suffering from outages. More details about `ServiceResolver` can be found at `https://developer.hashicorp.com/consul/docs/connect/config-entries/service-resolver`.

Consul Service Mesh also has provision for gateways to manage traffic from outside the mesh. It supports three kinds of gateway:

- **Mesh gateways** are used to enable and secure communication between data centers. It acts as a proxy providing Ingress to Service Mesh while at the same time securing the traffic using mTLS. Mesh gateways are used to communicate between Consul Service Mesh instances deployed in different data centers and/or Kubernetes clusters. A good hands-on exercise on mesh gateways is available at `https://developer.hashicorp.com/consul/tutorials/kubernetes/kubernetes-mesh-gateways`.
- **Ingress gateways** are used to provide access to services in the mesh to clients outside the mesh. The client can be outside the mesh but in the same Kubernetes cluster, or completely outside the cluster but within or beyond the network perimeter of the organization. You can read more about Ingress Gateway at `https://developer.hashicorp.com/consul/docs/k8s/connect/ingress-gateways`.
- **Terminating gateways**, like ServiceEntry in Istio, are used to allow workloads within the mesh to access services outside the mesh. To use a terminating gateway, users need to also use a configuration called `ServiceDefault`. This is where the details about the external service are defined, and it is referred to by the terminating gateway. You can read more about terminating gateways at `https://developer.hashicorp.com/consul/docs/k8s/connect/terminating-gateways`.

Finally, Consul Service Mesh also provides comprehensive observability of the mesh. The sidecar proxies collect and expose data about the traffic traversing the mesh. The metrics data exposed by the sidecar proxies are then scraped by Prometheus. The data includes Layer 7 metrics such as HTTP status code, request latency, and throughput. The Consul control plane also provides some metrics such as config synchronization status, exceptions, and errors, like the Istio control plane. The tech stack for observability is also like Istio; like Istio, Consul also supports integration with various other observability tools, such as Datadog, to get insight into Consul Service Mesh health and performance. You can read more about Consul Service Mesh observability at `https://developer.hashicorp.com/consul/tutorials/kubernetes/kubernetes-layer7-observability`.

I hope this section provided a brief but informative rundown of how Consul Service Mesh operates, what the various constructs in Consul Service Mesh are, and how they operate. I am sure you must have noticed the similarities between Consul Service Mesh and Istio: they both use Envoy as a sidecar proxy, and the Consul service discovery chain closely resembles Istio virtual services and destination rules; Consul Service Mesh gateways are very similar to Istio gateways. The main difference is how the control plane is implemented and the use of an agent on each node of the cluster. Consul Service Mesh can run on VMs to provide benefits of Service Mesh on legacy workloads. Consul Service Mesh is backed by HashiCorp and is tightly integrated with HashiCorp's other products, including HashiCorp Vault. It is also offered as a freemium product. There is also an Enterprise version for organizations needing enterprise support and a SaaS offering called HCP Consul that provides a fully managed cloud service to customers who want one-click mesh deployments.

Uninstalling Consul Service Mesh

You can use consul-k8s to uninstall Consul Service Mesh using the following commands:

```
% consul-k8s uninstall -auto-approve=true -wipe-data=true
..
    Deleting data for installation:
    Name: consul
    Namespace consul
 ✓ Deleted PVC => data-consul-consul-server-0
 ✓ PVCs deleted.
 ✓ Deleted Secret => consul-bootstrap-acl-token
 ✓ Consul secrets deleted.
 ✓ Deleted Service Account => consul-tls-init
 ✓ Consul service accounts deleted.
```

You can uninstall consul-k8s CLI using Brew on macOS:

```
% brew uninstall consul-k8s
```

Gloo Mesh

Gloo Mesh is a Service Mesh offering from Solo.io. There is an open source version called Gloo Mesh and an enterprise offering called Gloo Mesh Enterprise. Both are based on Istio Service Mesh and claim to have a better control plane and added functionality on top of open source Istio. Solo.io provides a feature comparison on its website outlining the differences between Gloo Mesh Enterprise, Gloo Mesh Open Source, and Istio, which you can access at `https://www.solo.io/products/gloo-mesh/`. Gloo Mesh is primarily focused on providing a Kubernetes-native management plane through which users can configure and operate multiple heterogeneous Service Mesh instances across multiple clusters. It comes with an API that abstracts the complexity of managing and operating multiple meshes without the user needing to know the complexity under the hood caused by multiple Service Meshes. You can find details about Gloo Mesh at `https://docs.solo.io/gloo-mesh-open-source/latest/getting_started/`. This is a comprehensive resource on how to install and try Gloo Mesh. Solo.io has another product called Gloo Edge, which acts as a Kubernetes Ingress controller as well as an API gateway. Gloo Mesh Enterprise is deployed along with Gloo Edge, which provides many comprehensive API management and Ingress capabilities. Gateway Gloo Mesh Enterprise adds support for external authentication using OIDC, OAuth, API key, LDAP, and OPA. These policies are implemented via a custom CRD called ExtAuthPolicy, which can apply these authentications when routes and destinations match certain criteria.

Gloo Mesh Enterprise provides WAF policies to monitor, filter, and block any harmful HTTP traffic. It also provides support for data loss prevention by doing a series of regex replacements on the response body and content that is logged by Envoy. This is a very important feature from a security point of view and stops sensitive data from being logged into the log files. DLP filters can be configured on listeners, virtual services, and routes. Gloo Mesh also provides support for connecting to legacy applications via the SOAP message format. There are options for building data transformation policies to apply XSLT transformation to modernize SOAP/XML endpoints. The data transformation policies can be applied to transform request or response payloads. It also supports special transformations such as via Inja templates. With Inja, you can write loops, conditional logic, and other functions to transform requests and responses.

There is also extensive support for WASM filters. Solo.io provides custom tooling that speeds up the development and deployment of web assemblies. To store WASM files, solo.io provides WebAssembly Hub, available at `https://webassemblyhub.io/`, and an open source CLI tool called wasme. You can read more about how to use Web Assembly Hub and the wasme CLI at `https://docs.solo.io/web-assembly-hub/latest/tutorial_code/getting_started/`.

As Gloo Mesh and other products from Solo.io are closely integrated with the Enterprise Service Mesh offering, you get a plethora of other features, and one such feature is a global API portal. The API portal is a self-discovery portal for publishing, sharing, and monitoring API usage for internal and external monetization. When using a multi-heterogeneous mesh, users don't need to worry about managing observability tools for every mesh; instead, Gloo Mesh Enterprise provides aggregated metrics across every mesh, providing a seamless experience of managing and observing multiple meshes.

In enterprise environments, it is important that multiple teams and users can access and deploy services in the mesh without stepping on each others' toes. Users need to know what services are available to consume and what services they have published. Users should be able to confidently perform mesh operations without impacting the services of other teams. Gloo Mesh uses the concept of workspaces, which are logical boundaries for a team, limiting team Service Mesh operations within the confines of the workspace so that multiple teams can concurrently use the mesh. Workspaces provide security isolation between configurations published by every team. Through workspaces, Gloo Mesh addresses the complexity of muti-tenancy in Enterprise environments, making it simpler for multiple teams to adopt Service Mesh with config isolation from each other and strict access control for safe muti-tenant usage of the mesh.

Gloo Mesh is also integrated with another Service Mesh based on a different architecture than Istio. The mesh is called Istio Ambient Mesh, which, rather than adding a sidecar proxy per workload, adds a proxy at the per-node level. Istio Ambient Mesh is integrated with Gloo Mesh, and users can run their sidecar proxy-based mesh along with their per-node proxy Istio Ambient Mesh.

Gloo Enterprise Mesh, with integration with Solo.io products such as Gloo Edge, makes it a strong contender among Service Mesh offerings. The ability to support multi-cluster and multi-mesh deployments, multi-tenancy via workspaces, strong support for authentication, zero-trust networking, and mature Ingress management via Gloo Edge makes it a comprehensive Service Mesh offering.

Kuma

Kuma is an open source CNCF sandbox project donated to CNCF by Kong Inc. Like Istio, Kuma also uses Envoy as the data plane. It supports multi-cluster and multi-mesh deployments, providing one global control plane to manage them all. At the time of writing this book, Kuma is one single executable written in GoLang. It can be deployed on Kubernetes as well as on VMs. When deployed in non-Kubernetes environments, it requires a PostgreSQL database to store its configurations.

Let's start by downloading and installing Kuma, followed by hands-on exercises on this topic:

1. Download Kuma for your operating system:

```
% curl -L https://kuma.io/installer.sh | VERSION=2.0.2 sh
-
INFO Welcome to the Kuma automated download!
INFO Kuma version: 2.0.2
INFO Kuma architecture: arm64
INFO Operating system: Darwin
INFO Downloading Kuma from: https://download.konghq.com/
mesh-alpine/kuma-2.0.2-darwin-arm64.tar.gz
```

2. Install Kuma on minikube. Unzip the download file and, in the unzipped folder's `bin` directory, run the following commands to install Kuma on Kubernetes:

```
% kumactl install control-plane | kubectl apply -f -
```

This will create a namespace called `kuma-system` and install the Kuma control plane in that namespace, along with configuring various CRDs and admission controllers.

3. At this point, we can access Kuma's GUI using the following command:

```
% kubectl port-forward svc/kuma-control-plane -n kuma-
system 5681:5681
```

Open `localhost:5681/gui` in your browser and you will see the following dashboard:

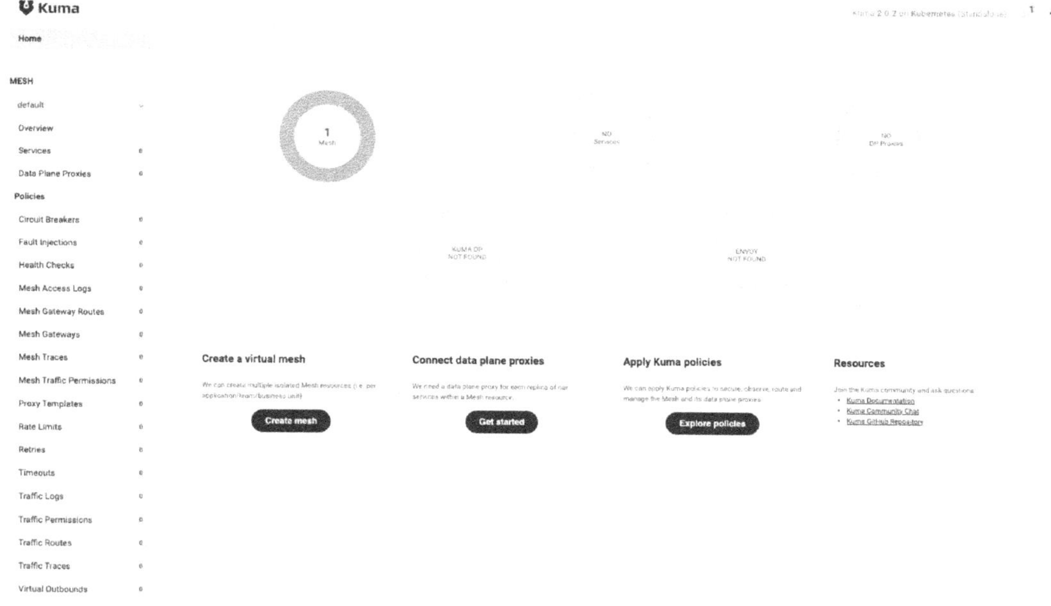

Figure A.7 – Kuma dashboard

The Kuma GUI provides comprehensive details about the mesh. We will use this to check the configuration as we build policies and add applications to the mesh. On the home page of the GUI, you will notice that it shows one mesh called **default**. A mesh in Kuma is a Service Mesh that is logically isolated from other Service Meshes in Kuma. You can have one Kuma installation in a Kubernetes cluster, which can then manage multiple Service Meshes perhaps for each team or department deploying their apps in that Kubernetes cluster. This is a very important concept and a key differentiator of Kuma from other Service Mesh technologies.

Deploying envoydemo and curl in Kuma mesh

The deployment file is available at `AppendixA/Kuma/envoy-proxy-01.yaml`. The noticeable difference compared to Istio in the deployment file is the addition of the following label, which instructs Kuma to inject its sidecar proxies into `envoydummy`:

```
kuma.io/sidecar-injection: enabled
```

The following commands will deploy the `envoydummy` and `curl` applications:

```
% kubectl create ns appendix-kuma
namespace/appendix-kuma created
% kubectl apply -f AppendixA/Kuma/envoy-proxy-01.yaml
configmap/envoy-dummy created
service/envoydummy created
deployment.apps/envoydummy created
serviceaccount/envoydummy created
```

After a few seconds, check whether the Pods have been deployed and the sidecars injected using the following commands:

```
% kubectl get po -n appendix-kuma
NAME                          READY   STATUS    RESTARTS   AGE
curl                          2/2     Running   0          71s
envoydummy-767dbd95fd-tp6hr   2/2     Running   0          71s
serviceaccount/curl created
pod/curl created
```

The sidecars are also called **data plane proxies** (**DPPs**), and they run along with every workload in the mesh. DPP comprises a data plane entity that defines the configuration of the DPP and a kuma-dp binary. During startup, kuma-dp retrieves the startup configuration for Envoy from the Kuma control plane (kuma-cp) and uses that to spawn the Envoy process. Once Envoy starts, it connects to kuma-cp using XDS. kuma-dp also spawns a core-dns process at startup.

It is worth noticing that installing Kuma and deploying an application has been a breeze. It is very simple, and the GUI is very intuitive, even for beginners.

Using the GUI, let's check the overall status of the mesh.

From **MESH** | **Overview**, you can see newly added DPPs:

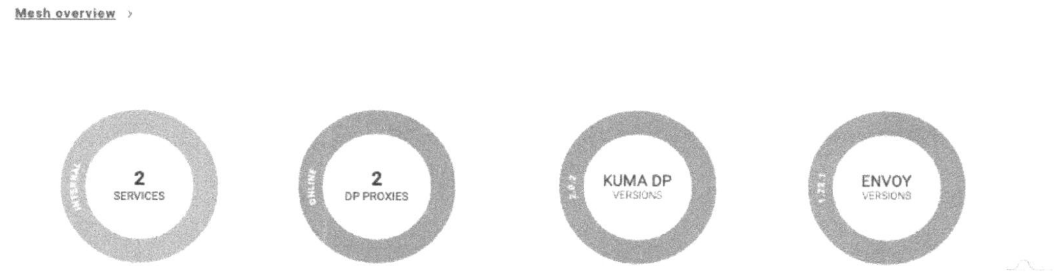

Figure A.8 – Mesh overview in the Kuma GUI

From **MESH** | **Data Plane Proxies**, you can find details about the workloads:

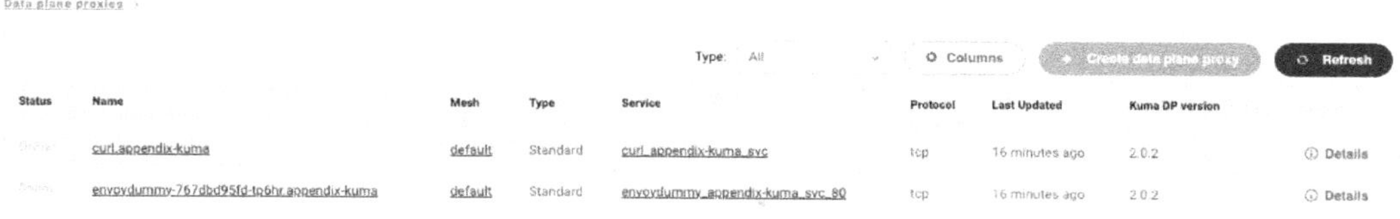

Figure A.9 – Data plane proxies

Now that we have installed apps, we will perform some hands-on exercises with Kuma policies to get experience with Kuma.

We will start by accessing the `envoydummy` service from the `curl` Pod:

```
% kubectl exec -it pod/curl -n appendix-kuma -- curl http://
envoydummy:80
V1----------Bootstrap Service Mesh Implementation with Istio--
--------V1%
```

The output is as expected. By default, Kuma allows traffic in and out of the mesh. By default, all traffic is unencrypted in the mesh. We will enable mTLS and deny all traffic in the mesh to establish zero-trust networking. First, we will delete the policy that allows all traffic within the mesh using the following command:

```
% kubectl delete trafficpermissions/allow-all-traffic
```

`allow-all-traffic` is a traffic permission policy that allows all traffic within the mesh. The previous command deletes the policy, thereby restricting all traffic in the mesh.

Next, we will enable mTLS within the mesh to enable secure communication and let `kong-dp` correctly identify a service by comparing the service identity with the DPP certificate. Without enabling mTLS, Kuma cannot enforce traffic permissions. The following policy enables mTLS in the default mesh. It makes use of an inbuilt CA, but in case you want to use an external CA then there are also provisions to provide externally generated root CA and key. Kuma automatically generates certificates for every workload with SAN in SPIFEE format.

```
apiVersion: kuma.io/v1alpha1
kind: Mesh
metadata:
  name: default
spec:
  mtls:
    enabledBackend: ca-1
    backends:
    - name: ca-1
      type: builtin
```

In the config file, we have defined that this policy applies to the `default` mesh. We have declared a CA named `ca-1` of the `builtin` type and have configured it to be used as the root CA for mTLS by defining `enabledBackend`. The configuration file is available at `AppendixA/Kuma/enablemutualTLS.yaml`. You can apply the configuration using the following commands:

```
% kubectl apply -f AppendixA/Kuma/enablemutualTLS.yaml
```

After enabling mTLS, let's try to access `envoydummy` from `curl`:

```
% kubectl exec -it pod/curl -n appendix-kuma -- curl http://
envoydummy:80
curl: (52) Empty reply from server
command terminated with exit code 52
```

The output is as expected because mTLS is enabled and there is no `TrafficPermission` policy allowing the traffic between `curl` and `envoydummy`.

To allow traffic, we need to create the following `TrafficPermission` policy:

```
apiVersion: kuma.io/v1alpha1
kind: TrafficPermission
mesh: default
```

```
metadata:
  name: allow-all-traffic-from-curl-to-envoyv1
spec:
  sources:
    - match:
        kuma.io/service: 'curl_appendix-kuma_svc'
  destinations:
    - match:
        kuma.io/service: 'envoydummy_appendix-kuma_svc_80'
```

Note that the `kuma.io/service` fields contain the values of the corresponding tags. Tags are sets of key-value pairs that contain details of the service that the DPP is part of and metadata about the exposed service. The following are tags applied to DPP for `envoydummy`:

```
% kubectl get dataplane/envoydummy-767dbd95fd-tp6hr -n
appendix-kuma -o json | jq '.spec.networking.inbound[].tags'
{
  "k8s.kuma.io/namespace": "appendix-kuma",
  "k8s.kuma.io/service-name": "envoydummy",
  "k8s.kuma.io/service-port": "80",
  "kuma.io/protocol": "http",
  "kuma.io/service": "envoydummy_appendix-kuma_svc_80",
  "name": "envoydummy",
  "pod-template-hash": "767dbd95fd",
  "version": "v1"
}
```

Similarly, you can fetch the value of the `curl` DPP. The configuration file is available at `AppendixA/Kuma/allow-traffic-curl-to-envoyv1.yaml`. Apply the configuration using the following command:

```
% kubectl apply -f AppendixA/Kuma/allow-traffic-curl-to-
envoyv1.yaml
trafficpermission.kuma.io/allow-all-traffic-from-curl-to-
envoyv1 created
```

After applying the configuration, test that you can access `envoydummy` from `curl`:

```
% kubectl exec -it pod/curl -n appendix-kuma -- curl http://
envoydummy:80
```

```
V1----------Bootstrap Service Mesh Implementation with Istio--
--------V1%
```

We have just experienced how you can control traffic between workloads in the mesh. You will find this very similar to `ServiceIntentions` in Consul Service Mesh.

Traffic management and routing

Now we will explore traffic routing in Kuma. We will deploy version v2 of the `envoydummy` service and route certain requests between version v1 and v2.

The first step is to deploy version v2 of `envoydummy`, followed by defining traffic permission to allow traffic between the `curl` Pod and the `envoydummy` v2 Pod. The files are at `AppendixA/Kuma/envoy-proxy-02.yaml` and `AppendixA/Kuma/allow-traffic-curl-to-envoyv2.yaml`. Apply the configuration, and once you have applied both files, test that `curl` is able to reach both v1 and v2 of the `envoydummy` Pod:

```
% for ((i=0;i<2;i++)); do kubectl exec -it pod/curl -n
appendix-kuma -- curl http://envoydummy ;done
V2----------Bootstrap Service Mesh Implementation with Istio--
--------V2
V1----------Bootstrap Service Mesh Implementation with Istio--
--------V1
```

Next, we will configure the routing by using a Kuma policy called `TrafficRoute`. This policy allows us to configure rules for traffic routing in the mesh.

The policy can be split into four parts to make it easier to understand:

1. In the first part, we are declaring the `TrafficRoute` policy. The basic usage of the policy is documented at `https://kuma.io/docs/2.0.x/policies/traffic-route/`. Here, we are declaring that the policy applies to the default mesh and to any request in the mesh originating from `curl_appendix-kuma_svc` with a destination of `envoydummy_appendix-kuma_svc_80`:

```
apiVersion: kuma.io/v1alpha1
kind: TrafficRoute
mesh: default
metadata:
  name: trafficroutingforlatest
spec:
  sources:
    - match:
```

```
          kuma.io/service: curl_appendix-kuma_svc
    destinations:
      - match:
          kuma.io/service: envoydummy_appendix-kuma_svc_80
```

2. Next, we are configuring any request with a prefix of `'/latest'` to be routed to the DPP with the tags that are highlighted under `destination`:

```
conf:
   http:
   - match:
       path:
         prefix: "/latest"
     destination:
       kuma.io/service: envoydummy_appendix-kuma_svc_80
       version: 'v2'
```

3. Then, we are configuring request with a prefix of `'/old'` to be routed to the data plane with the tags that are highlighted under `destination`:

```
   - match:
       path:
         prefix: "/old"
     destination:
       kuma.io/service: envoydummy_appendix-kuma_svc_80
       version: 'v1'
```

4. Finally, we are declaring the default destination for requests that do not match any of the paths defined in previous parts of the config. The default destination will be the DPP with the tags highlighted in the following code:

```
     destination:
       kuma.io/service: envoydummy_appendix-kuma_svc_80
```

The configuration file is available at `AppendixA/Kuma/trafficRouting01.yaml`. Apply the configuration and test the following scenarios:

- All requests with `'/latest'` should be routed to version v2:

```
% for ((i=0;i<4;i++)); do kubectl exec -it pod/curl -n
appendix-kuma -- curl http://envoydummy/latest ;done
```

```
V2----------Bootstrap Service Mesh Implementation with
Istio----------V2
V2----------Bootstrap Service Mesh Implementation with
Istio----------V2
V2----------Bootstrap Service Mesh Implementation with
Istio----------V2
V2----------Bootstrap Service Mesh Implementation with
Istio----------V2
```

- All request with '/old' should be routed to version v1:

```
% for ((i=0;i<4;i++)); do kubectl exec -it pod/curl -n
appendix-kuma -- curl http://envoydummy/old ;done
V1----------Bootstrap Service Mesh Implementation with
Istio----------V1
V1----------Bootstrap Service Mesh Implementation with
Istio----------V1
V1----------Bootstrap Service Mesh Implementation with
Istio----------V1
V1----------Bootstrap Service Mesh Implementation with
Istio----------V1
```

- All other requests should follow the default behavior:

```
% for ((i=0;i<4;i++)); do kubectl exec -it pod/curl -n
appendix-kuma -- curl http://envoydummy/xyz ;done
V2----------Bootstrap Service Mesh Implementation with
Istio----------V2
V2----------Bootstrap Service Mesh Implementation with
Istio----------V2
V1----------Bootstrap Service Mesh Implementation with
Istio----------V1
V1----------Bootstrap Service Mesh Implementation with
Istio----------V1
```

The request routing works as expected, and it is similar to how you would configure the same behavior using Istio. Now, let's look at the load balancing properties of Kuma Mesh. We will build another traffic routing policy to do weighted routing between version v1 and v2 of `envoydummy`. Here is a snippet of the configuration available at `AppendixA/Kuma/trafficRouting02.yaml`:

```
  conf:
    split:
      - weight: 10
        destination:
          kuma.io/service: envoydummy_appendix-kuma_svc_80
          version: 'v1'
      - weight: 90
        destination:
          kuma.io/service: envoydummy_appendix-kuma_svc_80
          version: 'v2'
```

After applying the configuration, you can test the traffic distribution using the following command:

```
% for ((i=0;i<10;i++)); do kubectl exec -it pod/curl -n
appendix-kuma -- curl http://envoydummy/xyz ;done
```

The traffic should be distributed between the two versions in approximately a 1:9 ratio. You can perform traffic routing, traffic modification, traffic splitting, load balancing, canary deployments, and locality-aware load balancing using a `TrafficRoute` policy. To read more about `TrafficRoute`, please use the comprehensive documentation available here: https://kuma.io/docs/2.0.x/policies/traffic-route.

Kuma also provides policies for circuit breaking, fault injection, timeout, rate limiting, and many more things. A comprehensive list of Kuma policies is available here: `https://kuma.io/docs/2.0.x/policies/introduction/`. These out-of-the-box policies make Kuma very easy to use with a very shallow learning curve.

In the hands-on example so far, we have been deploying all workloads in the default mesh. We discussed earlier that Kuma allows you to create different isolated meshes, allowing teams to have isolated mesh environments within the same Kuma cluster. You can create a new mesh using the following configuration:

```
apiVersion: kuma.io/v1alpha1
kind: Mesh
metadata:
  name: team-digital
```

The configuration is available in `AppendixA/Kuma/team-digital-mesh.yaml`. Apply the configuration using the following command:

```
% kubectl apply -f AppendixA/Kuma/team-digital-mesh.yaml
mesh.kuma.io/team-digital created
```

Once you have created the mesh, you can create all the resources within the mesh by adding the following annotations to the workload deployment configurations:

```
kuma.io/mesh: team-digital
```

And add the following to the Kuma policies:

```
mesh: team-digital
```

The ability to create a mesh is a very useful feature for enterprise environments and a key differentiator of Kuma compared to Istio.

Kuma also provides built-in Ingress capabilities to handle north-south traffic as well as east-west traffic. The Ingress is managed as a Kuma resource called a gateway, which in turn is an instance of kuma-dp. You have the flexibility to deploy as many Kuma gateways as you want, but ideally, one gateway per mesh is recommended. Kuma also supports integration with non-Kuma gateways, also called delegated gateways. For now, we will talk about built-in Kuma gateways and, later, briefly discuss delegated gateways.

To create a built-in gateway, you first need to define `MeshGatewayInstance` along with a matching `MeshGateway`. `MeshGatewayInstance` provides the details of how a gateway instance should be instantiated. Here is an example configuration of `MeshGatewayInstance`, which is also available at `AppendixA/Kuma/envoydummyGatewayInstance01.yaml`:

```
apiVersion: kuma.io/v1alpha1
kind: MeshGatewayInstance
metadata:
  name: envoydummy-gateway-instance
  namespace: appendix-kuma
spec:
  replicas: 1
  serviceType: LoadBalancer
  tags:
    kuma.io/service: envoydummy-edge-gateway
```

In the config, we are setting that there will be `1 replica` and a `serviceType` of `LoadBalancer`, and we have applied a tag, `kuma.io/service: envoydummy-edge-gateway`, which will be used to build the association with `MeshGateway`.

In the following configuration, we are creating a `MeshGateway` named `envoydummy-edge-gateway`. The configuration is available in `AppendixA/Kuma/envoydummyGateway01.yaml`:

```
apiVersion: kuma.io/v1alpha1
```

```
kind: MeshGateway
mesh: default
metadata:
  name: envoydummy-edge-gateway
  namespace: appendix-kuma
spec:
  selectors:
  - match:
      kuma.io/service: envoydummy-edge-gateway
  conf:
    listeners:
      - port: 80
        protocol: HTTP
        hostname: mockshop.com
        tags:
          port: http/80
```

The `MeshGateway` resource specifies the listeners, which are endpoints that accept network traffic. In the configuration, you specify ports, protocols, and an optional hostname. Under `selectors`, we are also specifying the `MeshGatewayInstance` tags with which the `MeshGateway` configuration is associated. Notice that we are specifying the same tags we defined in the `MeshGatewayInstance` configuration.

Next, we will define `MeshGatewayRoute`, which describes how a request is routed from `MeshGatewayInstance` to the workload service. An example configuration is available at `AppendixA/Kuma/envoydummyGatewayRoute01.yaml`. Here are some snippets from the file:

- Under `selectors`, we are specifying the details of the gateway and the listener to which this route should be attached. The details are specified by providing the tags of the corresponding gateway and listeners:

```
spec:
  selectors:
    - match:
        kuma.io/service: envoydummy-edge-gateway
        port: http/80
```

- In the `conf` part, we provide Layer 7 matching criteria for the request, such as the path and HTTP headers, and the destination details:

```
conf:
  http:
```

```
      rules:
        - matches:
            - path:
                match: PREFIX
                value: /
          backends:
            - destination:
                kuma.io/service: envoydummy_appendix-
kuma_svc_80
```

- And last but not least, we allow traffic between the edge gateway and the envoy dummy service by configuring `TrafficPermission` as described in following snippet. You can find the configuration at `AppendixA/Kuma/allow-traffic-edgegateway-to-envoy.yaml`:

```
kind: TrafficPermission
mesh: default
metadata:
  name: allow-all-traffic-from-curl-to-envoyv1
spec:
  sources:
    - match:
        kuma.io/service: 'envoydummy-edge-gateway'
  destinations:
    - match:
        kuma.io/service: 'envoydummy_appendix-kuma_
svc_80'
```

With traffic permission in place, we are now ready to apply the configuration using the following set of commands:

1. Create `MeshGatewayInstance`:

```
% kubectl apply -f AppendixA/Kuma/
envoydummyGatewayInstance01.yaml
meshgatewayinstance.kuma.io/envoydummy-gateway-instance
created
```

2. Create `MeshGateway`:

```
% kubectl apply -f AppendixA/Kuma/envoydummyGateway01.
yaml
meshgateway.kuma.io/envoydummy-edge-gateway created
```

3. Create `MeshGatewayRoute`:

```
% kubectl apply -f AppendixA/Kuma/
envoydummyGatewayRoute01.yaml
meshgatewayroute.kuma.io/envoydummy-edge-gateway-route
created
```

4. Create `TrafficPermissions`:

```
$ kubectl apply -f AppendixA/Kuma/allow-traffic-
edgegateway-to-envoy.yaml
trafficpermission.kuma.io/allow-all-traffic-from-curl-to-
envoyv1 configured
```

You can verify that Kuma has created a gateway instance using the following commands:

```
% kubectl get po -n appendix-kuma
NAME                         READY   STATUS    RESTARTS   AGE
curl                         2/2     Running   0          22h
envoydummy-767dbd95fd-
br2m6                     2/2     Running   0          22h
envoydummy-gateway-instance-75f87bd9cc-
z2rx6   1/1     Running   0          93m
envoydummy2-694cbc4f7d-
hrvkd                    2/2     Running   0          22h
```

You can also check the corresponding service using the following command:

```
% kubectl get svc -n appendix-kuma
NAME                      TYPE    CLUSTER-IP      EXTER-
NAL-IP   PORT(S)          AGE
envoydummy          Clus-
terIP      10.102.50.112   <none>        80/TCP          22h
envoydummy-gateway-instance   LoadBal-
ancer   10.101.49.118   <pending>     80:32664/TCP   96m
```

We are now all set to access `envoydummy` using the built-in Kuma gateway. But first, we need to find an IP address through which we can access the Ingress gateway service on minikube. Use the following command to find the IP address:

```
% minikube service envoydummy-gateway-instance --url -n
appendix-kuma
http://127.0.0.1:52346
```

Now, using http://127.0.0.1:52346, you can access the envoydummy service by performing curl from your terminal:

```
% curl -H "host:mockshop.com" http://127.0.0.1:52346/latest
V1----------Bootstrap Service Mesh Implementation with Istio--
--------V1
```

You have learned how to create a MeshGatewayInstance, which is then associated with MeshGateway. After the association, kuma-cp created a gateway instance of the built-in Kuma gateway. We then created a MeshGatewayRoute that specifies how the request will be routed from the gateway to the workload service. Later, we created a TrafficPermission resource to allow traffic flow from MeshGateway to the EnvoyDummy workload.

Kuma also provides options for using an external gateway as Ingress, also called a delegated gateway. In a delegated gateway, Kuma supports integrations with various API gateways, but Kong Gateway is the preferred and most well-documented option. You can read more about delegated gateways at https://kuma.io/docs/2.0.x/explore/gateway/#delegated.

Like Istio, Kuma also provides native support for both Kubernetes and VM-based workloads. Kuma provides extensive support for running advanced configurations of Service Mesh spanning multiple Kubernetes clusters, data centers, and cloud providers. Kuma has a concept of zones, which are logical aggregations of DPPs that can communicate with each other. Kuma supports running Service Mesh in multiple zones and the separation of control planes in a multi-zone deployment. Each zone is allocated its own horizontally scalable control plane providing complete isolation between every zone. All zones are then also managed by a centralized global control plane, which manages the creation of and changes to policies that are applied to DPPs and the transmission of zone-specific policies and configurations to respective control planes of underlying zones. The global control plane is a single pane of glass providing an inventory of all DPPs across all zones.

As mentioned earlier, Kuma is an open source project that was donated by Kong to CNCF. Kong also provides Kong Mesh, which is an enterprise version of Kuma built on top of Kuma, extending it to include capabilities required for running critical functionality for enterprise workloads. Kong Mesh provides a turnkey Service Mesh solution with capabilities such as integration with OPA, FIPS 140-2 compliance, and role-based access control. Coupled with Kong Gateway as an Ingress gateway, a Service Mesh based on Kuma, additional enterprise-grade add-ons and reliable enterprise support makes Kong Mesh a turnkey Service Mesh technology.

Uninstalling Kuma

You can uninstall Kuma Mesh using the following command:

```
% kumactl install control-plane | kubectl delete -f -
```

Linkerd

Linkerd is a CNCF graduated project licensed under Apache v2. Buoyant (`https://buoyant.io/`) is the major contributor to Linkerd. Out of all Service Mesh technologies, Linkerd is probably one of the earliest, if not the oldest. It was initially made public in 2017 by Buoyant. It had initial success, but then it was criticized for being very resource hungry. The proxy used in Linkerd was written using the Scala and Java networking ecosystem, which uses the **Java Virtual Machine** (**JVM**) at runtime, causing significant resource consumption. In 2018, Buoyant released a new version of Linkerd called Conduit. Conduit was later renamed Linkerd v2. The Linkerd v2 data plane is made up of Linkerd2-proxy, which is written in Rust and has a small resource consumption footprint. Linkerd2- proxy is purpose built for proxying as a sidecar in Kubernetes Pods. While Linkerd2-proxy is written in Rust, the Linkerd control plane is developed in Golang.

Like other open source Service Mesh technologies discussed in this *Appendix*, we will discover Linkerd by playing around with it and observing how it is similar to or different to Istio. Let's start by installing Linkerd on minikube:

1. Install Linkerd on minikube using the following command:

```
% curl --proto '=https' --tlsv1.2 -sSfL https://run.
linkerd.io/install | sh
Downloading linkerd2-cli-stable-2.12.3-darwin...
Linkerd stable-2.12.3 was successfully installed
Add the linkerd CLI to your path with:
  export PATH=$PATH:/Users/arai/.linkerd2/bin
```

2. Follow the suggestion to include linkerd2 in your path:

```
export PATH=$PATH:/Users/arai/.linkerd2/bin
```

3. Linkerd provides an option to check and validate that the Kubernetes cluster meets all the prerequisites required to install Linkerd:

```
% linkerd check --pre
```

4. If the output contains the following, then you are good to go with the installation:

```
Status check results are √
```

   If not, then you need to resolve the issues by going through suggestions at `https://linkerd.io/2.12/tasks/troubleshooting/#pre-k8s-cluster-k8s%20for%20hints`.

5. Next, we will Install Linkerd in two steps:

   I. First, we install the CRDs:

   ```
   % linkerd install --crds | kubectl apply -f -
   ```

   II. The, we install the Linkerd control plane in the `linkerd` namespace:

   ```
   % linkerd install --set proxyInit.runAsRoot=true |
   kubectl apply -f -
   ```

6. After installing the control plane, check that Linkerd is fully installed using the following commands:

   ```
   % linkerd check
   ```

   If Linkerd is successfully installed, then you should see the following message:

   ```
   Status check results are √
   ```

That complete the setup of Linkerd! Let's now analyze what has been installed:

```
% kubectl get pods,services -n linkerd
NAME                            READY   STATUS    RESTARTS   AGE
pod/linkerd-destination-86d68bb57-
447j6       4/4      Running   0          49m
pod/linkerd-identity-5fbdcccbd5-
lzfkj         2/2      Running   0          49m
pod/linkerd-proxy-injector-685cd5988b-
5lmxq    2/2      Running   0        49m

NAME    TYPE         CLUSTER-IP       EXTER-
NAL-IP   PORT(S)     AGE
service/linkerd-dst            ClusterIP
   10.102.201.182   <none>      8086/TCP   49m
service/linkerd-dst-headless         Clus-
terIP   None            <none>       8086/TCP   49m

service/linkerd-identity        ClusterIP   10.98.112.229
<none>        8080/TCP   49m

service/linkerd-identity-headless         Clus-
terIP   None            <none>       8090/TCP   49m
```

```
service/linkerd-policy                ClusterIP   None
<none>        8090/TCP   49m

service/linkerd-policy-validator      ClusterIP   10.102.142.68
<none>        443/TCP    49m

service/linkerd-proxy-injector        ClusterIP   10.101.176.198
<none>        443/TCP    49m

service/linkerd-sp-validator          ClusterIP   10.97.160.235
<none>        443/TCP    49m
```

It's worth noticing here that the control plane comprises many Pods and Services. The `linkerd-identity` service is a CA for generating signed certificates for Linkerd proxies. `linkerd-proxy-injector` is the Kubernetes admission controller responsible for modifying Kubernetes Pod specifications to add linkerd-proxy and proxy-init containers. The `destination` service is the brains of the Linkerd control plane and maintains service discovery and identity information about the services, along with policies for securing and managing the traffic in the mesh.

Deploying envoydemo and curl in Linkerd

Now let's deploy envoydummy and curl apps and check how Linkerd performs Service Mesh functions. Follow these steps to install the application:

1. Like most Service Mesh solutions, we need to annotate the deployment descriptors with the following annotations:

   ```
   annotations:
     linkerd.io/inject: enabled
   ```

   The configuration file for the `envoydummy` and `curl` apps along with annotations is available in `AppendixA/Linkerd/envoy-proxy-01.yaml`.

2. After preparing the deployment descriptors, you can apply the configurations:

   ```
   % kubectl create ns appendix-linkerd
   % kubectl apply -f AppendixA/Linkerd/envoy-proxy-01.yaml
   ```

3. That should deploy the Pod. Once the Pod is deployed, you can check what has been injected into the Pods via the following commands:

   ```
   % kubectl get po/curl -n appendix-linkerd -o json | jq
   '.spec.initContainers[].image, .spec.initContainers[].
   name'
   ```

```
"cr.l5d.io/linkerd/proxy-init:v2.0.0"
"linkerd-init"
% kubectl get po/curl -n appendix-linkerd -o json | jq
'.spec.containers[].image, .spec.containers[].name'
"cr.l5d.io/linkerd/proxy:stable-2.12.3"
"curlimages/curl"
"linkerd-proxy"
"curl"
```

From the preceding output, observe that Pod initialization was performed by a container named `linkerd-init` of the `cr.l5d.io/linkerd/proxy-init:v2.0.0` type, and the Pod has two running containers, `curl` and `linkerd-proxy`, of the `cr.l5d.io/linkerd/proxy:stable-2.12.3` type. The `linkerd-init` container runs during the initialization phase of the Pod and modifies iptables rules to route all network traffic from `curl` to `linkerd-proxy`. As you may recall, in Istio we have `istio-init` and `istio-proxy` containers, which are similar to Linkerd containers. `linkerd-proxy` is ultra-light and ultra-fast in comparison to Envoy. Being written in Rust makes its performance predictable and it doesn't need garbage collection, which often causes high latency during garbage collection passes. Rust is arguably much more memory safe than C++ and C, which makes it less susceptible to memory safety bugs. You can read more about why `linkerd-proxy` is better than envoy at `https://linkerd.io/2020/12/03/why-linkerd-doesnt-use-envoy/`.

Verify that `curl` is able to communicate with the `envoydummy` Pod as follows:

```
% kubectl exec -it pod/curl -c curl -n appendix-linkerd -- curl
http://envoydummy:80
V1----------Bootstrap Service Mesh Implementation with Istio--
--------V1%
```

Now that we have installed the `curl` and `envoydummy` Pods, let's explore Linkerd Service Mesh functions. Let's start by exploring how we can restrict traffic within the mesh using Linkerd.

Zero-trust networking

Linkerd provides comprehensive policies to restrict traffic in the mesh. Linkerd provides a set of CRDs through which policies can be defined to control the traffic in the mesh. Let's explore these policies by implementing policies to control traffic to the `envoydummy` Pod:

1. We will first lock down all traffic in the cluster using following:

   ```
   % linkerd upgrade --default-inbound-policy deny --set
   proxyInit.runAsRoot=true | kubectl apply -f -
   ```

We used the `linkerd upgrade` command to apply a `default-inbound-policy` of `deny`, which prohibits all traffic to ports exposed by workloads in the mesh unless there is a server resource attached to the port.

After applying the policy, all access to the `envoydummy` service is denied:

```
% kubectl exec -it pod/curl -c curl -n appendix-linkerd
-- curl --head http://envoydummy:80
HTTP/1.1 403 Forbidden
content-length: 0
l5d-proxy-error: unauthorized request on route
```

2. Next, we create a server resource to describe the `envoydummy` port. A Server resource is a means of instructing Linkerd that only authorized clients can access the resource. We do that by declaring the following Linkerd policy:

```
apiVersion: policy.linkerd.io/v1beta1
kind: Server
metadata:
  namespace: appendix-linkerd
  name: envoydummy
  labels:
    name: envoydummy
spec:
  podSelector:
    matchLabels:
      name: envoydummy
  port: envoydummy-http
  proxyProtocol: HTTP/1
```

The configuration file is available at `AppendixA/Linkerd/envoydummy-server.yaml`. The server resource is defined in the same namespace as the workload. In the configuration file, we also define the following:

- `podSelector`: Criteria for selecting the workload
- `port`: Name or number of the port for which this server configuration is being declared
- `proxyProtocol`: Configures protocol discovery for inbound connections and must be one of the following: unknown, `HTTP/1`, `HTTP/2`, `gRPC`, `opaque`, or `TLS`

Apply the server resource using the following command:

```
% kubectl apply -f AppendixA/Linkerd/envoydummy-server.
yaml
server.policy.linkerd.io/envoydummy created
```

Although we have applied the server resource, the `curl` Pod still can't access the `envoydummy` service unless we authorize it.

3. In this step, we will create an authorization policy that authorizes `curl` to access `envoydummy`. The authorization policy is configured by providing server details of the target destination and service account details being used to run the originating service. We created a server resource named `envoydummy` in the previous step and, as per `AppendixA/Linkerd/envoy-proxy-01.yaml`, we are using a service account named `curl` to run the `curl` Pod. The policy is defined as follows and is also available at `AppendixA/Linkerd/authorize-curl-access-to-envoydummy.yaml`:

```
apiVersion: policy.linkerd.io/v1alpha1
kind: AuthorizationPolicy
metadata:
  name: authorise-curl
  namespace: appendix-linkerd
spec:
  targetRef:
    group: policy.linkerd.io
    kind: Server
    name: envoydummy
  requiredAuthenticationRefs:
    - name: curl
      kind: ServiceAccount
```

4. Apply the configuration as follows:

```
% kubectl apply -f AppendixA/Linkerd/authorize-curl-
access-to-envoydummy.yaml
authorizationpolicy.policy.linkerd.io/authorise-curl
created
```

Once the `AuthorizationPolicy` is in place, it will authorize all traffic to the Envoy server from any workload running using a `curl` service account.

5. You can verify the access between the `curl` and `envoydummy` Pods using the following command:

```
% kubectl exec -it pod/curl -c curl -n appendix-linkerd -
curl  http://envoydummy:80
V1----------Bootstrap Service Mesh Implementation with
Istio----------V1
```

Using `AuthorizationPolicy`, we have controlled access to ports presented as servers from other clients in the mesh. Granular access control, such as controlling access to an HTTP resource, can be managed by another policy called `s`.

We can understand this concept better via an example, so let's make a requirement that only requests whose URI start with `/dummy` can be accessible from `curl`; requests to any other URI must be denied. Let's get started:

1. We need to first define an `HTTPRoute` policy as described in the following code snippet:

```
apiVersion: policy.linkerd.io/v1beta1
kind: HTTPRoute
metadata:
  name: envoydummy-dummy-route
  namespace: appendix-linkerd
spec:
  parentRefs:
    - name: envoydummy
      kind: Server
      group: policy.linkerd.io
      namespace: appendix-linkerd
  rules:
    - matches:
      - path:
          value: "/dummy/"
          type: "PathPrefix"
        method: GET
```

The configuration is also available at `AppendixA/Linkerd/HTTPRoute.yaml`. This will create an HTTP route targeting the `envoydummy` server resource. In the `rules` section, we define the criteria for identifying requests that will be used to identify the HTTP request for this route. Here, we have defined to rule to match any request with the `dummy` prefix and the `GET` method. `HTTPRoute` also supports route matching using headers and query parameters. You can also apply other filters in `HTTPRoute` to specify how the request should be processed during the request or response cycle; you can modify inbound request headers, redirect requests, modify request paths, and so on.

2. Once we have defined `HTTPRoute`, we can modify the `AuthorizationPolicy` to associate with `HTTPRoute` instead of the server, as listed in the following code snippet and also available at `AppendixA/Linkerd/HttpRouteAuthorization.yaml`:

```
apiVersion: policy.linkerd.io/v1alpha1
kind: AuthorizationPolicy
metadata:
  name: authorise-curl
  namespace: appendix-linkerd
spec:
  targetRef:
    group: policy.linkerd.io
    kind: HTTPRoute
    name: envoydummy-dummy-route
  requiredAuthenticationRefs:
    - name: curl
      kind: ServiceAccount
```

The configuration updates `AuthorizationPolicy` and, instead of referencing the server (`envoydummy` configured in `AppendixA/Linkerd/authorize-curl-access-to-envoydummy.yaml`) as the target, the policy is now referencing `HTTPRoute` (named `envoydummy-dummy-route`).

Apply both configurations and test that you are able to make requests with the `/dummy` prefix in the URI. Any other request will be denied by Linkerd.

So far in `AuthorizationPolicy` we have used `ServiceAccount` authentication. `AuthorizationPolicy` also supports `MeshTLSAuthentication` and `NetworkAuthentication`. Here is a brief overview of these authentication types:

- `MeshTLSAuthentication` is used to identify a client based on its mesh identity. For example, the `curl` Pod will be represented as `curl.appendix-linkerd.serviceaccount.identity.linkerd.local`.
- `NetworkAuthentication` is used to identify a client based on its network location using **Classless Inter-Domain Routing** (**CIDR**) blocks.

Linkerd also provides retries and timeouts to provide application resilience when systems are under stress or suffering partial failures. Apart from support for usual retry strategies, there is also a provision for specifying retry budgets so that retries do not end up amplifying resilience problems. Linkerd provides automated load balancing of requests to all destination endpoints using the **exponentially weighted moving average** (**EWMA**) algorithm. Linkerd supports weight-based traffic splitting, which is useful for performing canary and blue/green deployments. Traffic splitting in Linkerd uses the

Service Mesh Interface (SMI) Traffic Split API, allowing users to incrementally shift traffic between blue and green services. You can read about the Traffic Split API at `https://github.com/servicemeshinterface/smi-spec/blob/main/apis/traffic-split/v1alpha4/traffic-split.md` and SMI at `https://smi-spec.io`. Linkerd provides a well-defined and documented integration with Flagger to perform automatic traffic shifting when performing canary and blue/green deployments.

There is a lot more to learn and digest about Linkerd. You can read about it at `https://linkerd.io/2.12`. Linkerd is ultra-performant because of its ultra-light service proxy build using Rust. It is carefully designed to solve application networking problems. The ultra-light proxy performs most Service Mesh functions but lacks in features such as circuit breaking and rate limiting. Let's hope that the Linkerd creators bridge the gap with Envoy.

Hopefully, you are now familiar with the various alternatives to Istio and how they implement Service Mesh. Consul, Linkerd, Kuma, and Gloo Mesh have lots of similarities, and all of them are powerful, but Istio is the one that has the biggest community behind it and the support of various well-known organizations. Also, there are various organizations that provide enterprise support for Istio, which is a very important consideration when deploying Istio to production.

Index

A

B

C

D

F

G

H

I

L

M

N

O

P

R

S

T

U

V

W

Packtpub.com

Subscribe to our online digital library for full access to over 7,000 books and videos, as well as industry leading tools to help you plan your personal development and advance your career. For more information, please visit our website.

Why subscribe?

- Spend less time learning and more time coding with practical eBooks and Videos from over 4,000 industry professionals
- Improve your learning with Skill Plans built especially for you
- Get a free eBook or video every month
- Fully searchable for easy access to vital information
- Copy and paste, print, and bookmark content

Did you know that Packt offers eBook versions of every book published, with PDF and ePub files available? You can upgrade to the eBook version at packtpub.com and as a print book customer, you are entitled to a discount on the eBook copy. Get in touch with us at customercare@packtpub.com for more details.

At www.packtpub.com, you can also read a collection of free technical articles, sign up for a range of free newsletters, and receive exclusive discounts and offers on Packt books and eBooks.

Other Books You May Enjoy

If you enjoyed this book, you may be interested in these other books by Packt:

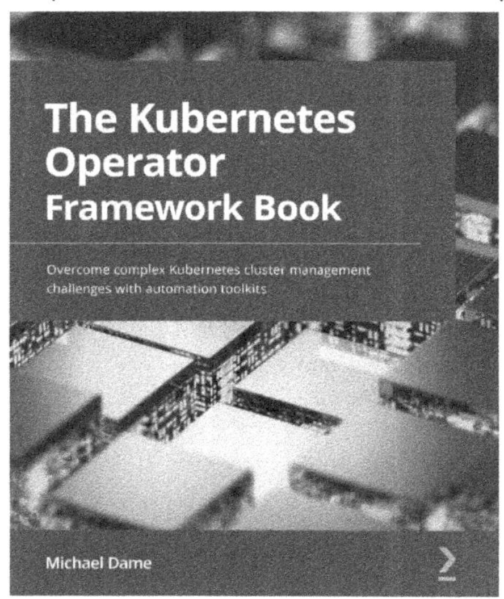

The Kubernetes Operator Framework Book

Michael Dame

ISBN: 978-1-80323-285-0

- Gain insight into the Operator Framework and the benefits of operators
- Implement standard approaches for designing an operator
- Develop an operator in a stepwise manner using the Operator SDK
- Publish operators using distribution options such as OperatorHub.io
- Deploy operators using different Operator Lifecycle Manager options
- Discover how Kubernetes development standards relate to operators
- Apply knowledge learned from the case studies of real-world operators

Learning DevOps - Second Edition

Mikael Krief

ISBN: 978-1-80181-896-4

- Understand the basics of infrastructure as code patterns and practices
- Get an overview of Git command and Git flow
- Install and write Packer, Terraform, and Ansible code for provisioning and configuring cloud infrastructure based on Azure examples
- Use Vagrant to create a local development environment
- Containerize applications with Docker and Kubernetes
- Apply DevSecOps for testing compliance and securing DevOps infrastructure
- Build DevOps CI/CD pipelines with Jenkins, Azure Pipelines, and GitLab CI
- Explore blue-green deployment and DevOps practices for open sources projects

Packt is searching for authors like you

If you're interested in becoming an author for Packt, please visit `authors.packtpub.com` and apply today. We have worked with thousands of developers and tech professionals, just like you, to help them share their insight with the global tech community. You can make a general application, apply for a specific hot topic that we are recruiting an author for, or submit your own idea.

Share Your Thoughts

Now you've finished *Bootstrapping Service Mesh Implementations with Istio*, we'd love to hear your thoughts! Scan the QR code below to go straight to the Amazon review page for this book and share your feedback or leave a review on the site that you purchased it from.

`https://packt.link/r/1803246812`

Your review is important to us and the tech community and will help us make sure we're delivering excellent quality content.

Download a free PDF copy of this book

Thanks for purchasing this book!

Do you like to read on the go but are unable to carry your print books everywhere?

Is your eBook purchase not compatible with the device of your choice?

Don't worry, now with every Packt book you get a DRM-free PDF version of that book at no cost.

Read anywhere, any place, on any device. Search, copy, and paste code from your favorite technical books directly into your application.

The perks don't stop there, you can get exclusive access to discounts, newsletters, and great free content in your inbox daily

Follow these simple steps to get the benefits:

1. Scan the QR code or visit the link below

`https://packt.link/free-ebook/9781803246819`

2. Submit your proof of purchase
3. That's it! We'll send your free PDF and other benefits to your email directly

www.ingramcontent.com/pod-product-compliance
Lightning Source LLC
Chambersburg PA
CBHW081503050326
40690CB00015B/2909

9781803246819